Stormwater Design for Sustainable Development

About the Author

Dr. Ronald Rossmiller, PhD, PE, has expertise ranging from broad conceptual planning to detailed design of water resources facilities and stormwater management projects. As a national expert in stormwater, Dr. Rossmiller teaches short courses and seminars throughout the United States. The former National Program Director for Stormwater Management at HDR Engineering, Inc., he currently manages his own consulting company, Stormwater Consultants.

Stormwater Design for Sustainable Development

Ronald L. Rossmiller

New York Chicago San Francisco
Athens London Madrid
Mexico City Milan New Delhi
Singapore Sydney Toronto

Cataloging-in-Publication Data is on file with the Library of Congress.

McGraw-Hill Education books are available at special quantity discounts to use as premiums and sales promotions, or for use in corporate training programs. To contact a representative please visit the Contact Us page at www.mhprofessional.com.

Stormwater Design for Sustainable Development

1 2 3 4 5 6 7 8 9 0   QVS/QVS   1 2 0 9 8 7 6 5 4 3

ISBN 978-0-07-181652-6
MHID 0-07-181652-6

The pages within this book were printed on acid-free paper.

Sponsoring Editor	**Proofreader**
Larry S. Hager	Eina Malik
Acquisitions Coordinator	**Indexer**
Bridget Thoreson	Robert Swanson
Editorial Supervisor	**Production Supervisor**
David E. Fogarty	Pamela A. Pelton
Project Manager	**Composition**
Shruti Chopra,	MPS Limited
MPS Limited	
	Art Director, Cover
Copy Editor	Jeff Weeks
Erica Orloff	

To my wife, Susan, who has been the loving
and solid foundation of our family
of seven for over fifty years

Contents

Preface

Numerous books have been written concerning aspects of low-impact development (LID) and the triple bottom line (TBL). However, they are somewhat general in nature. They discuss the concepts behind LIDs and describe ways in which these concepts can be implemented. The same is true of the TBL. Books describe and discuss its components: people, planet, and profit. Many discuss the planet which usually is a discussion of how to improve the environment.

I am a civil engineer, and engineers tend to be problem solvers. Rather than talking about what we should be doing, we go ahead and do something about it. Most times we improve the human condition. Sometimes, while trying to solve a problem, we create other problems.

Fortunately for me, early in my career, I read a quote by Herbert Hoover, a mining engineer who went on to become president of the United States. He said in part:

> *Engineering*! It is a great profession. There is the fascination of watching as a figment of the imagination emerges through the aid of science to a plan on paper. Then it moves to realization in stone or metal or energy. Then it brings jobs and homes to men. Then it elevates the standards of living and adds to the comforts of life. This is the engineer's high privilege.
>
> The great liability of the engineer compared to men of other professions is that his works are out in the open where all can see them. His acts, step by step, are in hard substance. The engineer simply cannot deny he did it. If his works do not work, he is damned.
>
> But the engineer himself looks back at the unending stream of goodness which flows from his successes with satisfactions that few professions may know. And the verdict of his fellow professionals is all the accolade he wants.

This book is my attempt to combine the many facets of LID, TBL, Federal Emergency Management Agency (FEMA), Environmental Protection Agency (EPA), and local and state reviewers as they impact upon a design team. The team changes a thought into an acceptable design with plans and specifications as the team develops, retrofits, or redevelops a parcel of land for some use that blends all of the above into a pleasing whole.

We all start out with a blank sheet of paper. On it we draw the outline of our site, pathways of swales and creeks, natural and manmade features that should be preserved, and rights-of-way and easement alignments for existing and future streets and utilities. Then we review planning and zoning ordinances plus state and federal regulations for flooding and pollution.

In the distant and recent past, we usually laid out straight streets at right angles to each other and then formed rectangular lots within the blocks. Little if any attention was paid to the natural topography. Natural drainageways were obliterated in the grading process to fit our preconceived arrangement of streets and lots. Drainage was relegated to streets and storm sewers. Pollutants were simply allowed to find their way into our creeks and streams. Overhead power and telephone lines contributed to sight pollution and were replaced after wind, rain, and snowstorm events. All of this required the use of resources, labor, machinery, and huge sums of money.

Hopefully, we have been learning from these past mistakes and are now developing, retrofitting, and redeveloping our land and water resources while incorporating the concepts of LID and the TBL. This book's contents are the result of my education, experiences, and continuous learning over the last 50-plus years. My original focus had been on the development of rural and urban drainage systems. In the past two decades, it has been on developing better ways to develop urban areas by doing it wisely and cheaply and blending all features of developments into safe, attractive, economic, and sustainable places in which to live, work, and play.

Ronald L. Rossmiller

Abbreviations

A	Area
AHW	Allowable headwater
AMC	Antecedent moisture condition
ASCE	American Society of Civil Engineers
BMP	Best management practice
BOD	Biological oxygen demand
C	Runoff coefficient
CADD	Computer aided drafting and design
CF	Channel flow
CFR	Code of Federal Regulations
cfs	cubic feet per second
CN	Curve number
CSP	Corrugated steel pipe
Cu	Copper
CWP	Center for Watershed Protection
D	Depth
d	Diameter
dB	Decibel
Dc	Critical depth
DCIA	Directly connected impervious area
DDF	Depth duration frequency
Dn	Normal depth
DNR	Department of Natural Resources
DOT	Department of Transportation
EGL	Energy grade line
EPA	Environmental Protection Agency
ET	Evapotranspiration
F	Infiltration

FAA	Federal Aviation Agency
FHWA	Federal Highway Administration
FIA	Flood Insurance Agency
fps	feet per second
F&W	Fish and Wildlife
GIS	Geographic information system
GW	Groundwater
HDPE	High-density polyethelene
HDS	Hydraulic Design Series
HEC	Hydraulic Engineering Circular
HEC	Hydrologic Engineering Center
HGL	Hydraulic grade line
HIG	Hole in the ground
ho	Greater of TW or (Dc + D)/2
HSG	Hydrologic soil group
H/V	Horizontal to vertical
HW	Headwater
I	Infiltration
Ia	Initial abstraction
IDF	Intensity duration frequency
Imp	Imperviousness
ISHC	Iowa State Highway Commission
KN	Kjeldal nitrogen
L	Length
LACFCD	Los Angeles County Flood Control District
Lca	Length to the center of a drainage area
LID	Low-impact development
LSo	Conduit length × conduit slope
MOP	Manual of Practice
N	Nitrogen
NFIP	National Flood Insurance Program
NIMBY	Not in my back yard
NRCS	Natural Resources Conservation Service
NRDC	Natural Resources Defence Council
NWS	National Weather Service
O&M	Operation and maintenance
OS	Open space
OSHA	Office of Safety and Health Administration

P	Phosphorus
P	Precipitation
Pb	Lead
PL	Public Law
PMP	Probable maximum precipitation
POTW	Publically owned treatment works
PVC	Polyvinyl chloride
Q	Flow rate, measured in cfs or inches depending on usage
RCB	Reinforced concrete box
RCP	Reinforced concrete pipe
ROW	Right of way
RP	Return period
S	Slope
SCF	Shallow concentrated flow
SCS	Soil Conservation Service
SF	Sheet flow
Sf	Friction slope
SLDI	Sustainable Land Development International
So	Channel or conduit slope
SRO	Surface runoff
SWM	Stormwater management
TBL	Triple bottom line: people, planet, profit
TKN	Total Kheldal nitrogen
TN	Total nitrogen
TP	Technical paper
TP	Total phosphorus
TSS	Total suspended solids
USCOE	United States Corps of Engineers
USDI	United States Department of the Interior
USDOT	United States Department of Transportation
USGS	United States Geological Service
UV	Ultraviolet
UWEX	University of Wisconsin Extension
VOC	Volatile organic compound
WP	Wetted perimeter
Zn	Zinc

CHAPTER 1

Introduction

1.1 Why This Book

1.1.1 Where We Were and Are Now

We have come a long way since abandoning nomadic ways and settling in small villages. These grew into towns, towns grew into cities, and cities grew into metropolises. Now we have megacities. Drainage was left to natural creeks, streams, and rivers. As villages grew, paths and streets were used to convey runoff. As towns grew, "night soil" was simply thrown out windows into streets. As cities grew, sewage and rainfall runoff were combined into small and large combined sewer systems. As cities grew, we separated sewage from stormwater runoff and treated this waste in sewage plants. Runoff was conveyed in storm sewers that emptied into creeks.

As cities grew, sewage plants went beyond primary to secondary and tertiary treatment, and their names changed to wastewater treatment plants. As cities grew and we became more prosperous, we decided that our way of living in urban areas was polluting our air, water, and land resources. An Environmental Protection Agency (EPA) study concluded that stormwater runoff contained many pollutants, so we developed some best management practices (BMPs). One BMP was construction of detention basins, large and small, all over the United States to control both the quantity and quality of stormwater runoff. Today, we have numerous BMPs classified as source and treatment controls plus dozens of managerial practices to control and remove pollutants from stormwater.

To aid us in design, computers and programs have replaced slide rules and calculators. They allow us to ask and answer many more "what happens if" questions much more quickly and easily—as long as we have the computer capability, programs, and data needed for their use.

1.1.2 For Whom Was This Book Written?

This leads me to for whom this book was written. It was written for those civil engineers, planners, and those in other disciplines that work for small- and medium-sized cities, counties, and consulting firms. They must follow the same federal, state, and local laws, rules, and regulations that large cities, metropolises, megacities, and national consulting firms must follow.

However, they do this with far fewer resources in terms of money, staff, computer and program capabilities, and expertise. Over the past decades, the power, capabilities, and sophistication of computers and programs have allowed us to develop answers to the degree of improvement to stormwater quantity and quality. We can do this—*if we* have the personnel, budget, and expertise to use them. Some programs are within the reach of smaller entities.

The EPA requires that BMPs be used to enhance runoff quality. The Federal Emergency Management Agency (FEMA) currently requires that we reduce flooding due to stormwater runoff. As long as approved BMPs are used and flooding is curtailed, current laws are satisfied. We may be leading toward requirements to quantify how much pollutants are reduced at some point in the future.

BMPs are defined by the EPA as:

1. Activity schedules, prohibitions of practices, maintenance procedures, and other physical, structural, and/or managerial practices to prevent or reduce pollution of Waters of the US.

2. BMPs include treatment systems, operating procedures, and practices to control site runoff, spillage or leaks, sludge or waste disposal, or drainage from raw material storage.

3. BMPs can also be defined as any program, technology, process, siting criteria, operating method, measure, or device, which controls, prevents, removes, or reduces pollution.

4. BMPs are classified as source or treatment controls. Source controls are operational practices that prevent pollution by reducing potential pollutants at their source and keeping them out of stormwater. They typically do not require construction. Treatment controls are methods of treatment to remove pollutants from stormwater that require construction.

1.1.3   Purposes of This Book

While this book's title is *Stormwater Design for Sustainable Development*, it is also written to get the point across that developing land for various uses is more than just arranging streets, locating buildings, handling stormwater runoff quantity and quality, and landscaping the site. In too many cases the result is drab, cookie-cutter subdivisions, uninteresting commercial and industrial areas, flooding, pollution, tens of thousands killed and injured in traffic accidents, along with the loss of possessions and heartache accompanying the above.

We do get it right in some locations. The ideas and methods in this book can be utilized by everyone, no matter their size, to design developments that are a credit to and a sustainable addition to their communities. The result is a development that blends the existing topography and drainage channels, curvilinear street layout, building locations, utilities, and both older and fresher types of BMPs into an attractive, pleasant, safe, economic, and sustainable whole.

This integrated approach to land development benefits every type of land use. Natural drainageways do not disappear during site grading because of some notion of where streets should be. Rather, existing land forms are preserved, and grading is reduced. Curvilinear streets replace rectangular grids, thereby reducing total street lengths by 15 to 50 percent with fewer intersections, saving lives and hundreds of thousands or millions of dollars in construction, operation, and maintenance costs. They also incorporate measures to reduce stormwater quantities and enhance stormwater quality requirements.

A main purpose of this book is to explain each portion of the design process in some detail for various stormwater management facilities for both water quantity and water quality control. These include BMPs that have been used for decades and newer, fresher BMPs. In certain larger developments with rolling to hilly terrain, there is still a place to use traditional detention basins that were constructed by the hundreds of thousands over the past decades. In addition, millions of culverts exist in the United States. A least-cost method of reducing flow rates and enhancing water quality is to retrofit suitable existing detention basins and culverts. They are already in place and deserve as much attention as BMPs known as rain gardens, bioswales, infiltration basins and trenches, greenroofs, porous concrete, permeable pavers, etc.

A second purpose is to discuss the merits of using curvilinear streets rather than the traditional grid pattern of rectangular streets and lots that have been used for centuries in our cities. This concept was developed in the 1980s and can reduce total street and utility lengths by 15 to 50 percent. Its use for new developments in single- and multi-family areas serve both sustainability and all portions of the triple bottom line (TBL) of people, planet, and profits (3Ps).

A third purpose is to show how each portion of a development can be blended into a single harmonius whole using the concepts of low-impact development and the 3Ps. This includes streets, utilities, lot sizes, building placement, on-site open space, and other BMPs. This is of major interest to buyers, renters, and developers. Developers because they want to sell their projects faster; buyers and renters because they want to live and work in areas that are attractive, economic, safe, colorful, and sustainable.

A fourth purpose is to do the above while meeting all current laws, rules, regulations, and ordinances regarding developments in urban and rural areas.

A final purpose is to do all of the above while allowing developers to use 100 percent of their sites for profitable construction. One thing developers hate is to "lose" a portion of their buildable land for other purposes to meet some local requirements. Doing away with a need to build almost all on-site "normal" detention basins is a distinct plus for the ideas and methods presented in this book. "Normal" in the sense of constructing the old land-consuming single-purpose detention basins we constructed in the past.

1.1.4   Sustainability and the Triple Bottom Line

Current societal thinking about new developments, retrofits, and redevelopment centers around sustainability and the TBL of people, planet, and profit.

> Sustainable land development is the art and science of planning, financing, regulating, designing, managing, constructing, and marketing conversion of land to other uses through team-oriented, multi-disciplinary approaches which balance the needs of people, planet, and profit—for today, and future generations. (Sustainable Land Development Today 2008)

The TBL is defined as a pattern of resource use that aims to meet human needs while preserving the environment so these needs can be met not only in the present but also for future generations. Sustainable development ties concerns for the carrying capacity of natural systems with the social challenges facing humanity. The following explanation of the 3Ps was taken from Sustainable Land Development International (2010).

> The "people" component refers to fair and beneficial practices toward meeting the various needs of communities. Social sustainability includes issues related to safety, quality of life, connectedness, and access to social, economic, and cultural opportunities.
>
> The "planet" aspect refers to sustainable environmental practices. These issues are the most commonly discussed component of sustainability. The "greening" of our

society through innovative products and constant mention by politicians and media has brought these concerns to the forefront. In the land development industry, it is focused on reduction of stormwater runoff impacts, improving water quality, and increasing energy efficiency in building construction.

The "profit" component is the bottom line shared by all commerce and includes more than just developer's financial needs. A good neighborhood design often represents a significant community economic benefit. In most cases a developer drives the project, with communities' development officials in tow. But, none of this happens without a project leading to a profitable situation for private developers.

1.2   The Book's Contents

This is a book about what and how to do it. "It" is the layout of developments in which we live, work, and play plus the hydrologic and hydraulic design of stormwater facilities needed to control both runoff quantity and quality from rainfall events. Not the way since the pilgrims came ashore, but using the sustainability concepts of low-impact development and the TBL of people, planet, and profit.

Recent years have seen exciting new ideas about how land should be developed. These ideas do away with cookie-cutter subdivisions, replacing notions of how streets should be laid out, how homes and buildings are placed on lots, and materials used to construct them. Land-consuming detention basins still have a place in certain locations along with fresher BMPs ideas.

I started as a design engineer for a county, then a state agency with an interest in stormwater management facilities of all types and sizes, then as a professor of civil engineering at a state university, at two private consulting firms, and as a decades-long teacher of short courses for another state university and other organizations. I was always asking two questions: Why are you doing it that way? Where did you get that number from?

A first portion of this book gives some background information on certain concepts of the why we are doing it that way. Books have been written on these concepts, but few follow through with details on how a development's components can be designed using these concepts so that when constructed, each blends with the others into an aesthetic, pleasant, safe, economic, and sustainable development. There are also some reminders that designers do not work in a vacuum. They must conform to federal, state, and local laws, rules, regulations, and ordinances. Their designs must also take into account the desires of politicians and the various publics.

The book's next portion is concerned with the several portions of the design process to develop stormwater management facilities for both water quantity and quality control. It answers the second question: Where did you get that number from? Designers must remain in control of their designs—not be at the mercy of people who write computer programs.

For example, rain is the driving force that determines facility sizes and shapes. We need to know its intensity, depth, duration, and/or time distribution of the rain, depending on method used. Where do we get these numbers from? Some information is given in Chap. 6 as to how we measure rain, the gages we use, how we intrepret the data, and how we use these data. But how good is this information? Do we simply accept the data from the old U.S. Weather Bureau and now National Oceanic and Atmospheric Administration (NOAA)? We usually do because they are the best source for rainfall data.

This portion includes methodologies, equations, variables, and their subtleties for each part of the design process. Simply plugging numbers into spreadsheets and computer programs is not good enough. We must remain in charge. The weir equation is $Q = CLH^{3/2}$, where L is weir length. Most books and programs state this. So for a 5-ft weir, we enter 5.0: wrong. L is the length of its vena contracta, the length to which it is reduced as water passes over the weir.

Another subtlety: a BMP's outlet could be a pipe or it could be a conduit plus a riser at its upstream end with various weir and orifice openings in its height to reduce ouflow rates for various storm magnitudes to conform with local ordinances. During passage of a flood, a weir is a weir during a hydrograph's first portion, can become an orifice during times when temporary pond depths are higher, then become a weir again during the hydrograph's recession limb. The depth-outflow relationship must, during each time step in a flood's passage, be correct for the relative heights of water in the temporary pond and inside the riser to use the proper equation and correct value of head in either the weir or orifice equation at that point in time. A poor depth-outflow relationship results in incorrect flow rates through the outlet and incorrect depths of water in a temporary pond whether it is a rain garden or a multiacre detention basin.

The final portion of this book is a series of appendices. Each appendix contains one or more ways in which a parcel of land can be developed for some type of land use that changes blank sheets of paper into a final set of calculations, design plans, and specifications. Most appendices are replete with figures, tables of calculations, and text that describe each step in the design process and present examples of developing correct depth-outflow relationships.

These newer ideas result in less acreage for streets, more room for on-site open space, and more or larger lot sizes. People live in attractive, safer, and sustainable surroundings. It is also true for office buildings, small shopping centers, malls, manufacturing locations, and institutional and governmental buildings. There is little flooding or pollutants moving off-site, resulting in a healthier and safer environment. Developers make a profit since costs are less, fewer resources are used, plus developments are more attractive and economical for buyers and lessees. Governments and utility companies are happy because operation and maintenance costs are less.

One task of agency staff is to approve plans of various consulting firms and others to ensure they meet the EPA's and the FEMA's requirements and ensure these plans result in communities in which we would like to live, work, and play. What we plan, design, approve, and construct today will exist for decades to come. Will our grandchildren bless us or curse us?

In past centuries, stormwater facilities consisted of ditches and creeks, then we added storm sewer systems. With the advent of FEMA and the EPA, we added facilities to control runoff quantity and quality to our repertoire. Today, we are blessed with better ways to accomplish development and stormwater management with less cost and fewer resources that are safer, more livable and attractive, and economical for developers, people, cities, and counties.

1.3 A Better Way to Develop Land

Chapter 3 begins the discussion of developing residential neighborhoods that get us away from myriads of cookie-cutter subdivisions with rectangular street grids and lots that exist, usually with no attention paid to the existing terrain. Developers and civil engineers apparently love to move dirt, rearrange a landscape, remove dirt, move dirt back so owners can grow grass, and spend hundreds of thousands or millions of dollars in the process. While rearranging the landscape, they remove many if not all of the natural swales that used to drain the site.

More millions are spent to construct storm sewers for a 2- to 10-year storm. However, most designers neglect or do not understand that inlets are the most important part of these systems. Small inlets only pick up 6 months of runoff while the other 1.5 or 9.5 years' worth flow down gutters to somewhere. Then a larger storm occurs, and designers find out, to their dismay, that gutter runoff ends up in owners' basements or first floors or the showroom floor of a local car agency. There is a better way to develop land that prevents future flooding problems, solves some current flooding problems, and reduces pollutant loads leaving a site.

Chapter 3 introduces the concept of coving and curvilinear streets. While this subject seems to be far removed from stormwater design, it has everything to do with developing land into an attractive, pleasant whole. Streets, buildings, sidewalks, utilities, and BMPs must be laid out and designed so they are sustainable and incorporate all facets of people, planet, and profit. Drainage is just one aspect of urban infrastructure and must be blended in as an attractive asset to an area.

1.4   Intent of This Book

Intent of this book is to help readers to understand not only the development of neighborhoods and other land uses, incorporating many BMPs for new and retrofitted developments, but also to appreciate and understand the following thoughts:

1. A team of people with various disciplines is needed to successfully design developments.

2. Look for fatal flaws that make your project infeasible as a first task in your investigations. Do not waste time and money on something that will never be constructed.

3. Have many meetings with reviewers and agencies you need permits from to ensure that you make investigations and calculations needed to get project approvals.

4. Learn to ask the question "Why and how am I using this method?" Do this while reflecting on why it is a good method to use, and make sure you understand and can use it correctly.

5. Learn to ask the question "Where did I get this number from, and is it a good one to use here?" for every input you make to a spreadsheet or computer program. Then write down why you used that number so you remember later when explaining your work to your boss or reviewer about why you did what you did. Finally, check the input for accuracy and check the output for reasonableness.

6. And last, do not design anything unless you understand what is to be done, methods and equations used, and have people skills to work with others, so you and other team members can bring a design to life as a pleasing, safe, and long-lasting addition to a community.

1.5   Spreadsheets

Spreadsheets for all components of the steps needed to design and/or retrofit projects using many kinds of BMPs are available for download at www.mhprofessional.com/sdsd. They have been used extensively in many types of situations and have had the bugs worked out of them. However, since we reach perfection asymtotically, there may still be a stray bug here and there.

Output is arranged in tables so you can include them in reports. None are locked— but be careful if you want to make any, other than cosmetic, changes, i.e., any changes other than removal of colors, changes in row height or column width, or heading names. Explanation sheets for each column and line of every spreadsheet are on the website, so, hopefully no instructions should be needed to successfully use them other than the explanations in this book.

1.6   Organization of the Book

The first five chapters provide a setting for what developments and what we call BMPs are about. Chapters 6 through 16 describe in great detail how to determine allowable water depths, develop inflow hydrographs for various components of a development; estimate storage volumes for above, surface, and underground locations; calculate outflow structure hydraulics for numerous types of facilities; and calculate routing curves and hydrograph routing for the above. Appendices A through H explore ways in which these development concepts and inclusion of BMPs into them for various projects can be designed through the use of worked examples. Appendix I is a list of source-control BMPs. Appendix J is a list of Manning's *n* values. References are included in Appendix K.

1.7   Some Final Thoughts

An interesting item about BMP hydrologic and hydraulic design is that the methods, equations, and calculations are all the same. It makes no difference if you are designing rain gardens, streets with porous pavements, greenroofs, parking lots, infiltration trenches, retrofitting an existing multiacre detention basin, or working on Boulder Dam. The calculations are all the same. The only difference is in the size of the numbers. Each can be thought of as a detention basin if we define it as a hydraulic structure that detains runoff, reducing flows, and enhancing its quality.

As you read this book, the design storms used are the 2-, 10-, and 100-year, 24-hour events. I did this deliberately to show what occurs on a site. Drainage facility sizes may seem large to some, but are economical in terms of long-term construction, operation, and maintenance costs, and blend into their surroundings as attractive amenities that advance sustainability concepts.

You could infiltrate every drop of rain onsite from a 100-year storm. There would be no need for a storm sewer system. However, with clayey soils, underground detention with thicker gravel layers is more expensive. Or you could design for a more frequent storm such as a 25- or 50-year event, but in these cases your design must handle the 100-year excess runoff in some manner so no one and no thing gets flooded and water quality concerns are still met.

Infiltrating all rain is not possible in water rights laws states. Users are entitled to certain water volumes from either or both surface and underground sources. If we infiltrate all water, no surface water is available. Our designs must preserve the historic split of surface and groundwater resources. Storm sewers leading to creeks are needed. However, during rarer storm events, our designs must prevent flooding. Also, prior to water in storm sewers reaching a stream, its quality must be improved. Remove pollutants on-site before runoff reaches inlets.

A few last thoughts:

1. Recognize that developers want to use 100 percent of their sites for profitable construction.

2. Local, state, and federal rules and regulations must be followed.

3. People want to live, work, and play in attractive, safe, affordable, and sustainable surroundings.

4. Developers, cities, counties, and utility companies want to spend the least amount possible to construct, operate, and maintain developments.

CHAPTER 2

Developing Low-Impact Developments

2.1   Low-Impact Development

The Environmental Protection Agency (EPA) defines low impact development (LID) as a development approach (or retrofit or redevelopment) that works with nature to manage stormwater as close to its source as possible. LID employs principles such as preserving and recreating natural landscape features, minimizing effective imperviousness, and creating functional, appealing site drainage that treats stormwater as a resource rather than a waste product. Using LID, water is managed to reduce built-area impacts and promote natural water movement within an ecosystem or watershed.

2.2   America Is Becoming Green

Citizens, officials, and politicians at all levels of government have joined the environmental movement. We now are beginning to have green streets, green buildings, greenroofs, the greening of America, green space, and green, as opposed to gray, infrastructure.

The two main differences between gray and green infrastructure are

1. Green infrastructure controls runoff and pollutants at their source (the point where rain falls) through the use of various best management practices (BMPs).

2. Gray infrastructure centralizes storage in one location (large on-site or regional detention basins). Green infrastructure distributes them in small areas throughout a site (biting off and controlling rainfall and pollutants in small chunks). Designing green developments and stormwater facilities this way is the key to sustainable, aesthetically attractive, livable, and lower short- and long-term development, operation, and maintenance costs.

Gray infrastructure includes the following to manage stormwater runoff: alleys, bridges, catch-basin inlets, creeks, culverts, curbs, ditches, driveways, detention basins, gutters, lakes, parking lots, rivers, sidewalks, storm sewers, streams, streets, and swales.

Green infrastructure includes the following to manage stormwater runoff: buffer strips and zones, catch-basin inserts, cisterns, adding or eliminating curbs and gutters, detention basins, dry wells, filter strips, gardens, greenroofs, infiltration basins and trenches, lawns, level spreaders, permeable pavers, porous asphalt and concrete pavements, rain barrels, rain gardens, retention basins, soil amendments, trees, tree boxes, underground basins, and vegetated bioswales.

The EPA sent a memo to personnel in October 2011 entitled "Achieving Water Quality Through Integrated Municipal Stormwater and Wastewater Plans." One part stated, "As you know, given the multiple benefits given with green infrastructure, EPA strongly encourages the use of green infrastructure and related innovative technologies, approaches, and practices to manage stormwater as a resource, reduce sewer overflows, enhance environmental quality, and achieve other economic and community benefits." Thus, the EPA encourages use of green infrastructure, and by inference, the concepts of sustainability, LID, and the triple bottom line (TBL).

LID incorporates five ideas:

1. Keep rainfall as close as possible to where it falls.

2. Infiltrate as much rain as possible.

3. Bite off rain in small chunks so BMPs on lots and developments are small and numerous.

4. After development, keep the site as close as possible to its natural hydrologic condition.

5. Use BMPs to keep pollutants on a site as much as possible.

The following five concepts can be used to implement LID's ideas as we develop, retrofit, or redevelop land for various uses to achieve sustainability and the TBL.

1. Maximize a development's attractiveness.

2. Minimize peak runoff rates and volumes leaving a site.

3. Maximize reduction of pollutants leaving a site.

4. Minimize construction, operation, and maintenance costs.

5. Use BMPs as an opportunity to educate local citizens of their environmental benefits

The concepts are not independent of each other. They are interdependent with each adding to the others so the total is not only equal to the sum of its parts but is greater than the whole, with each blending into the others making developments desirable locations in which to live or work. Their planning and design must comply with laws, rules, regulations, and ordinances.

These five concepts are elaborated on in the following sections.

2.3   Maximize Development's Attractiveness

Some of the ways to maximize a development's attractiveness are listed below:

1. Eliminate a grid pattern of streets. Instead, use curvilinear streets with uniform variation of front-yard setbacks. Reduce number of intersections and potential accidents.

2. Use a variety of home models and exteriors with each type identified on the plat.

3. Use a variety of trees and shrubs in both front and back yards.

4. Use rain gardens to add additional color to the development.

5. Use a series of pathways that augment wide sidewalks to serve their purpose of allowing people to mingle and wander throughout a development.

6. Incorporate stormwater management into a neighborhood's planning and design so residents and visitors are not unduly inconvenienced during large, rarer rainfall events. Blend these facilities in as normal, unobtrusive, and attractive components so residents and visitors may be unaware of their role in stormwater-runoff quantity and quality control.

7. Incorporate usable open space that allows residents/visitors to socialize with each other.

8. Maintain and enhance any of the site's natural features.

9. Place all utilities underground.

2.4 Minimize Runoff Rates and Volumes

Some of the methods to minimize runoff rates and volumes are described below. Other BMPs used to accomplish this are described in Chapter 5.

1. Retain rainfall on roofs by constructing greenroofs where possible.

2. Convey remaining roof runoff through downspouts into one or more of the following: infiltration trenches, lawns, grassed swales, and/or rain gardens.

3. Construct patios, driveways, sidewalks, walkways, and streets of porous concrete or permeable pavers in new developments. Use porous concrete in retrofitting or redevelopment projects when these need to be replaced. Convey remaining runoff onto grass and into rain gardens.

4. Convey lawn runoff to vegetated bioswales. Swales should have mild slopes (or intermittent berms if mild slopes are not possible) to slow runoff and allow it to infiltrate.

5. If soils are clayey, use soil amendments and gravel in and under BMPs. Small underground perforated pipes could be placed in the gravel to convey runoff.

6. With increased experience with porous concrete and asphalt pavements, cities and counties are using it for streets. Underlaying them with a layer of gravel will infiltrate all rain from a 100-year, 24-hour storm, regardless of soil type. This stores runoff until it infiltrates into the native soil. Reports and conference papers report success in their use. Future reports will verify their success and/ or indicate design and construction improvements needed.

2.5 Maximize Reduction of Pollutants

The previously discussed BMPs to minimize runoff rates and volumes also serve to trap pollutants onsite.

1. People use fertilizers, herbicides, and insecticides to nourish lawns and gardens. Reducing chemicals to quantities needed and using them correctly reduces amounts available.

2. Slowing runoff velocities and then temporarily storing it allows more infiltration to take place and more pollutants to be trapped. Use of infiltration trenches, lawns, porous and permeable pavements, swales, rain gardens, and other BMPs will all do this.

3. If a storm-sewer system is included in the design, then catch-basin inserts can be used to trap sediment, grease, oil, heavy metals, and urban litter.

4. In commercial and industrial areas, numerous other source and treatment control BMPs can be used to reduce pollutants leaving a site.

5. During public works projects, such as constructing a new water line, the excavated soil stockpiled adjacent to the trench should be covered to prevent the excavated material from washing into a storm sewer during rainfall events.

2.6   Minimize Construction, Operation, and Maintenance Costs

Developers, cities, counties, and utility companies want to minimize their construction, operation, and maintenance costs. These can be done in the following ways.

1. Minimize total length of streets. In addition to reducing the cost of the street itself, cost of sidewalks and all utilities also will be reduced by minimizing their lengths.

2. Resist the temptation to rearrange existing landscapes. Keep grading to a minimum to reduce its cost. Leave much of the site alone. Moving, stockpiling, and moving topsoil again onto lots is minimized. This minimizes soil compaction and gives a better chance to grow lawns.

3. Use methods that keep rainfall as close as possible to where it falls. Using on-site BMPs, infiltrate as much rain as possible so runoff is minimized, thus reducing runoff rates and volumes, and allowing use of smaller drainage-system sizes. These range from swales to storm-sewer systems to detention basins of every type either required or recommended by local ordinances.

4. Using/applying these methods could eliminate any need for land-consuming detention basins in some developments, freeing up land for more profitable construction and open space.

5. Infiltrating rainfall will keep many pollutants on-site by trapping them in various BMPs.

6. Provision of open space for passive and active recreation provides more opportunities for on-site reduction of peak-runoff rates and volumes and enhancing runoff water quality.

2.7   Educational Opportunities

While some BMPs can be unobstrusive and unnoticed by citizens, such as porous pavements and underground detention, others are highly visible to the public, such as rain gardens and greenroofs on gazebos and shelters, and other park buildings. Greenroofs are being used more and more on commercial, industrial, and municipal buildings due to their reductions in rainfall runoff and reductions in energy use for heating and cooling needs.

BMPs, such as rain barrels, infiltration trenches and basins, vegetated swales, and tree boxes can be also used as educational tools for their improvements in reduction of runoff peaks, volumes, and pollutants. Several signs in many lands uses with diagrams, pictures, and words serve to educate the public as to why these BMPs are being used and their benefits.

2.8   Realistic Approach

Certain items should be investigated before a parcel of land is designed and developed. Are there any fatal flaws that would prohibit its development? Check to see if there are any financial or legal encumbrances of which developers and designers should be aware. These could include existing mortgages or fees owed, as well as development fees imposed by a local jurisdiction for plan checking and inspection. Some communities impose a per-acre cost for stormwater improvements in a watershed for regional basins and "improvements" to creeks. Are there existing or proposed easements that will impact where streets and buildings can be located? These could include utility companies, such as water, sewer, gas, and power. Space and access to utility corridors could be made part of open-space areas. Are there future rights-of-way (ROWs) for major streets and/or highways? These all impact street layouts and connections to them.

2.8.1   Evolution of a Development

We start out with a blank sheet of paper. First, draw the site boundaries. Make sure they are accurate. Second, draw lines showing street ROWs bordering one or more sides or within the site. Third, draw lines for alignments and ROWs of all minor or major swales or creeks originating within and outside it. Fourth, draw lines representing existing easements bordering or within the site boundaries. Fifth, on a separate piece of paper show the parcel's topography to the same scale.

Overlay these two sheets so we can view how topo influences our choice of how and where we place internal curvilinear streets. What does the site's topo look like: flat or rolling or hilly? Undeveloped land is drained naturally by existing slopes, swales, and creeks. We do not need to move one shovel of dirt to have it drain. Do any drainage ways qualify as wetlands? Many jurisdictions require certain widths of buffer zones surrounding wetlands. If some drainageways are creeks or streams rather than swales, then some jurisdictions require that the 100-year floodplains plus be dedicated to or an easement be given to the local jurisdiction for their operation and maintenance. Are there cultural or historic or scenic sites within the parcel of land that should be preserved and made part of open space within the site?

Once we plan to develop a site, we need to develop streets to access every portion of it. Lines are drawn so property owners can identify exactly what piece of land he/she owns. We can do this by spending a lot of money, or we can spend a lot less. We can end up with a bland, ugly development or end up with an attractive, pleasant, and safe place in which to live, work, and/or play. The choice of one or the other is the result of decisions we make along the way.

2.8.2   Some Suggestions

So how should we move forward to achieve sustainable developments and the TBL? The following are some suggestions.

1. Leave existing swales, creeks, and streams alone. Place them in their own ROWs with widths such that they also serve as corridors for people-pathways to meander through the site.

2. Drain lots to existing waterways. In some portions of a site, this could be done without moving any dirt. Other portions may need some regrading. Not moving dirt saves money for the profit portion. Using existing waterways is a gain for the planet and people components.

3. Use the concept of biting rain off in small chunks using various on-site BMPs to handle both runoff quantity and quality. Doing this with other BMPs could reduce the size of a storm-sewer system. This is a factor in the profit component— less design and construction costs and less need for long-term system operation and maintenance costs. Natural drainage adds color and variability to a site, a plus for the people and planet side of sustainability.

4. Using curvilinear streets can reduce their total lengths by 15 to 50 percent. This also reduces lengths of all utilities and private systems, such as the internet and cable TV. The reduced ROW needs can be translated into more and/or larger average-sized lots and/or more land for open-space uses. These are pluses for all aspects of sustainability.

5. Streets and sidewalks/pathways should have two different and independent alignments. Both are curvilinear and both provide access to all parts of a site. Just as streets have the purpose of leading people to places people want to go both within and away from a development, sidewalks/paths also lead people to places they want to walk or bike to within a development. Narrow sidewalks on both sides in cookie-cutter, grid-type subdivisions lead to nowhere. They are not well-used because where people want to go are not within walking distance.

6. Use whatever BMPs the team decides to use on the site. These should meet both present, and probably future, federal, state, and local requirements for reductions in peak-flow rates and volumes and enhancements in runoff water quality.

7. The only time I recommend using a land-consuming detention basin, with a pond plus extended detention for a frequent water-quality storm event plus water plants is if the existing terrain is sufficiently rolling so existing storage volumes needed to reduce the 100-year, 24-hour storm runoff peak by 50 to 95 percent are available without moving one shovel of dirt.

8. Six inches of soil on greenroofs retain two inches of rainfall. They add to a new building's cost, but this is offset by individual and community-wide benefits. Greenroofs can be colorful, add to an area's birds, insects, and butterflies, reduce heating and cooling costs by 20 to 30 percent, and eliminate a drab view of shingles and metal roofs. They add oxygen and reduce an area's heat island effect by reducing roof temperatures from 175° F to less than 100° F. The use of greenroofs also reduces the sizes of other BMPs needed. They add to all aspects of the TBL.

2.9 Balanced-Layered Approach to Sustainability

Layers involved in a balanced development are safety, environmental, economic, aesthetic, architectural, landscape, and the realistic approach described above. Use them to create developments that include several integrated layers including safety;

spaciousness; quality of life; connectedness; lower-cost BMPs that reduce flow rates, volumes, and runoff pollutants; other sustainable environmental practices; economic homes or buildings to buy; profits to developers; and economic benefits to communities. Elements of a sustainable development include its local streets, sidewalks, lot sizes, front- (coving) and side-yard setbacks, driveways, homes or buildings, gathering places (open space), architecture, rain gardens, lawns, shrubs, and trees.

2.9.1 Local Streets

Streets within residential areas should never be laid out in a grid pattern. The following details are from Harrison (2008) and are discussed in detail in Chap. 3.

1. Streets should be curvilinear with no intervening tangents. Limit speeds to 25 ± 5 mph. Do this by keeping curve radii from 200 to 300 ft. With higher speeds and smaller radii, drivers are uncomforable because of centrifical forces. Curves exceeding 180° should be common.

2. Street widths should be 28 ft wide and in cul-de-sacs 26 ft wide within 50-ft ROWs. On single-lane streets, widths should be reduced to 20 ft.

3. Street layout should be rational, allowing drivers/pedestrians to maintain their sense of direction so they do not get lost. Developments for them both should be about connectivity and flow. Streets and sidewalks lead to locations where people want to go. Cul-de-sacs are okay as long as emergency vehicles and pedestrian connectivity are not restricted.

4. Flow is defined as the ability to enter and safely traverse a development with a minimum number of stops and turns while experiencing a sense of space and place along the way.

5. Curvilinear streets reduce the number of intersections. This increases safety because there are fewer opportunities for unwanted interactions between drivers, pedestrians, and cyclists to occur, i.e., accidents, deaths, and personal property damage.

6. An important consideration for developers, cities, and counties is curvilinear streets reduce street lengths needed to access lots by 15 to 50 percent.

7. Consider the following regarding cul-de-sacs:

   a. Cul-de-sac radii are set by a fire truck's turning radius.
   b. Make this radius much larger than required.
   c. Use narrower one-way roadways around cul-de-sac ends.
   d. Only pave the roadway. This eliminates some impervious areas.
   e. Develop their inner portions as open space or landscaped gathering places with BMPs. Plant them with grass, shrubs, and trees plus benches and lighting to encourage use.

2.9.2 Sidewalks

Sidewalks should have the following features:

1. Sidewalks and pathways should safely lead users to all portions of a development.

2. Alignments of sidewalks should be independent of street alignments.

3. Place sidewalks on one side of streets. They and pathways lead people to places they want to go, i.e., shopping, picnic tables, passive and active recreational areas, and peaceful arbors.

4. Widths should vary from 5 to 8 ft wide depending on the volume of traffic they carry.

5. Sidewalk alignments should be meandering with large radii on reverse curvatures.

6. Pathways at a cul-de-sac's end and other pathways should continue to all portions of a site so people have access. Use 20-ft wide ROWs for paths, grass, shrubs, trees, benches, and lighting.

2.9.3   Lot Sizes

Local jurisdictions have zoning ordinances that dictate minimum size of lots in certain areas. Neighborhood lots with curving streets should have *average* lot sizes greater than the minimum required. This allows local officials to agree to both smaller and larger lots to follow the dictates of the existing terrain and proposed street layout.

2.9.4   Front-Yard Setbacks (Coving)

From Harrison (2008, page 88) (Rick Harrison, Rick Harrison Site Design, Published by SLDI): "By definition, a cove is an indent in the shoreline of a body of water. In land planning, a cove is defined as the breaking of the parallel relationship of house and curb, replacing the standard home setback pattern with one that sets each home along a fluid curving line, unique from the street pattern. This forms an indent into the building setback—a cove." Chapter 3 is devoted to this concept of coving and the use of curvilinear streets.

2.9.5   Side-Yard Setbacks

Side-yard setbacks should be 5 to 10 ft depending on lot width. However, homes are not parallel to side-lot lines so space between homes appears larger. Also with coving, homes are staggered and at various angles so no one looks into a neighbor's windows. This enhances a feeling of privacy.

2.9.6   Driveways

Today's homes typically have two- or three-car garages accessible from the street. The space is used for cars, storage, or workspace. This means that driveways will be 20- to 30-ft wide depending on a garage's entrance orientation. To reduce the amount of paving needed, the longer driveways should vary from 10-ft wide at the curb to their maximum widths one car length from the garage. This allows owners' and visitors' cars to park in the driveway and not along the curbs. Cars in driveways should not interfere with sidewalks.

2.9.7   Gathering Places: Open Space

Because of reduced acreage needed for street ROWs, some of this area can be used for more or larger lots. The remainder should be utilized for usable open space, such as walkways, park-like areas, small ponds for detention if needed, and recreational uses. Depending on a neighborhood's size, these recreational areas could include tennis courts, basketball hoops, horseshoe pits, picnic tables, gazebos, and shelters. These open spaces should be scattered throughout the neighborhood and accessible by streets and walkways to all residents and visitors.

2.9.8   Architecture

The curvilinear street pattern, when done correctly, ensure that side and rear yards are not visible from a street. Only the homes' fronts are seen. Thus, the first impression of prospective buyers, residents, and visitors is one of spaciousness, landscaping, and front views of homes. Homes are positioned with careful consideration of the view; they rarely face another home front directly. As one travels along a street, there is an ever-changing vista to behold. The blandness of cookie-cutter subdivisions along straight streets is gone and replaced by an attractive, ever-changing view of various home frontages and landscaping.

Exteriors of homes should be varied. Developers have a number of models to build and a variety of exteriors for them. Some have front porches, and some do not.

The city of Parker, Colorado (Stoll and Rossmiller, 2003), has passed an ordinance requiring the number of models and their variety of exteriors to be clearly identified on a plat. Each lot has its model and exterior identified. The variety of exteriors and models must be selected so their number and location is not repeated too often along a single street. With effective landscaping, this approach provides an attractive and ever-changing view of homes along a curving street.

Home orientation can reflect views, sun exposure in various seasons, and prevailing winds. Home interiors and room arrangements take advantage of views, sun, and newer construction materials. "Green buildings" take advantage of them in building costs and energy efficiency. Extensive and intensive greenroofs reduce annual heating and cooling costs by 20 to 30 percent.

2.9.9   Rain Gardens

One popular BMP that is useful throughout the United States are rain gardens. They are small detention basins using amended soil, plants, flowers, and shrubs. They are 6-in to 1-ft deep and vary in size from small to 2,500 sq ft and can be any shape. A rain garden is an attractive, functional landscaped area that mimics the hydrologic function of a forest. It is designed to capture and filter stormwater from roofs, driveways, sidewalks, lawns, parking lots, and roadways. It collects runoff and allows it to soak into the ground (Rain Garden Network, 2008a).

It is a garden designed to withstand extremes of moisture and pollutant concentrations, such as nitrogen and phosphorus. As in other bioretention cells, soils are amended with a very porous planting media (Urban Design Tools, 2011a), minimally to a depth of 6 in and ideally to depths of 2 to 4 ft. Soil is excavated and material imported. Media should be clean and weed seed free. They are sized from 3 to 30 percent of their drainage areas.

2.9.10   Lawns, Shrubs, and Trees

There are numerous species of flowers, grass, plants, shrubs, and trees used in urban areas of the United States. Many cities/counties have developed lists for species that will do well in their areas. Prairie grasses in some areas are preferable to Kentucky blue grass because of increased infiltration capabilities. Vegetable and/or flower gardens increase water use and add color and beauty to an area. Larger shrubs and trees (UW-Extension, 2008b) provide shade and habitat for birds and some animals. Species of trees can be used along streets with or without tree boxes. They can also be used among lawns in all types of land use to provide more variety and color to an area.

On many lots, short natural or manmade stone walls can add additional variety and interest to homes' frontal areas by creating small differences in elevation between grass and shrubbery.

2.9.11   Other BMPs

Many other BMPs can be included in single-family neighborhoods and other land uses. These include buffer strips and zones along adjacent creeks and streams, cisterns and rain barrels, eliminating curbs and gutters and using swales instead, greenroofs, infiltration trenches, porous asphalt and concrete pavements, permeable pavements, rain barrels, tree boxes, and underground detention basins.

2.10   Summary

This chapter has briefly discussed the concepts and components of LID and the TBL that should be incorporated into various types of development projects so they become locations in which people want to live, work, and play as well as becoming community assets. In addition, the chapter included brief discussions of the elements that are a part of every development. It is up to the design team to blend these elements into a pleasing whole that is safe, attractive, economic, and sustainable.

Also, the team needs to blend a series of on-site BMPs into each lot and project in such a manner that they are unobtrusive and blend in as attractive, natural features of each lot and project. These BMPs are required by EPA to achieve their stormwater quality regulations and as required by the Federal Emergency Management Administration (FEMA) to achieve their stormwater quantity regulations. These BMPs reduce the rate and volume of stormwater runoff and pollutants exiting a site.

The next three chapters begin the story of how the above objectives can be achieved.

CHAPTER 3

Coving and Curvilinear Streets

3.1 Introduction

This chapter is a departure from the title of this book, but it is an important chapter in implementing all portions of the triple bottom line (TBL) in single- and multi-family developments. Developers can use 100 percent of their sites and save large sums of money in construction costs. Homeowners and renters will be safer and happier in these neighborhoods for reasons explored in this chapter.

For the last few centuries, almost all residential development in the United States consisted of straight streets at right angles forming rectangular blocks with rectangular lots. This street and lot pattern was relieved when cul-de-sacs were used. Subdivision and building codes resulted in minimum front- and side-yard setbacks with homes and buildings set parallel to each other with the same setback. Little or some attention was paid to how streets, homes, and buildings fit existing terrains. Only in few areas were curved streets used. That view has been changing and hastened by the introduction of the TBL.

3.2 A Better Approach

Another way to design residential areas was introduced by Harrison (2008) Rick Harrison, Rick Harrison Site Design, Published by SLDI). He uses coved front yards, curvilinear streets with no tangents, wide sidewalk alignments separate from streets, homes at angles to others, creating a feeling of openness and estate-size lots. He showed how these areas could be attractive, safer, and economic places to live.

Sidewalks/pathways lead throughout neighborhoods. There is a reason for sidewalks to exist, rather than constructing them on both sides of every street. In the myriad subdivisions we have constructed in the past, there was little reason to use sidewalks. Why? Because we built cities that relied on horses, cars, buses, and trucks to buy groceries, get a haircut, shop for clothes, go to work, go to school, and to worship.

3.3 Coving (Front-Yard Setbacks)

The following thoughts and ideas paraphrase the attributes of coving (curvilinear streets and variable front yard setbacks) that change the characteristics and feelings about residential developments from subdivisions into neighborhoods. The results are

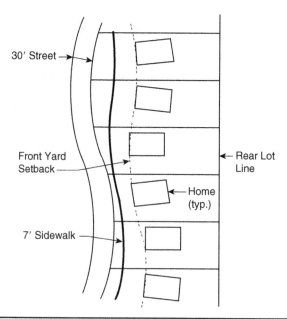

Figure 3.1 Coved front yard setback and sidewalk.

neighborhoods that are safer, more attractive, more affordable, with a greater feeling of privacy that result in locations that are much more pleasant in which to live and play. This concept was developed and then presented in a book by Harrison (2008) (Rick Harrison, Rick Harrison Site Design, Published by SLDI).

Coving is a varying of front yard setbacks, with its spacious meandering of openness, that gives a feeling of larger estate-sized lots. Setbacks vary from minimum to maximum in a rhythmic fashion. Coving allows sidewalks separate from streets and increases front yard sizes. Subdivision ordinances need no revision because they refer to minimum front yard setbacks. Front yard coving and winding streets gives a spacious feeling to neighborhoods as residents drive or walk along. (See Fig. 3.1.) Other than pushing some houses further back on their lots than they might be sited, there is no structural difference between a coved plat and a conventional one. But there is a world of difference in the finished product. Coving puts a street scape in motion. Instead of homes and walks paralleling streets, they meander to form unique shapes independent of each other. When done correctly, coving reduces street lengths up to 50 percent. Less street rights-of-way (ROW) means more land for lots and parks.

3.4 Benefits of Coving and Curvilinear Streets

The coving concept was developed by Rick Harrison, a Minneapolis-based urban designer in the late 1980s. He listed numerous advantages to curving neighborhood streets and coving front yards (Harrison, 2008) (Rick Harrison, Rick Harrison Site Design, Published by SLDI) such as the following:

1. Land is freed up to be devoted to larger lots and communal open-space uses.

2. The dollar cost of streets, curbs and gutters, and all utilities is reduced.

3. By setting homes at angles to a side-yard lot line, distance between homes appear to be and will be larger, while side-yard setbacks conform to ordinances.

4. By varying locations of homes on lots, owners can take better advantage of views.

5. By varying the front setback, no one looks directly into a neighbor's windows.

6. By varying the front setback, a much more open feel is given to the neighborhood.

7. Curved streets have a more pleasant and aesthetic appearance to drivers and pedestrians.

8. The open feeling allows for wider, meandering sidewalks.

9. Curving streets slow vehicles down and have a traffic-calming effect.

10. Curving streets allow better conformity and preservation of the natural topography.

11. Curved streets are adaptable to preserving features such as rocks, trees, and land forms.

12. Individual properties gain aesthetic value from the meandering setback lines.

13. Fewer internal intersections within a neighborhood promote safety and fewer accidents. Street patterns avoid four-way intersections, making the neighborhood safer.

14. Lower-development costs allow more affordable housing than traditional subdivisions and brings the price within the range of more prospective buyers.

15. Varying the front setbacks enhances individuality among home sites and adds privacy.

16. Coving a street layout creates an open environment for its future residents.

17. Coving allows walkways to meander instead of being parallel to streets. See Fig. 3.1.

18. This new design has curved lines meandering throughout and gives a feel of openness.

19. Curving streets and lot layouts allows preservation and use of natural swales for drainage.

20. Lesser roadway lengths and fewer intersections decreases the number of stopping and starting operations, reducing the amount of zinc on streets from tires and brake linings, releasing less exhaust into the atmosphere, and using less gasoline.

21. Shorter street and utility lengths have cost savings that are hundreds of thousands or millions of dollars depending on development size. Fewer resources are needed for construction. Long-term operation/maintenance costs by jurisdictions and utility companies are also reduced.

22. Placing utilities underground has two effects. Initial installation cost of power and telephone service is increased, but repair of lines is not needed because of wind, rain, and ice storms. Customers do not suffer loss of heat and spoiled food. Repair personnel do not spend time in hazardous weather conditions. Eliminating poles and lines eliminates sight pollution.

23. More land devoted to pervious areas allows more infiltration, reduced runoff, and enhanced water quality.

24. An important consideration is money saved by fewer resources used for accidents. There are fewer lost hours at work, fewer medical bills, fewer repair bills, fewer repairs to cars and bicycles, fewer police and towtrucks needed.

3.5   Traditional Subdivision

Figure 3.2 depicts a 186.3-ac traditional subdivision with straight streets and rectangular lots. Streets exist at top, bottom, and both sides. The 24,150 ft of street are 30-ft wide in 60-ft ROWs that use 33.3 ac. On the west side, 6.5 ac are devoted to open space and recreational activities. The other 146.5 ac are divided into 499 lots, 80 by 160 ft. Four-ft wide sidewalks are located in street ROWs on both sides. Straight streets encourage higher speeds and 18 intersections increase accident potential between vehicles, cyclists, and pedestrians. Lots slope from rear to front. A storm-sewer system is designed for a 5-year storm event. When a 100-year storm occurs, runoff is assumed to be street flow to somewhere. No provisions are made for runoff quality.

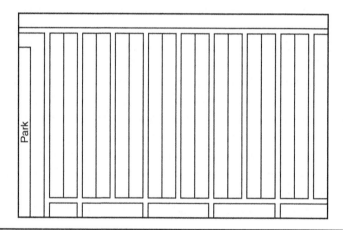

FIGURE 3.2   Single-family traditional subdivision.

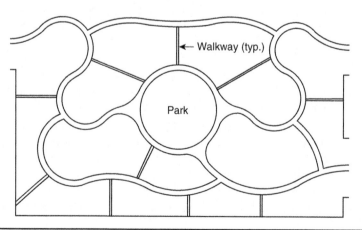

FIGURE 3.3   Single-family residential neighborhood.

3.6   Newer Neighborhood Approach

Figure 3.3 shows the newer approach. The change is lot widths and depths vary but still average 12,800 sq ft. Total street length is reduced to 15,000 ft, a 38 percent reduction. Streets occupy 20.7 ac instead of 33.3 ac. Open space and park occupy 16.2 ac instead of 6.5 ac. The other 149.4 ac contain 509, not 499 homes. Intersections are reduced from 18 to 11.

Reduced street length of 9,150 ft and reduced utility lengths represents a reduction of millions of dollars in construction costs plus more millions in long-term operation and maintenance costs. Using streetside vegetated swales rather than curbs and gutters saves an additional $10.00 \times (24,150 - 15,000) \times 2$ or \$183,000. Use of porous concrete and rain gardens reduce storm-sewer sizes or make a 5-year system adequate for a rarer storm—if inlets are long enough.

3.7   Newer Neighborhood Concept Applications

Three examples give an indication of how a different approach to street arrangement rather than a typical grid arrangement assists in attaining the objectives stated in Chaps. 1 and 2. Using this approach creates attractive and pleasant neighborhoods to live in, reduces development costs, reduces peak-flow rates and volumes, and reduces pollutants leaving a site.

Each example is accompanied by three figures. A site layout shows existing streets, swales, and ridge lines. A second figure is the site developed with a traditional street pattern, rectangular lots, and no open space. Topography is not a concern in street lay-out, and existing swales are obliterated by grading. Houses have the same front-yard setbacks and are parallel to each other.

The third figure assigned to each is the site using curvilinear streets. Some usable open space is created. Existing topography and drainage swales are incorporated into a neighborhood, reducing grading needed. Street lengths are reduced with commensurate savings in pavement, curbs and gutters, and utility costs. Homes are coved with uniformly varying front-yard setbacks. This coving along with curvilinear streets yields a more open and spacious feeling to these neighborhoods. Each curve brings a new vista into view. Homes are not parallel to each other, adding to a feeling of privacy.

Sidewalks are not parallel to streets; rather their alignments are independent of street alignments, and walkways give pedestrian access to all parts of these neighborhoods. Curvilinear streets act as traffic-calming devices, and a reduced number of intersections adds to a neighborhood's safety. Curvilinear streets reduce traffic speeds, and fewer intersections reduce stopping and starting needs, thus conserving resources and creating fewer pollutants.

While the three examples are all single-family developments, these ideas can be applied to other land-use types. Designs for these other land uses are explored in the appendices.

3.8   Altoona Heights

3.8.1   Traditional Grid Streets

This rectangular site contains 80 acres with existing streets, swales, and ridge lines as shown in Fig. 3.4. Figure 3.5 depicts a traditional subdivision. Each lot is 75 by 150 ft (11,250 sq ft), similar to existing lots in Altoona. Street ROWs are 60 ft. The site is on a crest, and slopes are generally east at 1.5 percent. Beginnings of small swales start but are eliminated during grading operations. A park is placed in the northwest corner and is the only open space.

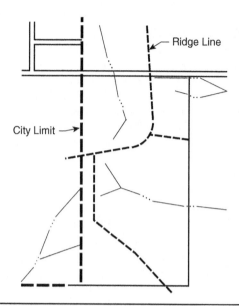

FIGURE 3.4 Existing streets, ridge lines, and swales.

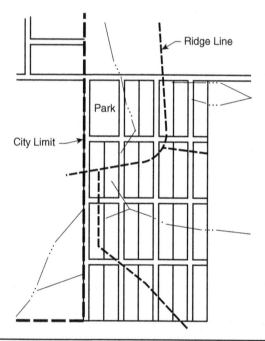

FIGURE 3.5 Altoona Heights—traditional subdivision.

The 223 lots occupy 57.6 ac. Street length is 11,800 ft within 16.2 ac. The park has 6.2 ac. Four-ft sidewalks exist on both sides of streets. Few lots are near the park so its use may be reduced. Swales are only starting to form, so their loss is not great. Storm-sewers outlet into swales at the sites's sides. Streets allow higher speeds, and 15 intersections increase accidents.

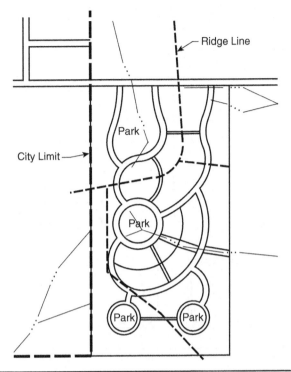

FIGURE 3.6   Altoona heights—curvilinear streets.

3.8.2   Curvilinear Streets

Streets total 8,700 ft, occupying 12.0 ac in 60-ft ROWs, and lead to four parks; two parks in the southern end are cul-de-sac interiors totaling 2.1 ac, and two parks accompanying existing swale areas occupy 7.8 ac for a total of 9.9 ac of open space. Curvilinear streets shown in Fig. 3.6 allow 229 lots to be formed on the remaining 58.1 acres with an average area of 11,050 sq ft, six more than the traditional subdivision and about 1.5 times the open space.

Sidewalks on one side plus pathways lead residents to open spaces scattered thoughout. They allow residents and visitors to engage in social interactions and recreational activities. Existing swales are preserved in place. Street length is reduced from 11,800 ft to 8,700 ft or 26 percent. This similarly reduces costs of pavement, curbs and gutters, and all utilities. Curvilinear street alignments are traffic calming and intersection reduction from 15 to 10 reduces accidents.

3.9   Walnut Creek Highlands

3.9.1   Traditional Grid Streets

This rectangular, 68-ac site is adjacent to Walnut Creek and Greenbelt Park as shown in Fig. 3.7 with its existing streets, swales, and ridge lines. A few existing homes along the south side bordering an existing street are removed. A ridge line along the southwest side has land sloping to the east and south. Swales that begin to form are eliminated by grading operations.

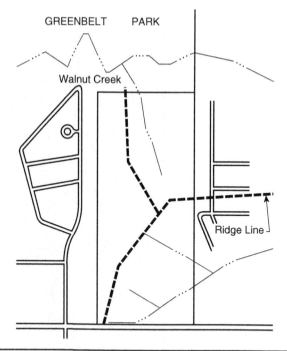

FIGURE 3.7 Walnut Creek—existing streets, ridge lines, and swales.

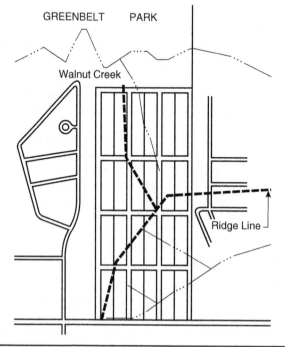

FIGURE 3.8 Walnut Creek—traditional subdivision.

The 167 lots are similar to others, 80 by 160 ft, and occupy 49.0 ac. See Fig. 3.8. Street ROWs are 60 ft wide, length of 13,800 ft, using 19.0 ac with 4-ft sidewalks on both sides. Streets allow higher speeds and 20 intersections encourage accidents. Storm sewers handle frequent event runoff. No provision is made for 100-year run-off other than water runs down streets to somewhere. There is no open space and no good way to visit Greenbelt Park.

3.9.2   Curvilinear Streets

The 178 lots with an average size of 12,800 sq ft cover 52.3 ac as shown in Fig. 3.9, 11 more lots than the traditional subdivision. Street length is 6,100 ft or 8.4 ac. Open space within the neighborhood is 7.3 ac, preserving existing swales and providing play activities in a park, as well as pathways and sidewalks to Greenbelt Park along Walnut Creek.

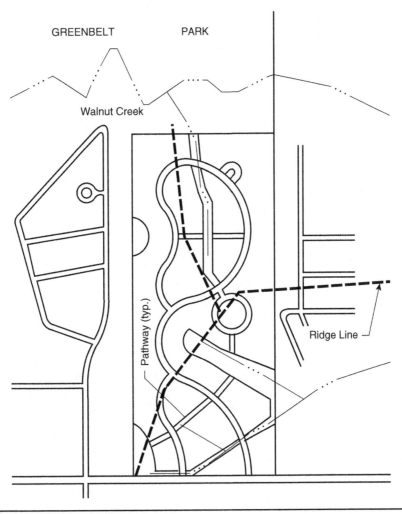

Figure 3.9   Walnut Creek—curvilinear streets.

Streets totaling 6,100 ft is a reduction of 56 percent from the traditional grid, resulting in huge savings for street, curb and gutter, and utility costs in construction, operation, and maintenance costs. Streets are a traffic-calming device with just 7 intersections instead of 20, further reducing chances for accidents. Slower speeds and reduced stopping and starting at fewer intersections results in fewer pollutants being generated. This pattern and coving of homes lends a more spacious feeling to the area, making it a more livable, attractive, and affordable place to live.

3.10 Santiago Creek

3.10.1 Traditional Grid Streets

This somewhat-rectangular site contains 67.4 ac located on a drainage divide between three watersheds as shown in Fig. 3.10. Eight on-site swales begin and flow in all four directions. Santiago Creek exists to the south. Land slopes are about 2.5 percent.

Figure 3.11 has 8,500 ft of streets in 60-ft ROWs using 11.7 ac and sidewalks. The other 55.7 ac contain 190 lots of 12,800 sq ft. Utilities are contained within streets and easements. The 16 intersections and street alignments encourage higher speeds, increasing

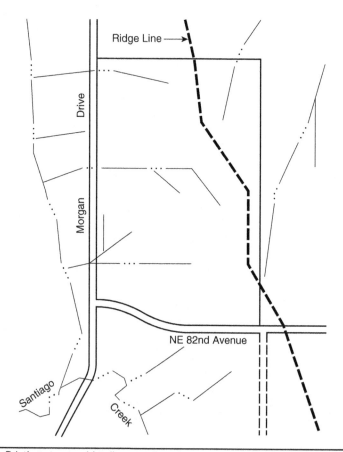

FIGURE 3.10 Existing streets, ridge lines, and swales.

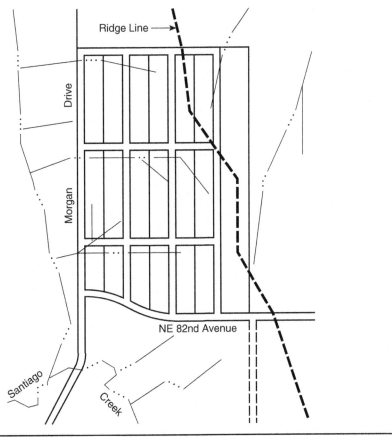

FIGURE 3.11   Santiago Creek—traditional subdivision.

accident potential. A storm-sewer system conveys frequent runoff events. Runoff from a 100-year event is conveyed by it and streets. There is no on-site open space.

3.10.2   Curvilinear Streets

Figure 3.12 has a street length of 6,900 ft and includes 9.5 ac in 60-ft ROWs. There are 3.0 ac of open space in three areas. The remaining 54.9 ac are occupied by 190 lots with an average size of 12,600 sq ft. This provides the same number of homes but some on-site open space. Curvilinear streets act as traffic-calming devices and only 7 intersections exist, rather than 16 as in the traditional subdivision. Reduced speeds and fewer stopping and starting operations reduces amounts of gasoline used and zinc and hydrocarbons deposited on streets and into the atmosphere. A combination of these two make this a safer and less-expensive place in which to live. Pathways and sidewalks lead most residents and visitors to open space areas. On-site BMPs reduce pollutant loads from this site using curvilinear streets.

Overall street length is reduced 19 percent. This is a moderate decrease in pavement, curb and gutter, and all utilities costs. Street grades and a little lot grading lead runoff to swales' beginnings in their own ROWs. These ROWs and other pathways lead residents and their visitors from all portions of the neighborhood to the open-space areas.

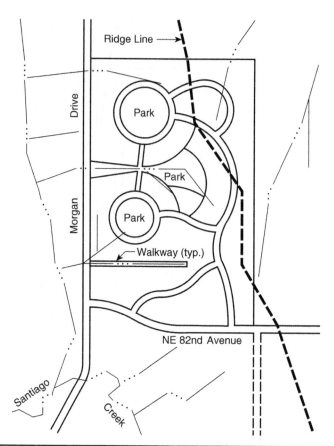

Figure 3.12 Santiago Creek—curvilinear streets.

3.11 Summary

This chapter described the many advantages of coving and curvilinear streets in creating more livable, pleasant, safer, economical, and sustainable neighborhoods. Three examples illustrated how these concepts could be used when developing plans for new single-family developments, demonstrating the benefits to both developers and buyers. Also, because of shorter lengths of streets and utilities, the long-term operation and maintenance costs for cities, counties, and utility companies are greatly reduced. Development of these attractive neighborhoods adds to communities' appeal to prospective inhabitants.

CHAPTER 4

Planning

4.1 Planning a New or Retrofitting an Existing Development

All successful developments meet the desires and requirements of owners, developers, local jurisdictions, and all interested parties. They must address the needs of future inhabitants, be they human, bird, wildlife, and/or fish.

Amount of preliminary planning to be done is based on a project's size and scope and results of meetings with an owner, developer, and others. These determine the size and make-up of a design team. If a project is a fast food restaurant, a single-family subdivision of only a few homes, or a multifamily development of only two or three acres, then a limited number of people and disciplines are needed.

If it involves a large development for some land use, then a design team has numerous people and disciplines. Make sure the budget includes money for numerous meetings with developer's representatives, various agency personnel involved in project approval, meetings with individuals and subgroups, public hearing presentations, and meetings with individuals and neighborhood groups. Long experience has shown that no matter how many meetings a firm plans and budgets for, more meetings and additional budget will be needed during the design and approval process.

However, whether a development is large or small, there are opportunities to create various best management practices (BMPs) within it. These include streets, driveways, patios, decks, lawns, parking lots, rain gardens, vegetative swales, greenroofs, infiltration trenches, rain barrels, cisterns, and "normal" detention basins. "Normal" is the one or several acres of land-consuming surface detention basins.

4.2 Fatal Flaw Analysis

Fatal flaws are whatever prevents a project from being built. This analysis should be the first item to investigate project feasibility. Do this before any time and money on data gathering and preliminary design is spent. Following is a partial list of items that could become fatal flaws.

1. An ancient Native American campground is discovered during a soils investigation. Preservationists require it be left alone. While I was working for the Iowa State Highway Commission (ISHC), this site was found in a proposed scenic highway alignment. The highway was moved away from a bluff overlooking a river.

2. Site soil borings for a new mobile home park were found with smoke coming out of the holes. Further investigation revealed this had been an old garbage dump site. Refuse was still smoldering.

3. A check of records found a proposed 100-ft rights-of-way (ROW) for a proposed major power line diagonally through the site of a proposed apartment complex.

4. ROW for existing large water-distribution pipes was found under the site for a new football stadium on the UCLA campus. While working for the Los Angeles County Flood Control District (LACFCD), we found these same pipes under a street we were going to use for a three-barrel box conduit as our flood channel; each barrel measured 10-ft wide by 10-ft high. We moved the alignment to an adjacent street.

5. Inadequate or unavailable funding has killed many projects. While working for the ISHC, I designed a triple 15-ft by 12-ft box culvert in a deep gully that ran diagonally under the intersection of a state and county highway. I proposed a single tapered inlet, doing this by buying a 160-ac farm, creating a 40-ac lake plus a recreation area, then turning it over to the county for a park plus funds for operation and maintenance. No flat water recreation existed within the county. The farmer and county thought this was a good idea. A lawyer in the state attorney general's office said this would be an unlawful use of highway funds. So instead of saving a half-million dollars, taxpayers spent over $2 million on a long triple-box culvert.

6. A major sewer line was found under a perfect site for a regional detention basin to include a large permanent pond to be used for flood control, water-quality enhancement, and water-based recreation. Another less desirable site had to be found.

7. Not in my back yard (NIMBY). Vocal opposition has been the death knell to many projects.

4.3   Proposed Land Use

Another item to be considered in planning is a site's proposed land use and existing land uses surrounding the site. Existing zoning must be considered. Changing the zoning to a proposed use can be a long and costly operation.

Single-family areas consist of lots on a few to a hundred acres or estate-sized lots on several acres. Homes are built around golf courses as part of the development. Multi-family areas can range from a few duplexes to large apartment complexes covering tens of acres with various types of recreational facilities. Commercial uses include neighborhood or major shopping centers and office parks for one or several tenants, private hospitals, universities, etc.

Industrial uses include parks with several tenants for light-manufacturing purposes to one manufacturing concern devoted to heavy industry. Municipal uses are governmental buildings such as city halls, post offices, maintenance facilities, schools, parks, hospitals, airports, water and wastewater treatment plants, power-generation facilities, etc. Open-space use can range from a city or regional park; to a few swingsets, sandboxes, wading pools, and horseshoe pits; to tennis, badminton, and basketball courts, gazebos, club houses, and golf courses.

In addition, if a city or county participates in the National Flood Insurance Program (NFIP), owners are eligible for reductions in their flood insurance premiums of 5 to 50 percent depending on a community's level in the program. Homeowners save over $100 while other property owners save thousands in annual premiums.

4.4   Information Needed

Numerous data are available at local, state, and federal levels that impinge on the successful completion of these plans or plans for a single development. Much data is needed before a project can be planned. The larger the proposed development, the longer the list. The following list is contained in a book published by McGraw-Hill for the American Society of Civil Engineers (ASCE) and the Water Environment Federation (WEF) (1992).

4.4.1   Topographic Information

Several types of topographic information are needed to assist designers and reviewers in the development of plans for a project.

1. City, county, U.S. Geological Survey (USGS), or other topographic mapping.
2. Aerial photographs.
3. Vegetation maps.
4. Soils maps.
5. Property surveys and ownership maps.
6. Field survey investigations to determine the following:
   a. Drainage basin areas and existing and projected land-use characteristics.
   b. Drainage basin boundaries to confirm interpretation of maps.
   c. Typical overland flow paths, swales, channels, and major drainageways.
   d. Ground and drainageway slopes and lengths.
   e. Typical channel cross sections.
   f. Sites potentially suitable for detention storage.
   g. All relevant drainage and flood control facilities, such as culverts, bridges, etc.
   h. Properties that have actually sustained flood damage in the past.
   i. Mapping needed for final design of local drainage systems.

4.4.2   Survey and Boundary Data

The following information is needed so that designers and reviewers are assured that plans for a proposed project do not conflict with the property rights of others.

1. Land boundaries and corners.
2. Benchmarks.
3. Aerial photographs and ground control.
4. Locations of utilities.
5. Existing streets, alleys, railroads, power lines, canals, schools, parks, and other physical features that will influence project feasibility and siting.

6. Existing ROWs and easements, along with their characteristics.

7. Drainage installation impediments, such as protected habitat, historic sites, cultural sites, etc.

4.4.3   Soils and Geologic Data

Many types of information concerning various soil and geologic characteristics of the project site and surrounding area are needed by the designers and reviewers to ensure that the constructed project is safe.

1. Infiltration characteristics and permeability of basin soils.

2. Excellent sources of soils and geologic data include Natural Resources Conservation Service (NRCS) soils reports and university files.

3. Other standard soil characteristics, such as gradation, density, and classification.

4. Bed and bank samples from drainageways and laboratory evaluations.

5. Soil strength properties for assessments of stability and foundations for drainage structures.

6. Data needed to assess suitability of dam sites if detention ponds are envisioned.

7. Groundwater elevations on a seasonal basis.

8. Bedrock locations and elevations, if foundations and trenches are involved.

9. Other geologic characteristics required by the hydraulic or structural designer.

10. Local geologic maps or reports that could influence projects.

11. Any natural hazards that could affect the drainage system such as hurricanes, landslides, and earthquakes.

4.4.4   Hydrologic and Hydraulic Data

The following types of hydrologic and hydraulic information are needed by designers and reviewers to assist them in developing and approving the drainage aspects of the project. Local rainfall data, including records from local weather stations, intensity-duration-frequency (IDF) curves, hyetographs, and design storm distributions.

1. Historic stream flow, sewer flow, and precipitation data.

2. Channel and pipe characteristics including slope, roughness coefficients, vegetation, stability (erosivity), state of maintenance, amount of debris, etc.

3. Water "flow-line" elevations required for hydraulic grade-line analysis, also sub-basin slopes for analysis of time of concentration (Tc) and model input.

4. Existing hydraulic structures including storm drains, inlets, culverts, channels, embankments, bridges, dams, ponds, and other similar items.

5. Existing intentional and inadvertent storage areas and associated flood-routing data.

6. As much data on large historic floods in the vicinity of the study area as can be derived through library research, interviews with professionals and residents, and other sources.

7. Water-quality data.

4.4.5   Regulatory Data

The following types of regulatory information are needed by designers and reviewers to make sure that all their requirements have been accounted for in the development of the project.

1. Zoning ordinances and maps.

2. Floodplain zoning and requirements (is community participating in the NFIP?).

3. Subdivision regulations.

4. Building and health codes.

5. Water and sewer standards.

6. Erosion, grading, water quality, and environmental protection ordinances.

7. Maps that show designated wetlands, wildlife habitats, receiving stream standards and classifications, and other items pertaining to environmental protection.

8. Environmental Protection Agency (EPA) 208 plans (available from local governments).

9. Historic stream-flow, sewer-flow, and precipitation data.

10. Stormwater management policy and criteria documents, including representative stormwater master plans and final design drawings and specifications.

11. Site-specific or adjacent stormwater master plans.

12. Natural Resources Conservation Service (NRCS) PL-566 plans and stormwater National Pollution Discharge Elimination System (NPDES) permits.

13. Flood insurance studies and maps.

14. Regional flood studies from ungaged basins, normally prepared by the USGS.

15. Drainage reports/plans detailing functions and projected future changes for upstream or downstream facilities or projects, especially if the designer intends to rely on them.

4.5   Other Basic Information Needs

The foregoing types of information are needed by designers and reviewers to ensure that the physical and legal implications have been properly considered and incorporated into the project. The following four types of information are needed to ensure that all portions of the triple bottom line (TBL) have been addressed in the alternatives considered for the project and the alternative used in the final set of plans and specifications represents a solution to the collective, sometimes conflicting, priorities of the parties involved.

4.5.1   Political Considerations

1. Type of local government—mayor or city manager.
2. Organization of local departments.
3. Neighborhood and other special interest groups.
4. Decision makers—both elected and non-elected (those who make policy decisions and those who have influence in policy making).

4.5.2   Social Considerations

1. Number of different publics involved.
2. Environmental concerns.
3. NIMBY.
4. Local inhabitants—both upstream and downstream of sites.
5. Cost concerns.

4.5.3   Financial Considerations

1. Other competing needs for funds: fire, police, education, welfare, existing debt.
2. Orientation of decision makers.
3. Attitudes of public and special interest groups.
4. Available sources of funding.

4.5.4   Information on BMPs

1. See Chap. 5.

CHAPTER 5

Types of Best Management Practices

In terms of stormwater facilities, best management practices (BMPs) come in all sizes, shapes, and types. They can be basins we have been used to designing: large, with or without permanent ponds, and serving a single development or several developments or on a regional basis. Today and in the future we are and will also be designing basins that do not look like detention basins. Instead they will:

1. Look like lawns, rain gardens, bioswales, parking lots, greenroofs, streets, etc.

2. Be scattered throughout developments, each handling runoff from a small portion of a site.

3. Allow developers to use 100 percent of their sites at a lesser cost for useful features.

4. Lead to sustainable developments that are attractive and easier and cheaper to maintain.

5.1 Land-Consuming Detention Basins

Before these newer BMPs are discussed, there is still a place in BMP design for the larger detention basins we have been constructing for several decades. I say this for three reasons:

1. Hundreds of thousands of them exist throughout the United States. Many of them were designed for water quantity only. They can be retrofitted to serve both quantity and quality purposes.

2. New, larger single-family developments constructed in rolling to hilly terrain can utilize steeper areas, unsuitable for homes, for water-quantity and -quality detention basins.

3. There are many millions of culverts throughout the United States. Some of them with suitable terrain and land uses upstream of them could be retrofitted to utilize their existing storage volumes for runoff-quantity and -quality control. If only a small percentage of them were suitable, there could still be over a million sites to retrofit to implement Federal Emergency Management Agency (FEMA) and Environmental Protection Agency (EPA) requirements.

By constructing a riser at upstream ends of their outlets with various weirs and orifices, these basins can be used to meet requirements with freeboard. In hydraulics, freeboard is defined as the difference between an allowable water surface elevation and some adjacent location, such as a roadway elevation, home or building elevation, crop-land elevation, or any other object that should not be flooded. A freeboard(s) is (are) always accompanied by one or more flood return periods. Since these culverts already exist, there should be no objections from neighbors and other interest groups. No land needs to be acquired. The only cost is riser design and construction. At public hearings, a presenter can discuss the small costs and benefits (open space and recreation) that will accrue to the neighbors and community.

For new, larger residential developments, valleys can be used as open space and blocked off with a berm or fill for a street to create a detention basin with a permanent pond to hide deposited silt and provide a setting for passive- and active-recreational activities. This should be done only if sufficient storage is created to sustain a permanent pond; a layer for a water-quality storm event; an upper layer for a 10- and 100-year, 24-hour storm event; and freeboard. This should only be done if sufficient fill for the berm or street is available from excess excavation from other areas of the site for any needed street and lot grading. Hillsides should be left alone.

Using land for a detention basin with only a single use is a waste. The following example of a mall located in a southeastern city I saw decades ago emphasizes this point. Several acres were included under roof plus numerous acres of parking. Grate inlets in the lots were connected to a detention basin using thousands of feet of pipe. The basin was over two acres in size, was 12 ft deep, and had 1:2 (H/V) side slopes. It was lined with asphalt and surrounded by a 6-ft high chain-link fence. Urban litter stuck to the fence creating an ugly mess. Commercial land then was worth over $2 million an acre. For well over $4 million plus the cost of all the pipes, the owner got an ugly hole that stored water once in a while. The city required peak flows after development be no more than prior to development.

The 2+ acres could have been used for more stores. Instead of this ugly hole and all the pipe, other BMPs could have been used. These include greenroofs and porous pavement parking lots underlain with a layer of gravel to provide storage for most or all rainfall. If soil percolation rates were large with a deeper gravel layer, no pipes would be needed. Another BMP could be several small basins located throughout parking areas, depressed 4 to 5 ft below the surface with 4:1 side slopes, covering the bottom and sides with grass with a cobble low-flow channel plus trees to visually break up a sea of asphalt. Outlets would be sized so runoff from 1- to 2-year events were completely captured. Upper portions would reduce runoff rates from the 5- to 100-year events. Another solution is to bury lengths of corrugated steel pipes (CSPs) in various portions of the parking lots to provide temporary storage.

5.2   Purposes of BMPs

We use BMPs to meet FEMA and EPA requirements. We increase time of concentration (Tc), reduce runoff volumes, and reduce runoff rates to less than existing peak flows for all storms from a drizzle through a 100-year, 24-hour event. We improve its quality by retaining pollutants on-site.

5.3   Types of BMPs

5.3.1   EPA's Definition

The official EPA definition for BMPs is listed below.

1. Activity schedules, prohibitions of practices, maintenance procedures, and other physical, structural, and/or managerial practices to prevent or reduce pollution of Waters of the US.
2. BMPs include treatment systems, operating procedures, and practices to control site runoff, spillage or leaks, sludge or waste disposal, or drainage from raw material storage.
3. BMPs can also be defined as any program, technology, process, siting criteria, operating method, measure, or device, which controls, prevents, removes, or reduces pollution.
4. BMPs are classified as source or treatment controls. Source controls are operational practices that prevent pollution by reducing potential pollutants at their source and keep them out of stormwater. They typically do not require construction. Treatment controls are methods of treatment to remove pollutants from stormwater that require construction.

5.3.2   Source Controls

Examples of source controls include the many managerial practices required through passage of local ordinances. Data on many maintenance and planning BMPs found on the Internet include agricultural nonpoint source pollution, conservation easements, construction entrances, construction-site runoff control, development districts, omitting curbs and gutters, erosion control and land clearing, infrastructure planning, illicit discharge detection and elimination, in-line storage, narrower streets, open-space design, overlay districts, good housekeeping for municipal operations, post-construction management in new and redevelopments, protection of natural features, public education and outreach on stormwater impacts, public involvement/participation, redevelopment, street design and patterns, street sweeping, subdivision design, urban forestry, and vehicle washing and maintenance at construction sites (USEPA, 2009d). A longer list of these prepared for a Phase I National Pollutant Discharge Elimination System (NPDES) permit in the 1990s is included as App. I.

5.3.3   Treatment Controls

Numerous treatment-control BMPs have been developed over the years to meet EPA standards. Descriptions of these can be found in manuals at all levels of government, on the Internet, and in conference proceedings. Computer programs have been written to estimate their effectiveness in reducing pollutants. Reports on research studies of their effectiveness abound.

EPA and the American Society of Civil Engineers (ASCE) Water Resources Research Council have joined in an effort over the last several years to coordinate research and their reports to standardize what and how results are reported. Their objective is to determine how and why some BMPs are more effective than others in reducing pollutants. This information will be disseminated so all city, county, state, and federal agencies, districts, universities, and Native American tribes can implement the effective BMPs.

5.3.4   BMP Descriptions

The following descriptions of some of the commonly used treatment-control BMPs are included to provide a feel for them. Readers desiring greater detail on any aspects of these BMPs can consult the numerous manuals, guidelines, and articles available on the Internet. My interest is in their physical design and blending them in as attractive features of a development. Details concerning their hydrologic and hydraulic design are contained in Chaps. 6 through 16.

5.4   Greenroofs

5.4.1   Definition of a Greenroof

Greenroofs are buildings' roofs partially or totally covered with vegetation and a growing medium planted over a waterproofing membrane. It can include other layers, such as a root barrier and drainage/irrigation systems. Use a company that is expert in planning, design, construction, and maintenance of greenroofs. Three types of greenroofs are modular, extensive, and intensive. Amended soil depths range from a few inches to several feet depending on plant and tree types.

5.4.2   Discussion

Planning, design, construction, and maintenance of greenroofs should be left to companies specializing in this BMP. Greenroofs are accepted in the United States as a BMP with benefits beyond a desire to meet EPA and FEMA rules in regard to runoff quality and quantity. The Internet describes the variety and benefits of greenroofs as part of new, retrofitted, and redeveloped sites.

Greenroofs are being used more and more on commercial, industrial, and institutional buildings. Some homes and apartments are also being constructed with greenroofs. Existing homes should add greenroofs only if soil depths are just 2 to 3 in and total weights are small so existing beams and rafters do not need to be replaced. There are three main types: extensive, intensive, and modular; all are described in detail on the Internet.

Six inches of an amended soil will retain two inches of rain. Gravel, as one of the greenroof layers, acts as a detention basin. The outlet is one or more downspouts. These should lead the delayed runoff to rain gardens or other infiltration BMPs.

5.5   Catch Basin Inserts

5.5.1   Definition

Catch basins as inlets to storm sewers include various curb or grate inlets. They are used in the United States and are designed to remove trash, sediments, heavy metals, grease, oil, and/or hydrocarbons.

5.5.2   Variety of Manufactured Inserts

Baskets, trays, bags, or screens are attached inside a catch basin inlet. A variety of devices are available; the Internet lists dozens of them. Some products consist of more than one tray or mesh grates. Such products use more filters or fabric to capture finer sediment or oil. With two or more trays, the top one serves as an initial sediment trap while the underlying ones are media filters. They are proprietary, therefore installation and maintenance should follow manufacturers' advice. Many varieties are available from manufacturers. Some capture only certain types of pollutants; others capture a variety; some are limited in sizes available and quantity of pollutants captured. Some are cheap, and some are more expensive; some need weekly while others require only semi-annual maintenance; some are available locally; others are available nationwide. Maintenance and disposal of pollutants are simple for some and more involved for other types. Numerous examples of inserts are available on the Internet by typing in "catch basin inserts."

5.6 Lawns

5.6.1 Definition

A lawn is an aesthetic area planted with various types of grasses which are maintained at a low and consistent height. Other flowering plants add colors and an aesthetic aura to an area.

5.6.2 Discussion

Lawns are based on soil type and grass species. Soil types are classified as a hydrologic soil group (HSG) (USDA, 1986). A and B soils have high infiltration rates whereas C and D soils have low rates. Grass types used include Kentucky bluegrass, Bermuda grass, weeping lovegrass, buffalo grass, and native grass mixtures. Roots are usually a few inches long, but buffalo grass has roots 10 ft long; roots prevent soil erosion. Thick, lush grass should have a 2 percent slope to capture runoff and pollutants and to prevent wet areas. Some owners overfertilize their lawns, resulting in pollutants added to storm-sewer systems.

5.7 Vegetated Swales (Bioswales)

5.7.1 Definition

Bioswales are landscaped elements designed to remove sediment and pollutants from surface runoff. They have a swaled drainage course with gently sloped sides and are filled with vegetation, compost, and/or riprap. Upper layers of soil can be removed and replaced with an amended soil with increased infiltration rates to capture more runoff and pollutants (USDA, 2008).

5.7.2 Description

Bioswales are located in all areas, then lead to rain gardens and other basins, and they are planted with grasses, flowers, shrubs, and trees. They are constructed on mild slopes, no less than 2 percent to prevent low, wet spots from occurring. With steeper slopes, use cobble or aggregate berms at locations along the bioswale's length to slow runoff. These berms act as level spreaders when overtopped.

Excavate from 2 to 5 ft depending on the soil's infiltration rate and depth to bedrock or the seasonal high-water table. Deeper excavation can provide for more storage in soil and gravel layers. Then they are filled with an engineered media classified as sandy loam and consisting of 50 percent construction sand, 20 to 30 percent topsoil with less than 5 percent maximum clay content, and 20 to 30 percent organic leaf compost (Low Impact Development Center, Inc., 2011).

5.8   Infiltration Trenches

5.8.1   Definition

Infiltration trenches are used at ground surface to intercept overland flows (USEPA, 2009a), are 3- to 12-ft excavations lined with fabric and filled with stone to create reservoirs for some design storm. See Fig. 5.1. Runoff percolates through the trench's bottom and sides into subsoil. Pretreatment is done in order to remove suspended solids from runoff before entering a trench. Pretreatment includes grass filter strips and vegetated swales with check dams. Its design storm is usually a 1- to 2-year event to provide treatment of the "first flush" effect.

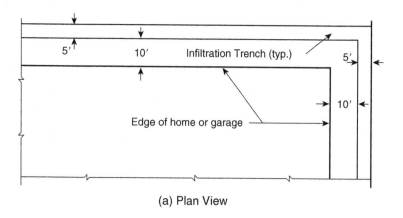

(a) Plan View

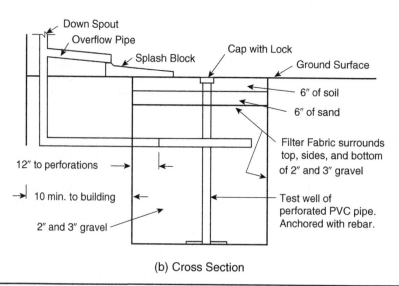

(b) Cross Section

FIGURE 5.1   Infiltration trench.

5.8.2   Advantages

Advantages of infiltration trenches include the following.

1. Trenches reduce volume of runoff from a drainage area.
2. Are effective for removing trace metals, nutrients, bacteria, and oxygen-demanding substances.
3. Reduce peak discharge rates by retaining the "first flush" of stormwater runoff and creating longer flow paths for runoff that increase time of concentration (Tc).
4. Provide groundwater recharge and baseflow in nearby streams.
5. Reduce size and cost of downstream stormwater control facilities and/or stormwater drainage systems by infiltrating stormwater in upland areas.
6. Appropriate for small sites of 10 acres or less.

5.8.3   Design

Infiltration trench failure can be avoided by using the following protective measures.

1. Careful site selection.
2. Treatment of sheet flow from a small drainage area to eliminate sediment.
3. Incorporation of pretreatment measures and a bypass for high-flow events.
4. Good construction techniques that prevent smearing, overcompaction, and operation of the trench during the construction period and performance of regular maintenance.

5.9   Rain Gardens (Rain Garden Network, 2008a)

5.9.1   Definition

Rain gardens are attractive and functional landscaped areas designed to capture and filter stormwater from roofs, driveways, sidewalks, patios, parking lots, and roadways. They collect water in bowl-shaped, vegetated areas and allow it to slowly soak into the soil. This reduces erosion potential and reduces pollutant amounts from surrounding areas.

Rain gardens are small gardens designed to withstand extremes of moisture and nutrient concentrations, particularly nitrogen and phosphorus, found in runoff. They are sited ideally close to a runoff's source and serve to slow stormwater as it runs off, giving it more time to infiltrate and less opportunity to gain momentum and erosive power.

5.9.2   Discussion

Rain gardens are gaining popularity because they are easy to construct, are attractive, and require only minimal maintenance once established. They are used in every land use to reduce peak runoff rates and improve runoff quality. They vary based on location: (1) they can be planted in a low point with shrubs, grasses, and plants (2) they can take the form of a dry river bed with pebbles, rocks, and plants; (3) if they are located on a slope, a depression is created by excavation and the material is used to form a berm around the lower end. Sites are overexcavated to include a deeper planting bed plus gravel to allow greater runoff storage. They duplicate the hydrology of a natural forest. Nutrients and water captured are used by plants for growth.

5.9.3 Soil Types

According to the Natural Resources Conservation Service (NRCS), soils can be classi-fied into four main hydrologic soil groups (HSG): A (sandy), B, C, and D (clay). Gardens with A and B soils can be constructed without an underdrain system. Gardens with C and D soils have an underdrain system with small perforated pipes wrapped in geofabric.

5.9.4 Benefits (Rain Garden Network, 2008b)

A rain garden will:

1. Filter runoff pollution to improve water quality.
2. Conserve water and recharge local groundwater.
3. Protect rivers and streams.
4. Remove standing water in yards to reduce mosquito breeding.
5. Increase beneficial insects that eliminate pest insects.
6. Reduce potential of home and building flooding.
7. Create habitats for birds and butterflies.
8. Survive drought seasons.
9. Reduce garden maintenance.
10. Enhance sidewalk appeal.
11. Increase garden enjoyment, add color, and beautify a neighborhood.

5.9.5 Size and Installation

Rain gardens range in size from small to over 2,500 sq ft to reduce runoff and pollution problems. They can be square, rectangular oval, kidney-shaped, or some free form. They are most effective when built to 3 to 30 percent of their tributary drainage area and have the following characteristics:

1. Gently sloping sides towards the center to prevent sudden drop-offs that lead to erosion.
2. Placed in a depression within a lawn, or in a location that receives roof runoff.
3. To avoid flooding improperly sealed foundations, construct them at least 10 ft away from structures, and direct water into them with a grassy swale, French drain, or gutter extension.

5.10 Perforated Pipe Surrounded by Gravel

5.10.1 Location and Construction

Perforated pipes surrounded by gravel are used with trenches near roof downspouts, bioswales, rain gardens, porous pavements, sidewalks, driveways, and patios where soil types have low-infiltration rates. Pipes are surrounded by filter fabric to prevent clogging, and are joined to storm sewers or swales.

They are useful in commercial and industrial land uses to intercept runoff from roof downspouts and impervious parking lots. Runoff from these areas into grass-lined swales underlain by perforated pipes and gravel depths reduces peak rates and volumes of runoff, traps some pollutants, and cools runoff water that is beneficial to fish in creeks, streams, and lakes. An open graded gravel layer is one to a few feet thick. Six to 12 in of 2- to 3-in gravel is placed below the pipe invert. Gravel depth depends on how much storage volume is desired.

5.11   Porous Pavements

5.11.1   Definition
Porous pavements are paved surfaces of porous asphaltic or portland cement with a stone reservoir below them that stores runoff before infiltrating runoff into the subsoil. Porous pavement often looks the same as a normal pavement but without "fines" that create void spaces to allow rapid infiltration (USEPA, 2009e) of tens of inches per hour.

5.11.2   Locations Where Used
Porous pavements can be used for streets, parking lots, sidewalks, driveways, patios, and playing surfaces for tennis, basketball, etc. In the past decades, their use for streets and parking lots in northern climates was restricted because of freeze-thaw concerns. But advances in their technologies, mixing, and construction, and ongoing familiarity of cities, counties, and contractors is increasing their use.

5.11.3   Layers
5.11.3.1   Surface Course
A surface course is open-graded asphalt concrete about 2 to 4 in thick. It consists of porous asphalt with little sand or dust with a pore space of about 16 percent, as compared to 2 to 3 percent for normal asphalt concrete. Strength and flow properties are similar to normal asphalt. A surface course is an open-graded portland cement concrete from 4 to 10 in thick depending on use. It consists of a concrete mix without fines (sand) with a pore space of 20 to 25 percent. Strength and flow properties are similar to normal concrete.

5.11.3.2   Filter Course
Filter course is a 1- to 2-in layer of half-inch crushed-stone aggregate. In addition to providing some infiltration, it provides stability for the next course during application of the surface layer.

5.11.3.3   Reservoir Course
Reservoir course is a base of 2- to 3-in stone of a depth determined by storage volume needed. Besides transmitting loads, it stores runoff water until it infiltrates. Where soils have low permeability, its thickness is increased. With soils primarily of clay or silt, infiltration capacity may be so slow that the soil is unacceptable as a subgrade, necessitating replacement by borrow material. If the material beneath is relatively impermeable, pipe drainage may have to be provided. A filter fabric is recommended between the reservoir course and the undisturbed soil.

5.11.4   Design Considerations

5.11.4.1   Slope

Slopes should not exceed 5 percent and should be as flat as possible. If low spots develop, then drop inlets should be installed to divert runoff into the reservoir course more quickly.

5.11.4.2   Reservoir Depth

This depth is such that it drains completely after a rare 100-year storm within 72 hours, allowing time for it to dry out and preserve capacity for the next storm. If the site has marginal soils for infiltration, then drain time could be 48 hours.

5.11.4.3   Effects of Frost

If frost penetrates deeper than the reservoir courses, and subgrade soil has potential for frost heaving; additional material should be added to the reservoir course to be below the frost zone.

5.11.5   Construction Considerations for Porous Asphalt Concrete

Several items are listed below in regards to the construction of porous asphalt pavement that must be taken into account to ensure a good final product.

1. Rope off the asphalt concrete area to keep heavy construction equipment off it so underlying soil is not compacted. Then only use equipment with tracks or oversized tires.

2. Install diversions to keep runoff from the site until construction is completed.

3. Construct the asphalt layer only when the air temperature is above 50°F and laying temperature is 230° to 260°F. Failure to follow this guideline could lead to premature hardening and subsequent loss of infiltration capacity.

4. Asphalt used in the mix is 85 to 100 percent penetration-grade since porous asphalt is subject to scuffing, occurring when front wheels of stationary vehicles are turned. It should be between 5.5 and 6.0 percent of the pavement's total weight. A lower limit assures adequate layers around the stones and an upper limit to prevent the mix from draining asphalt during transport.

5. Roll the asphalt when it is cool enough to withstand a 10-ton roller. Normally, only one or two passes are needed. More passes could reduce its infiltration capacity.

6. After rolling is completed, traffic should be kept off the porous pavement for a minimum of one day to allow proper hardening.

5.12   Permeable Pavements

5.12.1   Definition

Permeable is different from porous pavement since rain water passes around paver's edges (such as cobbles, bricks, or other manmade pavers) rather than passing through them (Urban Design Tools, 2011b). The soil's infiltration capacity is a key design variable for depth of base rock for stormwater storage and/or if an underdrain system is needed. Many permeable pavers are manufactured with many shapes and colors and are described on the Internet.

5.13   Sediment Basins

5.13.1   Definition

A sediment basin is a temporary pond prior to and after construction to capture eroded or disturbed soil that is washed off during storms, and is used to protect the water quality of a nearby storm sewer, channel, lake, or bay. The sediment settles before runoff is discharged.

5.13.2   Discussion

Sediment basins are separate basins or are located at upstream ends of detention basins. Their length is four times their width, narrower upstream and wider downstream. Flow meanders through them using berms, baffles, and islands. Their intent is to allow water to spend as much time as possible within the basin to increase settlement. A small outlet structure lengthens this time.

Sediment basins provide storage for captured sediment, which must be removed after construction is completed and grass and plants have become established to reduce post-construction erosion. This keeps detention volume intact in a downstream detention basin to capture runoff volume from its design storm event. Its outlet could include one or more small orifices to detain runoff for at least 48 to 72 hours.

5.14   Extended Detention Basins

5.14.1   Definition

They temporarily store runoff during rain events for a period of 48 to 72 hours to improve runoff water quality and are used within detention basins. They use a small outlet to delay passage of a water-quality storm event (6 month to 2 years) for the required time. All runoff is captured by either placing the next outlet or top of wall at a sufficient elevation above this outlet's invert at a dry detention basin's bottom or at the pond elevation of a wet detention basin.

5.14.2   Discussion

Three types of extended detention basin outlets are illustrated in Fig. 5.2. Figure 5.2a shows one outlet as a vertical CSP that is both perforated and unperforated. The lower portion is surrounded by geofabric and a cone of gravel. The top is open as an outlet for less frequent runoff events. Marshy plants use some runoff and nutrients for growth.

Figure 5.2b illustrates use of a small-diameter plastic or reinforced concrete (RC) pipe on an adverse slope as its outlet. The 3-ft dimension ensures that any vortex that forms does not penetrate to the pipe. This arrangement keeps all floatable debris and pollutants within the basin.

Figure 5.2c shows an underground perforated pipe. Runoff exits the sediment basin only after infiltrating. A berm is overtopped when runoff rate and volume exceeds the basin's capacity. A headwall's small outlet pipe has its invert elevation high enough so all runoff from this rain event is captured. A second outlet pipe is sized to attenuate the flow from a larger event.

Another outlet is simply a cobble or gravel berm between the sediment basin and the downstream detention basin. See Fig. 5.3. After a decade or two, the berm is removed, washed clean of the sediment trapped in it, and then replaced to return the outlet to its original capacity.

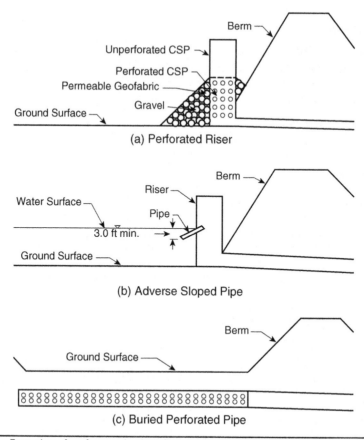

FIGURE 5.2   Examples of surface and subsurface outlets.

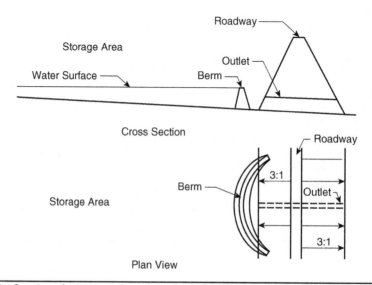

FIGURE 5.3   Creation of a berm upstream of an existing outlet structure.

5.15   Dry Detention Basins

5.15.1   Definition

These are installed in, or adjacent to, tributaries of rivers, streams, lakes, or bays designed to protect against flooding and downstream erosion by temporarily storing water. They are dry since they do not contain a permanent pool and thus drain dry after rainfall events. They are constructed above ground, on the surface, and below ground in vaults, gravel, or in a connected series of oversized pipes. They are used to reduce flows from 2- through 100-year events (USDOT, 2008).

5.15.2   Discussion

Figure 5.4 shows a basin with a vertical riser at the upstream end of its outlet conduit. The riser contains a number of weirs and orifices to contain and reduce flows from the 2- through 100-year storms plus improve runoff quality. The outlet conveys the 100-year reduced flow without water inside the riser rising enough to interfere with flow thru the weirs and orifices. Storage at various pool depths, together with outlets at various elevations, reduce flow rates to the degree desired. Outflow rates and volumes are a juggling act designed to serve intended functions.

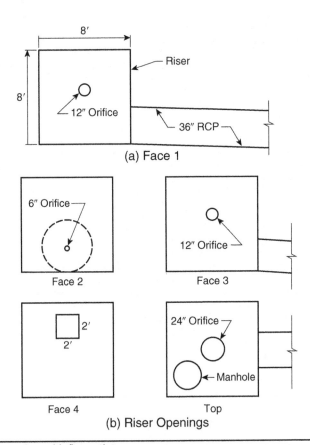

(a) Face 1

(b) Riser Openings

Figure 5.4   Square riser with five outlets.

An access road surrounds the basin for maintenance and for a walking/jogging trail. Some forms of recreation should be used in and surrounding the basin to make it more acceptable to people who live or work around it. It is grassed to enhance its attractiveness. Side slopes are at least 4:1 and vary around the perimeter. Its bottom is sloped from each side at a 2 percent slope to a low-flow cobble channel with a slope of 0.4 percent, that is used to prevent low spots.

5.15.3   Forebays

Every dry detention basin should have a separate or integral forebay at its upper end and whereever else runoff enters it as discussed previously to trap sediment and other pollutants.

5.16   Wet Detention Basins

5.16.1   Definition

Sometimes they are called wet ponds. See Fig. 5.5. This type of BMP is used to manage stormwater runoff to prevent flooding and downstream erosion, and improve water quality in a nearby storm sewer, channel, lake, or bay. It is an artificial lake with vegetation in and around its perimeter, and includes a permanent pond. These basins are also used for groundwater recharge, aesthetic improvement, recreation, or any combination of the above. Comments made above concerning dry detention basins apply to wet basins (USEPA, 2009c).

5.16.2   Difference from a Dry Detention Basin

The only difference between a wet detention basin and a dry detention basin is the wet basin has a permanent pond with benefits from both an aesthetic and a water-quality standpoint. The aesthetic difference is that the pool's surface hides the muddy bottom

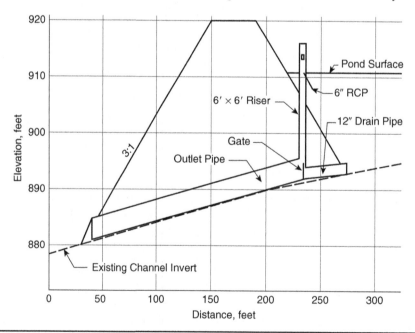

Figure 5.5   Outlet pipe and riser for a basin with a permanent pool.

that can exist in a dry detention basin. Sunlight reflecting off its surface is a pleasing addition to surrounding developments. For water quality, a pond provides a quiescent body of water that allows solids to settle. Water plants covering 25 percent of a pond's periphery provide nutrient uptake. Macro- and micro-invertebrates use pollutants for growth. Some plants grow best in 6 in of water, some prefer 12 in, while still others prefer 18 in of water. Initial grading of a detention basin requires care in its planning and construction.

Because someone may fall into a pond, a 10-ft safety bench is located around a pond's periphery 6 in below its water surface. Side slopes above the pool's surface should be a maximum of 4:1. As space permits, side slopes should vary from 4:1 to 10:1 or 20:1 to yield a more natural appearance. No fences are used; rather people should be invited in to enjoy it.

If fish are included, some portion of the pond should be 10 ft deep. Plants uptake some pollutants for growth and bugs on plants both above and below the water consume other pollutants. If the pond is large enough, it can be used for such recreational activities as swimming, boating, and fishing. A sandy beach can be constructed at the swimming area. Alongside the basin's top can be areas for other recreational areas and parking.

5.17   Buffer Zones

5.17.1   Definition

A buffer zone, or protective area, is an area along a waterway or water body where development is prohibited or limited. This reduces pollutants reaching it. A forested buffer by a stream aids bank stability and controls water temperatures. Its effectiveness depends on width, type(s) of vegetation, and pollutants. They are also useful for shielding differing land uses, e.g., a single-family area from a commercial or industrial area.

5.17.2   Discussion

Requirements are found in local and state guidelines. They range from simple to complex, narrow to wide widths, and loose to careful inspection. Some require residential lawns and pastures in agricultural areas not extend to creeks, and require that cattle are fenced off from them. Buffer zones have grasses, shrubs, and trees to prevent pollutants from washing into waterways.

5.17.3   Widths

The Wisconsin Department of Natural Resources (DNR) has developed regulations for minimum buffer widths (UW-Extension, 2008a). They are 75 ft for wetlands and streams in special interest areas and 50 ft for wetlands and streams susceptible to runoff contamination. They should be 10 percent of average-wetland width, but at least 10 to 30 ft for less susceptible wetlands, including those dominated by species such as reed canary grass.

5.17.4   Three-Zone Buffer System

The Center for Watershed Protection (CWP) has a three-zone buffer system (USEPA, 4/6/2009c), divided into three zones: streamside, middle core, and outer zone. Zones have four parts: function, width, vegetative target, and uses.

5.17.4.1 Streamside

The three-zone buffer system's function is to protect the physical integrity of the stream ecosystem; its width should be a minimum of 25 ft plus wetlands and critical habitat; undisturbed mature forest and reforest if grass. It has very restricted uses such as flood control, utility rights-of way, footpaths, etc.

5.17.4.2 Middle Core

Its function is to provide separation between upland development and the streamside zone; 50- to 100-ft wide depending on stream order, slope, and 100-year floodplain; managed forest with some clearing allowable; restricted uses, some BMPs, bike paths, and tree removal by permit.

5.17.4.3 Outer Zone

The outer zone is used to prevent encroachment and to filter backyard runoff; a 25-ft minimum setback to structures; forest is encouraged, but this zone is usually turf grass; unrestricted uses, e.g., residential uses including lawns, gardens, compost, yard wastes, and most stormwater BMPs.

5.17.5 Maintenance Considerations

An effective buffer-management plan should include establishment and management of allowable buffer zone uses. Boundaries must be well defined at all times. Without clear signs or markers defining it, boundaries become invisible. They should present an aesthetic view to surrounding residents, customers, and workers who must be well trained in the reasons for, used allowed in each zone, and types of maintenance that can and should be performed in each zone.

5.17.6 Design Factors

Buffer design can increase pollutant removal from runoff. See Table 5.1. (EPA, 4/6/2009f.)

Factors that Enhance Performance	Factors that Reduce Performance
Slopes less than 5 percent	Slopes greater than 5 percent
Contributing lengths <150 ft	Overland flow paths >300 ft
Water table close to surface	Groundwater far below surface
Check dams/level spreaders	Contact times less than five minutes
Permeable, but not sandy soils	Compacted soils
Growing season	Non-growing season
Long length of buffer or swale	Buffers less than 10 ft
Organic matter, humus, or mulch layer	Snowmelt conditions, ice cover
Small runoff events	Runoff events > then 2 year event
Entry velocity less than 1.5 ft/sec	Entry velocity more than 5.0 ft/sec
Swales that are routinely mowed	Sediment buildup at top of swale
Poorly drained soils, deep roots	Trees with shallow root systems
Dense grass cover, 6-in tall	Tall grass, sparse vegetative cover

TABLE 5.1   Factors Affecting Buffer Pollutant Removal Performance

CHAPTER 6

Precipitation

6.1   Summary of First Five Chapters

The first five chapters have described newer methods for planning and laying out developments to incorporate the concepts of sustainability and the triple bottom-line (TBL). Descriptions included street layouts, placement of homes and buildings, preserving existing natural drainage, data needs, and the numerous best management practices (BMPs) that have been used in the past and newer ones that assist in further achieving sustainability and the TBL.

These chapters have set the stage for this and the following several chapters that describe in some detail each step in the design process for sizing BMPs for runoff quantity and quality control. As discussed previously, this book is meant for those civil engineers, planners, and those in other disciplines who work for small- and medium-size cities, counties, and consulting firms whose end product is the development of our communities. The methods described and equations discussed in the following chapters were selected for their ease of use in keeping within the users' budget limitations, computer and software capabilities, and expertise.

The driving force for determining the sizing of all stormwater-management facilities is precipitation, the subject of this chapter. The following chapters each discuss one step in the design process. Within each chapter, one of the examples used is for a 126-acre residential area in Northeastern Iowa whose development began over 40 years ago. What I saw then was rolling to hilly terrain used for agricultural crops.

What was envisioned, planned, and constructed over a period of time is now single- and multi-family homes that I visited again a few years ago. The original gully and hillsides remain. A 4-lane divided circumferential city highway in the planning stage some 40 years ago is now in place. The gully receives drainage from the homes and apartments and exits through a 72-in reinforced concrete pipe (RCP) under the highway. When all of this was planned and constructed, the city was not yet involved with stormwater quality, only flood control.

6.2   Introduction

Precipitation comes in many forms: glaze, hail, rain, rime, sleet, and snow. In stormwater-drainage facility design, rainfall is the driving force for our calculations. We want to know its intensity, duration, total storm amount, and time distribution during a storm. We learn about these attributes by keeping records from previous events. Since rain is the starting point for calculations concerning water quantity and quality control, we want the best data for these variables. However, the more we know about how rainfall data is collected, recorded, and interpreted, the best we can say about the data is the numbers are our best estimate.

Errors creep in all the time: rain is spilled; 3.17 in is written down as 3.71; the wind blows; volunteers make mistakes or misjudge the accuracy of their observations; record lengths are too short for a decent statistical analysis; etc. Thus, when using these numbers and inputting them into a spreadsheet or computer program, we should use a conservative method and not skimp on facility sizes. Using this book's ideas plus your knowledge and experience, you can design facilities that will serve their intended purposes safely and faithfully for decades.

6.3   Rainfall Gaging Stations

We obtain precipitation records from rainfall gaging stations. The old U.S. Weather Bureau, now the National Oceanic and Atmospheric Administration (NOAA), established official gaging stations in major cities throughout the United States run by agency personnel. Some have been in continuous operation for decades; others have been active for a number of years and then become inactive or have been moved to another location.

Stations have a rain gage, an anemometer for wind speed, and a pan for evaporation. Rain is measured using one of three types: standard, weighing bucket, and tipping bucket (Wikipedia, 2011). A standard gage measures total rainfall amount but not its duration. The other two gages measure total rainfall, storm duration, and time distribution of rain throughout its duration.

6.3.1   Standard Gage

A standard rain gage is a cylinder 8 in in diameter and 2 ft high. A top funnel leads rain to a 20-in-high tube 2.53 in in diameter, an area one-tenth the 8-in opening's diameter. Thus, depth is magnified 10 times. A 1/10th-inch depth in a gage equals 1/100th-in of rain. The transparent tube is etched with lines each representing 1/100th-in of rain.

An inner tube measures 2 in of rain. If more rain occurs, excess rain overflows into the outer tube. The inner tube is removed and emptied. Then, cradling the gage under an arm, its contents are poured into the inner one. This operation is repeated until all rain is measured.

Care must be taken when pouring water from the cylinder into a 2.53-in tube to prevent spillage. Water weighs 62.4 lb/ft^3. Assume 10 in of rain falls in a rare storm event. The outer cylinder contains 8 in of water. Its weight is $0.7854 \times (8/12)^2 \times 8/12 \times 62.4$ equals 14.5 lb. Balancing this plus the gage's weight and carefully pouring water into the tube without spilling any is sometimes a problem.

6.3.2   Weighing Bucket

The weighing bucket consists of a storage bin with a pail that is weighed to record the rainfall amount that occurred. The bin contains a recording device using a pen on a rotating drum that rotates once every 24 hours. The pen makes an ink trace on a paper strip mounted on the drum. It has a grid with inches on its height and hours on its length. This gage measures rain, hail, and snow. Hail and snow are melted to record its weight, then converted into a rain depth.

6.3.3 Tipping Bucket

The tipping bucket consists of a funnel that collects and channels rain into a small seesaw-type container. After a pre-set amount of 1/100th of an inch falls, the lever tips, dumping the water, and sends a signal that is recorded on a strip of paper mounted on a drum that rotates once each 24 hours. This gage is not as accurate as the weighing-bucket gauge. During high-intensity rain periods, more rain falls while it is tipping and caught in the container, underestimating total rain. Some tipping gages incorporate weighing gauges so a more accurate amount is obtained.

6.3.4 Alter Shield

Rainfall events usually include wind that causes rain to fall at an angle. This results in depth being underestimated. Experiments have shown this to be the case. Higher-wind velocities increase underestimation. The old Weather Bureau included a figure in an article with a curve that showed wind velocity versus percent of underestimation (see Fig. 6.1). (Weather Bureau, 1934).

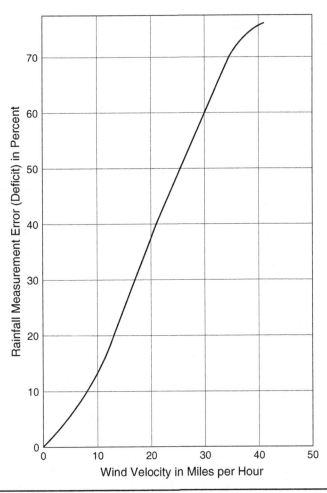

FIGURE 6.1   Wind induced rainfall measurement error.

At a wind velocity of 20 mph and an unshielded gage, depth is underestimated by about 38 percent. The rain gage used was an upright Hellman catch and was compared against a Koschmeider pit gage.

To reduce this error, a Weather Bureau employee invented the Alter Shield. It consists of a series of long, narrow triangular pieces of metal surrounding the rain gage. The effect is to somewhat negate the wind effect and obtain a better value of rainfall depth.

6.3.5   Volunteers

The average number of official rain-gage sites in the United States managed by NOAA personnel is just over four per state. This is very poor coverage when we consider that a rain gage's diameter is only 8 in. Numerous rainfall events occur but are not recorded because there is no gage within the area. To counteract this lack, thousands of volunteers throughout the United States record rain in gages and send monthly reports to NOAA offices located in their states. With volunteers' data included as well, coverage is about one gage every 200 sq mi. This is still a lot to ask of an 8-in diameter gage.

6.3.6   Gage Locations

They are placed in open areas containing no obstructions, such as buildings or trees, to block the rain. This also prevents water collected on roofs of buildings or leaves of trees from falling into a gage, resulting in inaccurate readings. It is better if the open area is surrounded by trees and/or buildings to cut down the wind effect, allowing a more accurate catch. A small meadow in a grove of trees makes an ideal location in rural areas. A small parking lot surrounded by buildings, a protected portion of a roof, or a lawn in a backyard are good locations in urban areas.

6.3.7   Areal Extent and Rainfall Depths in a Storm

A rain storm's areal extent is determined by sending agency personnel and other volunteers into a storm's area to measure rain in all types of containers. These can be the three types of gages mentioned previously, smaller plastic rain gages of various sizes used by gardeners and farmers, and pails, bins, paint cans, or jars containing rain. They are assigned portions of the overall storm area to collect data. Prior to going into the field, a short training session is held giving guidelines on how to fill out each portion of a form that includes the following items.

1. An accurate description of the location where the measurement was taken. If large-scale maps are available, then each position is shown on the map and numbered.

2. A description of the container and its location relative to its surroundings.

3. The measured rain depth in inches. Also, a narrative of whether the container held some water prior to the storm's occurrence.

4. Reasons for the observer's assessment of the data's accuracy (excellent, good, fair, or poor).

The best data is that gathered in rain gages measured by members of NOAA's network of volunteers. Next are those gages tended by gardeners and farmers who keep track of rainfall. Last would be containers that caught rain. Data are then compiled on a map showing each measurement followed by an e, g, f, or p. These data are then used to develop average rainfall depths in various portions of the storm's area using one of the following methods.

6.4   Average Rainfall over an Area

Three methods used to estimate depth over an area are arithmetic mean, Thiessen, and isohyetal.

6.4.1   Arithmetic Mean Method

This is the simplest method. All locations' depths are summed and divided by total number of locations. Its results are reliable in flat to gently rolling terrain. Results are better if locations are uniformly distributed and values are not too far from the mean. See Fig. 6.2a (Linsley, 1975).

6.4.2   Thiessen Method

This method allows for nonuniform gage distribution by using a weighting factor for each gage. Locations are plotted on a map, and connecting dashed lines are drawn. Solid perpendicular bisectors of these dashed lines form polygons around each gage. A polygon's boundary is the gage's effective area. These areas are estimated by planimetry and expressed as a percentage of total area. Average rain for the total area is computed by multiplying the gage's rainfall by its assigned percentage of area and totaling. See Fig. 6.2b (Linsley, 1975). It assumes linear rainfall variation between stations and assigns an area to the nearest station. This can be used for design, assuming no other rainfall data for that site's drainage area, or it can use rainfall data from the nearest NOAA gage. While this may not be best, it is the most we can do currently.

6.4.3   Isohyetal Method

This method for average rainfall is the most accurate. An isohyet is a line of equal rainfall. Gage amounts are plotted on a map with the decimal point assumed to be its location. Contours of equal rainfall are drawn by assuming linear interpolation between amounts. An area's rainfall is computed by averaging rain between two isohyets (taken as their average) by the area between isohyets, totaling these products, and dividing by total area. See Fig. 6.2c (Linsley, 1975).

6.5   Rainfall Intensity

Measured in inches per hour, obtained from statistical analyses of records, rainfall intensity is determined. A time period, 10 minutes, is selected and each storm examined to find the largest amount that fell in 10 minutes. All maxima are ranked in descending order and subjected to statistical analysis. Depths for six return periods are estimated from the data's mean, standard deviation, and skew. Depths are divided by duration used to obtain rainfall intensity, usually for the 2-, 5-, 10-, 25-, 50-, and 100-year events, and then plotted on duration-frequency paper.

Other durations are selected, analyzed, then plotted using the same procedure. Smooth curves and/or straight lines are drawn through same return-period points and are known as intensity-duration-frequency (IDF) curves. The IDF curves for Kansas City, Missouri, are shown in Fig. 6.3. They were published in Technical Paper No. 25 (Weather Bureau, 1955) for data collected up to about 1950 and then in HYDRO-35 (NOAA, 1977) using data gathered in the next 25 years.

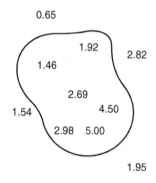

0.65

From Ray K. Lindsey, Max A. Kohler, Joseph L. H. Paulhus
Hydrology for Engineers, Second Edition

McGraw-Hill Book Company, 1975, p - 83

(1.46 + 1.92 + 2.69 + 4.50 + 2.98 + 5.00) / 6 = 3.09 in.

(a) Arithmetic Method

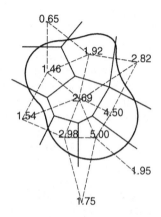

Observed precip. (in.)	Area* (sq. mi.)	Per cent total area	Weighted precipitation (in.) (col. 1 × col. 3)
0.65	7	1	0.01
1.46	120	18	0.26
1.92	109	17	0.33
2.69	120	18	0.48
1.54	20	3	0.05
2.98	92	14	0.42
5.00	82	12	0.60
4.50	76	12	0.54
2.82	32	5	0.15
	658	100	2.84

*Area of corresponding polygon within basin boundary

(b) Thiessen Method

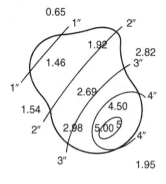

Isohyet (in.)	Area* enclosed (sq. mi.)	Net area (sq. mi.)	Average precip. in.	Precipitation volume (col. 3 × col. 4)
5	14	14	5.3	74
4	94	80	4.6	368
3	216	122	3.5	427
2	422	206	2.5	512
1	625	203	1.5	305
<1	658	33	0.8	26
				1712

Average = 1712 / 658 = 2.61 in.
* Within basin boundary

(c) Isohyetal Method

FIGURE 6.2 Areal rainfall averaging by three methods.

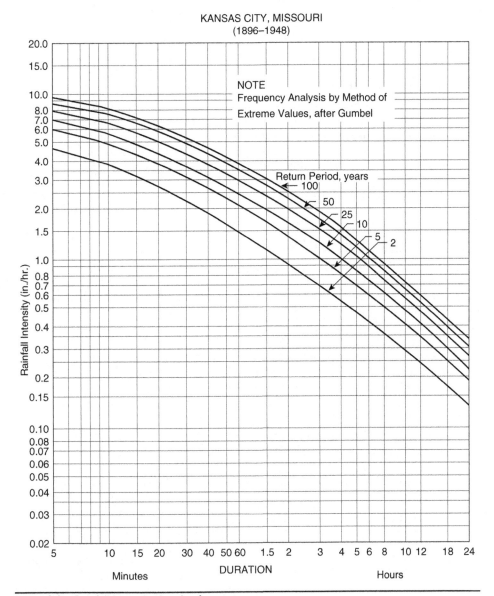

KANSAS CITY, MISSOURI
(1896–1948)

NOTE
Frequency Analysis by Method of
Extreme Values, after Gumbel

Return Period, years
← 100
50
25
10
5   2

Rainfall Intensity (in./hr.)

Minutes   DURATION   Hours

FIGURE 6.3   Rainfall intensity-duration-frequency curves.

6.6   Storm Duration

Storm duration is the time in minutes or hours rain fell between periods of zero rainfall. Short-duration storms can have high-rainfall intensities in them. Long-duration storms can also have some periods of high-intensity rainfall but also periods of low or moderate intensities. Longer storms usually have more rain depth than shorter storms and greater rain volumes. Some methods for estimating runoff hydrographs include storm duration as a variable.

6.7   Rainfall Depth

Depth and duration are used in many methods to develop inflow hydrographs and obtained from TP No. 40 (Weather Bureau, 1961). It includes U.S. maps listing rainfall depths for the 48 states with durations of 0.5-, 1-, 2-, 3-, 6-, 12-, and 24-hours and return periods of 1-, 2-, 5-, 10-, 25-, 50-, and 100-years but it is no longer in print. Maps for the Eastern two-thirds of the United States are contained in Appendix B of TR No. 55 (USDA, 1986). Maps of rainfall depths and durations for 21 Western states are found in NOAA Atlas 2 (1973) and 14 (2003).

6.8   Time Distribution of Rainfall

Rain does not fall at a uniform intensity during a storm. It drizzles, rains hard for a while, drizzles again, etc. I used various distributions of rainfall in my master's thesis as one variable in developing hydrographs. By changing from one distribution to another, I changed a hydrograph's peak flow rate by 50 percent. The NRCS (USDA, TR-55, 1986) has studied time distribution of rainfall during storms for U.S. rain gage data and found rain fell in four distinct patterns depending on location. Their locations are shown in TR 55, (USDA, 1986) on a map of the United States. See Fig. 6.4. Using a good time distribution of rain in your design storm event is crucial to obtaining peak flow rates, hydrograph shapes, and volumes of runoff.

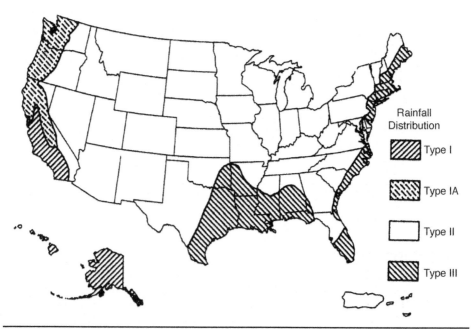

FIGURE 6.4   Areal extent of NRCS's synthetic rainfall distributions.

6.9   Variables Needed for Development and Other Projects

Three rainfall variables used for projects are depths, durations, and time distribution of rain. Intensities are not used for hydrograph development but are used in methods that estimate peak flow rates. Both rates and runoff volumes are included in hydrographs. The NRCS (USDA, 1986) method for developing hydrographs, one of about 50 methods, uses four variables.

1. Rainfall depth (P) in inches for various return periods for a 24-hour duration storm

2. Drainage area (A) in square miles

3. Time of concentration (Tc) in hours

4. Runoff curve number (CN), dimensionless

Their estimations are described in various chapters of TR-55 (USDA, 1986). This manual should be on every designer's shelf. The two most important variables to define hydrograph shape and magnitude are A and Tc. The most difficult to estimate is Tc.

6.10   Madison, Wisconsin, as a Case Study

Gaging station records in Madison, Wisconsin, show use of rain data for BMP design. It was active during the period June 1948 to December 1950 and January 1958 to May 1963, for a total of 3,286 days. If no rain or a trace fell, a zero was used. The largest rain recorded was 4.88 in. Rain occurred on 921 days or 28.0 percent. Total rain was 276.05 in, an average of 30.7 in per year.

Analyses of the records were made by the U.S. Weather Bureau. Intensities for durations and return periods are found in Technical Paper No. 25. TP-40 rainfall depths for durations and return periods are listed in Table 6.1. Table 6.2 is number of days rain occurred for 11 depth intervals plus total rain amounts for these same intervals for the nine year period.

For rain depths ≤0.10 in, there were 428 occurrences for a total of 15.0 in or 46.5 percent of the 921 events. For daily depths of rain >0.10 in and ≤1.00 in, there were 434 occurrences for a total of 146.0 in or 47.1 percent of the 921 events. For daily depths of

Duration Hours	Return Period, Years						
	1	**2**	**5**	**10**	**25**	**50**	**100**
0.5	1.0	1.2	1.4	1.6	1.8	2.0	2.2
1.0	1.2	1.4	1.7	2.0	2.2	2.5	2.8
2.0	1.4	1.7	2.1	2.4	2.7	3.0	3.3
3.0	1.6	1.9	2.3	2.6	2.9	3.2	3.6
6.0	1.8	2.1	2.6	3.1	3.5	3.8	4.2
12.0	2.1	2.5	3.1	3.5	4.1	4.5	5.1
24.0	2.4	2.9	3.5	4.0	4.6	5.2	5.9

TABLE 6.1   Rainfall Depths-Durations-Return Periods for Madison, inches

Depth Interval Inches	Number of Days	Cumulative Depth Inches
0.00–0.10	428	15.0
0.11–0.25	182	45.8
0.26–0.50	139	87.0
0.51–0.75	82	136.2
0.76–1.00	31	161.0
1.01–1.50	32	215.4
1.51–2.00	12	235.8
2.01–2.50	8	253.4
2.51–3.00	4	264.2
3.01–4.00	2	271.2
4.01–5.00	1	276.1
Total	921	276.1

TABLE **6.2**   Days of Precipitation and Cumulative Depths for Madison for the Period 6/1948–12/1950 and 1/1958–5/1963

rain >1.00 in and ≤4.88 in, there were 59 occurrences for a total of 115.1 in or 6.4 percent of the 921 events.

Another way of looking at the data is 46.5 percent of rainy days accounted for only 5.4 percent of total rain; 47.1 percent of rainy days accounted for 52.9 percent of total rain; and 6.4 percent of rainy days accounted for 41.7 percent of total rain that fell. Fifty-nine days of the 3,286 days accounted for nearly 42 percent of total rain. Thus $100 \times 921/3,286$ is about 28 percent. So 72 percent of the time, no rain fell during a 24-hour period or rain fell about one day in four. So $100 \times 59/3,286$ is 1.8 percent, <2.0 percent of the time was there more than 1 in of rain during a 24-hour period. Less than 2 percent of the days accounted for 42 percent of total rain.

Rain is a sometimes thing. Mostly, it is just a nuisance, especially if we forget an umbrella or we must jump across a puddle in a gutter. Only about 2 percent of the time, we *might* have problems during a rainfall event—and then again, we *might not* have any problems. This is why the general public complains at times, but they rarely get upset to where they demand action.

The general public thinks a 100-year storm occurs just once every 100 years. If that storm happened 5 years ago, it will be another 95 years before it happens again. They must understand it could happen anytime. It sometimes occurs twice in a single year as it did in Iowa in the 1970s.

CHAPTER 7

Drainage Area Estimation

7.1   Definition and Comments

Drainage area is defined as that area from which all runoff finds its way to a site in a watershed. Its estimation and time of concentration (Tc) are discussed in separate chapters because hydrograph shape and size are most sensitive to these two variables. Poor estimation of drainage area or Tc results in incorrect designs. After map delineation, watershed boundaries are verified in the field. This is more important in urban and urbanizing areas than in rural areas, especially if land slopes are flat. Grading operations on flat lands can change flow direction by 180 degrees.

Area is measured in acres or square miles. Runoff occurs under the action of gravity, so it flows downhill. More precisely, it flows to points of lesser energy. Water flows uphill if there is energy available—until it is lost due to friction and minor losses. Once this energy is gone, water forms a pond. If no outlet exists, it remains until the water either infiltrates and/or evaporates.

7.2   Steps in Watershed Delineation

There are six general steps in the watershed boundary delineation process.

1. Use recent topographic (topo) maps that include all areas contributing runoff to best management practice (BMP) sites. Revise existing contours to reflect final site grading. The revised map becomes part of the design plans with existing contours shown as dashed lines and revised contours as solid lines.

2. On a copy of the map, use a blue pencil or pen to show major and minor streams.

3. Locate closed contours and elevations that denote hilltops, saddles, and high points in streets.

4. Use a red pencil or pen to connect these hilltops and high points, which form a ridgeline around the watershed, forming the watershed boundary.

5. Use a planimeter or computer command to estimate the area in acres or square miles.

6. Verify this boundary in the field by driving and walking through and around the watershed after receiving permission from the owners.

Some designers do all steps. They like to see the drainage network and density, to assist in finding ridgelines, and in delineating sub-basins. Sub-basins are smaller areas within an overall watershed, areas of main stream tributaries, and/or BMP sites. Experienced designers omit steps two and three, simply drawing the watershed boundary after looking at the revised topo map.

7.3   Available Computer Tools

Computer tools such as computer-aided design (CAD), geographic information systems (GIS), digital terrain models, etc. are available to delineate watershed boundaries. However, a thorough understanding of how to delineate watersheds by hand is needed to correct errors and reviewing the computer output for accuracy. The following discussion concerning topographic maps will be useful in identifying features on topo maps useful in delineating watersheds, types of available maps, and other useful information.

7.4   Topographic Maps

Topo maps are obtained from a number of sources and include such items as contours (lines of equal elevation), roads, buildings, fences, culverts, bridges, and other man-made and natural features. Contour interval is the difference in elevation between two contour lines.

There are two main sources of topographic maps with United States Geological Survey (USGS) 7.5-minute quadrangle maps being the most common. Their scale is 1-in equals 2,000 ft, and their contour intervals can be 1, 2, 4, 5, 10, 25, or 50 ft. These multicolor maps have contour lines in brown, foliage in green, water features in blue, man-made objects in black, except for some highways that are red. Newly urbanized areas are shown in purple. City/county boundaries are shown in black. USGS 7.5-minute quadrangle maps are used for rural sites. The other topo maps are used more in urban settings. An exception to this is many sites in rural watersheds have areas of only a few acres so aerial topographic maps must be used.

Other topographic maps are derived from aerial mapping and have scales of 1-in equal to 10, 20, 40, 50, 100, 200, or 400 ft with contour intervals of 1, 2, 4, 5, or 10 ft. These maps are either black- or blue-line prints. They show more detail and are more up to date than USGS 7.5-minute quadrangle maps. Some communities and counties fly their areas every year or two.

7.5   Stream Network

On USGS quad sheets, perennial streams are shown as solid blue lines, and intermittent ones are blue as a short line, three dots, a short line, etc. Some tributaries have no lines. A channel is recognized by using contour lines depicted as vees. Water flows to points of lesser energy, so vees point upstream. This differentiates channels from sharp knolls in which vees point downhill to lower elevations. A channel is depicted in Fig. 7.1 as short vees and labeled with a 1.

Vees become rounded towards watersheds' upper ends. Channels become swales and then broader, shallower swales. Convex sides of rounded contour lines point upstream. A swale is labeled in Fig. 7.1 with a 2. Near a watershed's upper end, rounded contours become straight in some cases. These lines depict overland flow and are labeled in Fig. 7.1 with a 3.

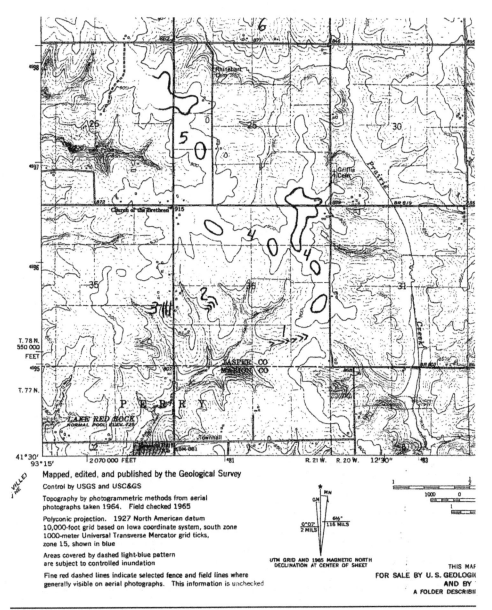

FIGURE 7.1 Stream and watershed elements.

7.6 Watershed Delineation in Rural Areas

7.6.1 Stream Network

A network is shown on a topo map using the following method. At a site, color in blue all perennial and intermittent streams (see Fig. 7.2). Then color in all major tributaries by drawing a line through the point of each vee, staying midway between

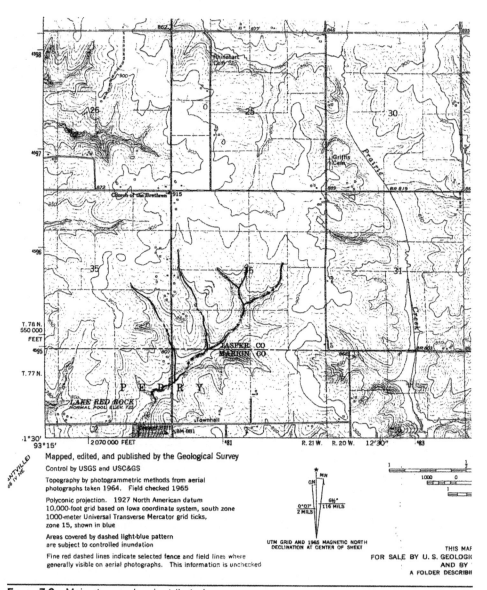

Mapped, edited, and published by the Geological Survey

Control by USGS and USC&GS

Topography by photogrammetric methods from aerial
photographs taken 1964. Field checked 1965

Polyconic projection. 1927 North American datum
10,000-foot grid based on Iowa coordinate system, south zone
1000-meter Universal Transverse Mercator grid ticks,
zone 15, shown in blue

Areas covered by dashed light-blue pattern
are subject to controlled inundation

Fine red dashed lines indicate selected fence and field lines where
generally visible on aerial photographs. This information is unchecked

UTM GRID AND 1965 MAGNETIC NORTH
DECLINATION AT CENTER OF SHEET

THIS MAP
FOR SALE BY U.S. GEOLOGIC
AND BY
A FOLDER DESCRIBII

FIGURE 7.2 Main stem and major tributaries.

contours forming these vees. As channel becomes swales, draw the line through a
rounded contour's most upstream point. See Fig. 7.3. Use this same method for each
minor tributary. See Fig. 7.4. Stop lines at their most upstream contour line. These
lines are identified because the next contour line's elevation is the same. Figure 7.4
should look like a trunk, branches, and twigs. Ends of the outermost twigs define a
watershed's shape. Perform this same procedure for adjacent stream systems surrounding your watershed. Do this for just those minor tributaries that lead water off
in another direction as shown in Fig. 7.4. Ridge lines are located between the ends of
these two sets of lines.

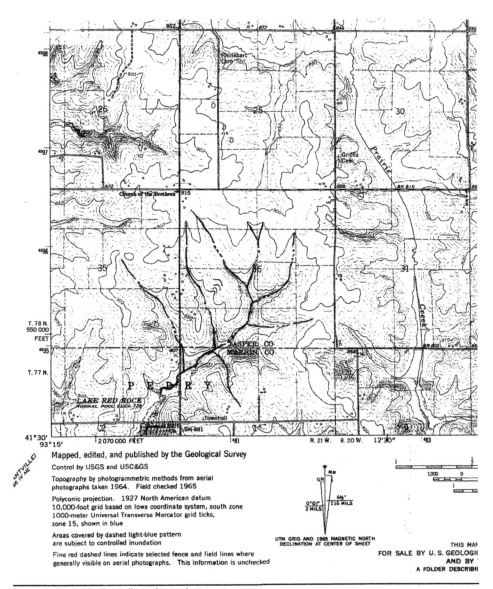

FIGURE 7.3   Add other tributaries to last contour.

7.6.2   Watershed Boundary

A ridgeline is located by a raindrop falling on one side of the ridgeline flows to a site while one falling on the other side of the ridgeline flows into another watershed. Topo map features also assist in drawing ridgelines. Find a contour that closes on itself, then all other contour lines inside it enclose smaller areas. If the innermost contour has the highest elevation, the contour lines represent a hill's shape, size, and height (signified by the number 4 in Fig. 7.1). A ridgeline passes through them.

At times, the contour lines form two tops, but lower contour lines close on themselves. Between two adjacent hilltops there is a lower elevation that is still higher than

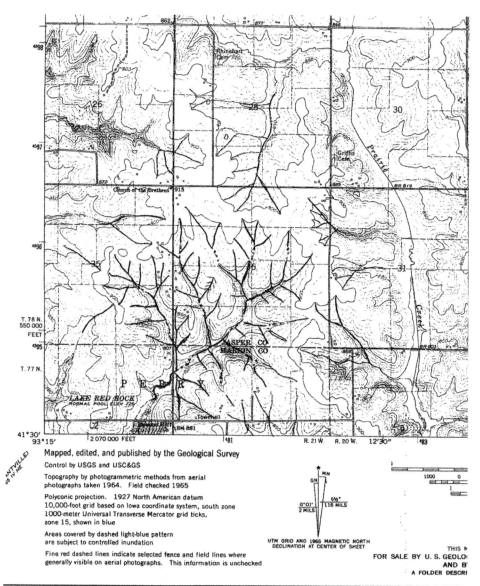

Figure 7.4   Add all minor tributaries to last contour.

adjacent contour lines. This is a saddle. Some topo maps show elevation and location of a saddle as a decimal point. On other topo maps, a saddle's location must be inferred from the position of adjacent contour lines. Ridgelines run through saddles. A saddle is labeled in Fig. 7.1 with a 5. Another clue for a ridgeline's location is a high point elevation in a street or highway. In some cases the decimal is the high point, but in other cases the elevation is shown adjacent to a road. In these cases, a high point is an "x" on the road or inferred to be midway between locations where two contour lines of the same elevation cross the road. A roadway's high point is labeled in Fig. 7.1 with a 6.

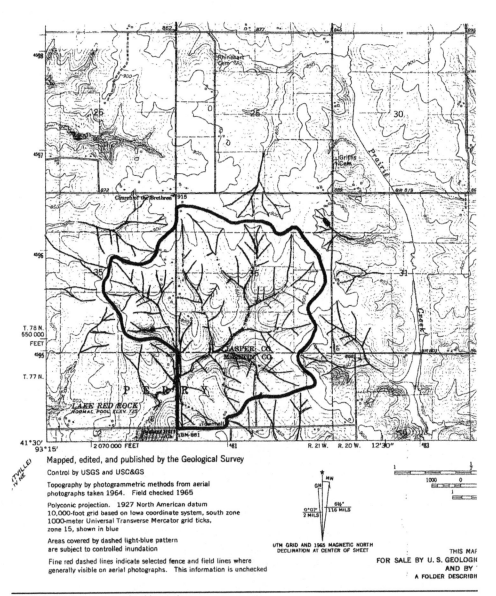

FIGURE 7.5   Draw watershed boundary.

A boundary is drawn in Fig. 7.5. Begin at a BMP site. Draw a line along a road's or berm's centerline until a higher elevation is reached. In some cases, usually small drainage areas, a ridge line meets a road at the site. Draw a line that follows the ridgeline around the watershed's periphery. Use hilltops, saddles, and high points on roads as guides. Draw your line through the middle of rounded contours. On some maps, elevations are shown on elongated hilltops. Your boundary is drawn through the decimal point. Continue the line until your site is reached.

7.7   Watershed Delineation in Urban Areas

The method used to delineate watershed boundaries in urban areas is similar to that used for rural areas. The only differences in their maps are its scale is larger and there is more detail.

7.7.1   Stream Network

Steps for outlining a stream network are the same: show the main stem and tributaries in blue, draw in other tributaries, and then draw in all minor tributaries. These steps are combined as shown in Fig. 7.6. In urban areas, it is especially important to drive through your entire watershed to ensure that all culverts have been located and shown on a map. Thus, if a tributary continues upstream of a street or highway, it will be shown correctly on your map. If there is no culvert under the roadway, then the road's center-line is the watershed boundary.

7.7.2   Watershed Boundary

A watershed boundary is drawn as shown in Fig. 7.6. Begin at your site. Draw a line along a road's or berm's centerline until a higher point is reached. Then continue using the same method used for rural areas. The boundary could follow the centerline of one or more streets.

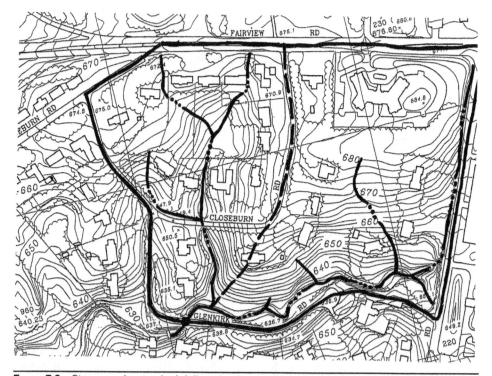

Figure 7.6   Stream and watershed delineation in an urban watershed.

7.8   Subarea Boundaries

Drawing lines for subarea boundaries is easy if a stream network is shown in blue on your topo map. Upstream end of twigs may be almost touching. General location of a boundary should be easy to see using the above clues for a ridgeline's location. Sub-basins are needed in larger watersheds for BMPs and if storage is to be utilized and hydrographs generated. Sub-basins are usually determined on a topographic basis, but in urban areas they may also need to be based on land uses and/or soil types in order to estimate more accurate runoff volumes.

7.9   Drainage Area Estimation

Area is estimated by using a planimeter or by a computer command if a watershed boundary is digitized in a geographic information system (GIS) or computer-assisted design and drafting (CADD).

7.10   Rural and Urban Areas

Methods described above are used to delineate watersheds. Field trips are needed to verify if the boundaries are correct. Much of it can be observed from local county roads so a windshield survey is adequate for the most part. For those areas not visible from the road, farm lanes are available after owners' permissions are obtained to enter their properties. Some walking is needed for that portion not visible from a lane. City streets afford easier access to watershed boundaries.

A detail that arises is whether runoff from a tributary is carried in a culvert or remains on its upstream side because no culvert is present. In this case, runoff from a tributary does not reach your site, so a watershed's size is reduced somewhat. Presence of a culvert at a location can only be determined by a field investigation because culverts may not be shown on maps.

7.11   Subtlety in Urban Areas

Subtleties are boundaries that may change during larger floods. This is because of the way water flows in streets. When it reaches an intersection, how grades and crowns are designed alter ways that water flows there. If intersecting streets' crowns are left intact, during a minor rain event water flows down a gutter and turns a corner at an intersection, because their crowns are higher than flow depth or, if a cross-pan is provided, water flows across an intersecting street. Some cities mark maps with arrows showing flow directions at streets and intersections.

However, during major storm events, such as a 25-, 50-, and 100-year event, water in streets is deeper because a system of inlets and storm sewers cannot completely intercept and convey these increased flows. Depths of flow can easily be higher than streets' crowns and water that turned a corner during more frequent rainfall events now flows across an intersection because crowns are not sufficiently high to turn runoff.

These increased depths may not affect a storm sewer much because they are already running full. But downstream culverts and channels may have more or less flow reaching them for which they were designed. This is usually not considered when delineating watersheds during design; however, it may be a reason that some culverts become inadequate and roads overtopped during major events. It would be prudent during design to determine whether or not this is possible.

7.12   Stream Network

Watersheds' stream patterns affect peak flows as shown in Fig. 7.7. Figure 7.7a is a watershed with a single main stem. Figure 7.7b is the same size but has four main stems. These stems bring runoff to a design site quicker because of a shorter Tc. Thus, the watershed shown in Fig. 7.7b yields a larger peak flow. In like manner, a stream network in Fig. 7.8b yields a larger peak flow than that in Fig. 7.8a because of its greater drainage density and its consequent reduction in Tc. Runoff reaches a channel quicker and once in a channel, runoff velocity is greater.

7.13   Detailed Drainage Area Delineation Summary

The following comments and guidelines are to be used to outline watershed boundaries.

1. Contours are lines of equal elevation
    a. They are usually solid lines. Dashed lines are halfway between solid contour lines.
    b. Every fifth contour line is a wider, darker line.
    c. Contour interval is elevation difference between two solid contour lines.
    d. Contour lines are usually somewhat parallel to each other.

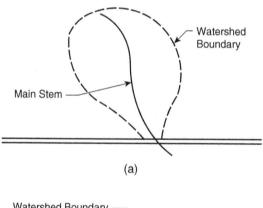

(a)

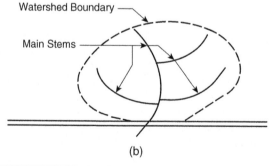

(b)

Figure 7.7   Effect of main stem on peak flow.

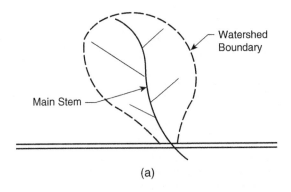

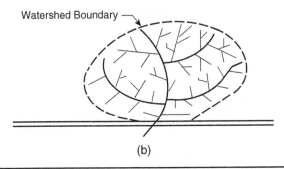

Figure 7.8   Effect of drainage density on peak flow.

      e. The closer these lines are to each other, the steeper is the land's slope.

      f. Small circles and ovals are interpreted as ridge tops.

      g. Small circles and ovals with short lines pointing towards the center are low points.

2. Step one in outlining a watershed tributary to some point on a stream is as follows:

      a. Beginning at a point of interest, draw in all perennial and intermittent streams.

      b. Streams can cross roadways if there is a bridge or culvert located at a crossing.

      c. Draw in remaining channel portions by following them upstream to the last contour.

      d. Draw in all tributaries to the main stream and all tributaries by following them upstream to a last contour.

      e. Result should be a fan-shaped series of lines that define all tributary streams and swales.

3. Step two in outlining a watershed tributary to some point on a stream is as follows:

      a. Locate small circles and ovals that are ridge tops.

      b. These are usually included within contour lines that close on themselves.

      c. When watercourse vees point towards each other, the divide is between the two points.

d. Water flows downhill quickly - pathways are perpendicular to contours.
- draw in light blue arrows that show these downhill pathways
- draw light blue lines in adjacent watersheds as well
- tails of these arrows point to a ridgeline, which is between these two tails

e. By mentally taking cross sections along a ridgeline, determine the drainage divide.
f. A ridgeline is usually taken as centered between a closed contour line.
g. A ridgeline can be along a saddle between two higher elevations.
h. If a culvert does not cross a roadway, the road itself is the watershed divide.
i. Using these comments, draw in the watershed divide (drainage area boundary).
j. Using these above comments, draw in any desired subwatershed boundaries.

7.13.1   Laying Out a Cross Section

The following six items are used to lay out a cross section of a channel and its floodplain. In item 2, a single straight line is usually not adequate to lay out a correct cross section. Natural drainageways have channels meandering within their floodplains; the floodplain itself also includes curves along it length. Water flows along these direction changes, so the series of lines forming a plan view of the cross section must be drawn perpendicular to the flow directions across the entire floodplain.

1. Draw a series of connected lines on a topo map to show a cross section's location.
2. Each portion of the overall line should be at right angles to channel flow or floodplains.
3. Lay a scale on the lines and measure distances between adjacent contour lines.
4. Use arithmetic graph paper and lay out suitable vertical and horizontal scales.
5. Plot these contour elevations and distances on this piece of graph paper.
6. Draw smooth curves and/or straight lines between these points.

7.13.2   Field Observations

Field observations are made to supplement information obtained from topo maps. These give answers to questions not answered by a topo map. These questions include:

1. Is a drainage divide sketched on a topo map correct?
2. Are there existing culverts at all locations where a stream crosses a road?
3. Are there any existing manmade features not shown on your topo map?

7.13.3   Subareas

This dendritic pattern of creeks, streams, and storm sewers in some cases, serves as one criterion for delineation of sub-basins, i.e., confluences of one channel with another. Other criteria for locating sub-basin boundaries include:

1. Locations where a hydrograph and/or peak flow rate is desired.
2. Downstream end of each existing pond and lake and existing and potential detention facility or retrofit locations where sufficient storage exists to attenuate peak flow rates.

3. Streets where existing or potential culvert and bridge sizes need to be checked.

4. Where land uses change.

5. Where soil types change.

6. Where channel and floodplain characteristics change.

7. To achieve a certain minimum size of subarea, e.g., 40 acres. This criterion is needed only for watershed master plan studies.

7.14   A Final Thought

In many cases, your site (watershed) is a single-family development or shopping center. It may be a self-contained site with no runoff draining into or through it. In these cases, your drainage area size is self-evident. However, even on these sites there could very well be a need for subareas because of locations of various BMPs locations with each needing to have their individual drainage areas identified.

CHAPTER 8

Time of Concentration Estimation

Developing time of concentration (Tc) estimates is not a trivial task. Selection of flow paths runoff takes to a best management practice (BMP) requires thought and experience. Each design, each Tc for each BMP, must be carefully done and documented. What your design hydrographs will look like depend in large part on your Tc estimates. The following discussions, equations, and examples set the stage for what I believe needs to be done each and every time. Examples in the appendices expand on this.

8.1 Definition

Tc is defined as time required for water to flow from the hydraulically most distant point in a watershed to the point of design. This definition means the longest pathway in time, not the longest in length. Rainfall intensity makes a difference. Rain falling on a pervious surface infiltrates until one of two events occur: (1) it exceeds infiltration rate and all abstractions have occurred, and (2) soil is saturated and its intensity is greater than minimum infiltration rate for that soil.

A watershed's antecedent moisture condition (AMC) also makes a difference. AMC is a measure of moisture amount stored in a soil profile at some point in time. Is a watershed completely saturated from previous rainfall or is it very dry because no rain has fallen in the last few weeks? At what state is the watershed between these two extremes when rain begins to fall?

No one has physically measured Tc. Thus, any Tc is just an estimate. If two methods yield close Tc values, all we know is that two methods yielded similar results, not that we have a good answer. Many methods estimate Tc, yielding estimates that can be an order of magnitude apart, even using the same variable values. If one method yields 5 minutes, another could estimate 50 minutes. I estimate Tc using the Natural Resources Conservation Service (NRCS) (USDA, 1986) method of velocities in three flow paths.

8.2 Types of Equations

Researchers have developed many Tc equations: some in minutes, hours, or days; some for flow velocity in feet per second (fps) and combined with lengths to obtain Tc. Some are derived for specific U.S. regions; others were derived for rivers and are not used for BMPs. I estimate Tc using velocities. However, even here there is a wide disparity in results depending on how designers obtain the variables.

8.3   Components of Tc

NRCS (USDA, 1986) states Tc is made up of three components: sheet flow (SF), shallow concentrated flow (SCF), and channel flow (CF). Tc is total flow times in these pathways. CF is a minor portion of Tc because velocities in SF and SCF are less than CF. SCF velocities range from 1 to 4 fps; CF velocities usually range from 2 to 15 fps. SF velocities are less than 0.1 fps.

8.4   Overland Flow or Sheet Flow

Sheet flow is flow over plane surfaces: lawns and parking lots, pastures, some culti-vated areas, and fallow land. Depths are in fractions of an inch. Except for parking lots, roughness element height is more than flow depth. A few equations are available to estimate time of overland or sheet flow. Each equation yields different results using the same input values. You need to use your judgment as to which to use or use that method required by a local entity.

8.4.1   Overland Flow Equations

The Federal Aviation Administration (FAA) (1970) and Kerby (Civil Engineering Maga-zine, 1959), Eq. (8.1) and Eq. (8.2), respectively, are examples of overland flow equations.

$$To = 1.8(1.1 - C) \, L^{0.5} \, S^{0.33} \qquad (8.1)$$

where To = overland flow time, minutes
C = runoff coefficient in the rational formula
L = length of overland flow path, feet
S = slope of overland flow path, percent

$$To = 0.667 \, (N \, L / S^{0.5})^{0.467} \qquad (8.2)$$

where N = Manning's roughness coefficient
L = length of overland flow path, feet
S = slope of overland flow path, feet/foot

Overland flow time is also estimated by Eq. (8.3). This equation is the kinematic wave approach by Ragan and Duru (1972, with permission from ASCE).

$$To = 0.931 \, (n \, L)^{0.6} / (i^{0.4} \, S^{0.3}) \qquad (8.3)$$

where To = overland flow time, minutes
n = Manning's roughness coefficient
L = length of overland flow path, feet
i = rainfall intensity, inches/hour
S = land slope, feet/foot

8.5   Shallow Concentrated Flow

Gutters and minor swales are SCF with depths measured in inches. SCF velocities for paved and unpaved surfaces can be estimated from Eq. (8.4) and Eq. (8.5), respectively. They are from *Urban Hydrology for Small Watersheds; Technical Release No. 55* (TR-55) (USDA, 1986).

$$Vp = 20.3 \ S^{0.5} \tag{8.4}$$

$$Vu = 16.1 \ S^{0.5} \tag{8.5}$$

where V = average flow velocity, feet per second
       S = watercourse slope, feet/foot

If flow is in a paved curb and gutter section, then Eq. (8.6) can be used to estimate flow velocity. This equation is taken from HEC No. 12 (Federal DOT, 1984).

$$V = 1.12 \ S^{0.5} \ D^{0.67} / n \tag{8.6}$$

where S = longitudinal slope of street, feet/foot
       D = depth of flow at curb face, feet
       n = Manning's roughness coefficient

8.6   Channel Flow

Flow in underground conduits; flow in streets greater than curb height; manmade and natural channels and floodplains are examples of channel flow. Flow depths are measured in inches or feet. Manning's equation, Eq. (8.7) is used to estimate channel-flow velocity (Manning, 1891). If a channel is a pipe or box, assume it is flowing full. If it is an open channel, assume it is flowing bank full.

$$V = 1.49 \ R^{0.667} \ S^{0.5} / n \tag{8.7}$$

where R = hydraulic radius, feet, A/WP
       A = area of flow, square feet
       WP = wetted perimeter of flow, feet
       S = channel slope, feet/foot
       n = Manning's roughness coefficient

8.7   Estimation of Tc

Time of flow in each of these three pathways is estimated by using Eq. (8.8).

$$T = L/60 \ V \tag{8.8}$$

where T = time of flow, minutes
       L = length of flow path, feet
       V = average velocity of flow, feet per second

Total Tc is estimated by adding flow times in each pathway. Do not use total length and an average slope if any of these three flow paths contains more than one slope. Instead, use length and slope of each portion, using an appropriate equation as many times as necessary.

8.8   Example of Tc Calculations

My flow-path interpretations may be different from yours but no cause for alarm. Designers make decisions based on their backgrounds and experiences. Your interpretations may easily be different from those of a checker or local reviewer. Your task as a designer is to convince them that your flow pathways are a better representation of the topography and definition of Tc.

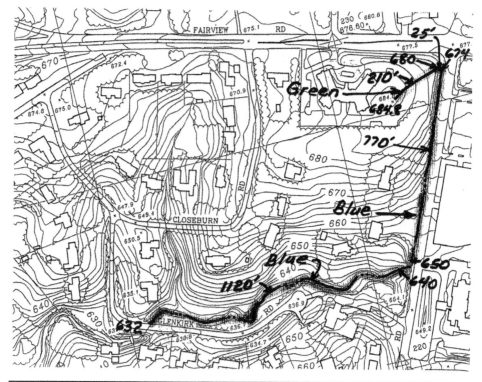

FIGURE 8.1 Tc Pathways for Example 8.1.

8.8.1 Example 8.1

A developed urban area watershed is tributary to a culvert under Glenkirk Road as shown in Fig. 8.1. In this case, only sheet flow and channel flow are present. The SF path is divided into two segments because two slopes are shown by contour lines being different distances apart. Channels are also separated into two flow paths, one for a ditch parallel to Park South Road and one for an open channel leading to a culvert. Contour lines are not the same distance apart in these two segments, so different slopes are used for each segment.

Flow path elevations and lengths are shown on Fig. 8.1. Manning's n values are assumed to be 0.24 for sheet flow, 0.035 for a ditch, and 0.045 for a channel; both have trapezoidal shapes. Ditch dimensions are a 1-ft bottom width with 1:1 side slopes, and 1-ft deep. Channel dimensions are a 3-ft bottom width with 2:1 side slopes, and a 2-ft depth.

Slopes in each flow segment are estimated as follows.

$$S = (684.8 - 680)/210 = 0.0229 \text{ ft/ft}$$
$$S = (680 - 674)/25 = 0.2400 \text{ ft/ft}$$
$$S = (674 - 650)/770 = 0.0312 \text{ ft/ft}$$
$$S = (640 - 632)/1120 = 0.00714 \text{ ft/ft}$$

Flow time in overland flow path 1

$To = 0.42 \ (NL)^{0.8} / ((P2)^{0.5} \ S^{0.4})$

$To = 0.42 \ (0.24 \times 210)^{0.8} / ((3.5)^{0.5} \ (0.0229)^{0.4})$

$To = 0.42 \ (23.01) / (1.871 \ (0.2208))$

$To = 23.4$ min

Flow time in overland flow path 2

$To = 0.42 \ (NL)^{0.8} / ((3.5)^{0.5} \ S^{0.4})$

$To = 0.42 \ (0.24 \times 25)^{0.8} / ((3.5)^{0.5} \ (0.2400)^{0.4})$

$To = 0.42 \ (4.19) / (1.871 \ (0.5650))$

$To = 1.7$ min

Flow time in channel path 1

$A = (d/2) \ (b1 + b2) = (1/2) \ (1 + (1 + 1 + 1)) = 2.0 \ ft^2$

$WP = b1 + 2 \ (d) \ (z^2 + 1^2) = b1 + 2 \ (1) \ (1^2 + 1^2)^{1/2} = 1 + 2.83 = 3.83 \ ft$

$V = 1.49 \ R^{2/3} \ S^{1/2} / n = 1.49 \ (2.0/3.83)^{2/3} \ (0.0312)^{1/2} / 0.035 = 4.88 \ fps$

$Tch = L/(60 \ V) = 770/(60 \times 4.88) = 2.6$ min

Flow time in channel path 2

$A = (d/2) \ (b1 + b2) = (2/2) \ (3 + (4 + 3 + 4)) = 14.0 \ ft^2$

$WP = b1 + 2 \ (d) \ (z^2 + 1^2) = b1 + 2 \ (2) \ (2^2 + 1^2)^{1/2} = 3 + 8.9 = 11.9 \ ft$

$V = 1.49 \ R^{2/3} \ S^{1/2} / n = 1.49 \ (14/11.9)^{2/3} \ (0.00714)^{1/2} / 0.045 = 3.11 \ fps$

$Tch = L/(60 \ V) = 1120/(60 \times 3.11) = 6.0$ min

Time of concentration

$Tc = 23.4 + 1.7 + 2.6 + 6.0 = 33.7$ min

This result is also shown in Table 8.1, a spreadsheet for estimating Tc.

Worksheet 2				
Overland (Sheet)	**1**	**2**	**3**	**Total**
Pathway Length, ft*	210.0	25.0	0.0	
Upstream Elevation, ft*	684.8	680.0	0.1	
Downstream Elevation, ft*	680.0	674.0	0.0	
Pathway Slope, ft/ft	0.02286	0.24000	10.00000	
Manning's n*	0.24	0.24	0.24	
2-yr, 24-hr Rainfall, in.*	3.5	3.5	3.5	
Flow Velocity, fps	0.15	0.25	0.23	
Travel Time, min	23.4	1.7	0.0	25.1

TABLE 8.1 Time of Concentration for Example 8.1

Worksheet 2				
Shallow Concentrated	1	2	3	Total
Pathway Length, ft*	0.1	0.1	0.1	
Upstream Elevation, ft*	0.1	0.1	0.1	
Downstream Elevation, ft*	0.0	0.0	0.0	
Pathway Slope, ft/ft	1.00000	1.00000	1.00000	
Equation Coefficient*	16.1	20.3	20.3	
Flow Velocity, fps	16.1	20.3	20.3	
Travel Time, min	0.0	0.0	0.0	0.0
Channel	1	2	3	
Pathway Length, ft*	770.0	1120.0	0.1	
Upstream Elevation, ft*	674.0	640.0	0.1	
Downstream Elevation, ft*	650.0	632.0	0.0	
Pathway Slope, ft/ft	0.03117	0.00714	1.00000	
Bottom Width, ft*	1.0	3.0	1.0	
Flow Depth, ft*	1.0	2.0	1.0	
Side Slope, H:V*	1.0	2.0	1.0	
Area, sq ft	2.00	14.00	2.00	
Wetted Perimeter, ft	3.83	11.90	3.83	
Manning's n*	0.035	0.045	0.035	
Flow Velocity, fps	4.87	3.12	27.61	
Travel Time, min	2.6	6.0	0.0	8.6
Gutter	1	2	3	
Pathway Length, ft*	0.1	0.1	0.1	
Upstream Elevation, ft*	0.1	0.1	0.1	
Downstream Elevation, ft*	0.0	0.0	0.0	
Pathway Slope, ft/ft	1.0	1.0	1.0	
Street Cross Slope, ft/ft*	0.02	0.02	0.02	
Flow Depth, ft*	0.25	0.25	0.25	
Flow Top Width, ft	12.5	12.5	12.5	
Manning's n*	0.016	0.016	0.016	
Flow Velocity, fps	27.78	27.78	27.78	
Travel Time, min	0.0	0.0	0.0	0.0
Total Travel Time, Min	26.1	7.7	0.0	33.7

An asterisk in the first column means a value must be input in the appropriate columns if different from the value existing in that column.

TABLE 8.1   Time of Concentration for Example 8.1 (*Continued*)

A large majority of total Tc is sheet-flow time. In many situations, designers and manuals fail to take this into account. Many manuals state that time to a storm sewer inlet is 5, 10, or 15 minutes. In rural situations, designers and manuals use this same language and ignore time of sheet flow. This results in a shorter Tc, higher rainfall intensities, larger peak-flow rates, larger drainage structures, and many more dollars spent. Remember, jurisdictions have their favorite method for estimating Tc. They may require that you use it—unless you can persuade them to use the three flow paths: sheet flow, shallow concentrated flow, and channel flow.

8.9   Detailed Tc Estimation

The following rules are used to develop data needed to estimate Tc for each subarea:

1. General
   a. Use the TR-55 (USDA, 1986) method of three flow paths.
   b. Determine the pathway that has the longest hydraulic travel time as per Tc's definition.
   c. Do not include buildings in a Tc pathway. Flow does not go through buildings. Assume the pathway is around it, even when such a pathway is not shown on a topo map.
   d. Elevations to estimate Tc should be taken from top maps for the three flow paths.
   e. Assume any storm sewer system shown on a topographic map is the complete system.
   f. Tc paths are measured to a detention pond's upper end when the pond is being modeled. If it is not modeled, ignore it and use the contour lines on a topo map.

2. Sheet flow
   a. Sheet flow is flow over parking lots, lawns, and open-space areas.
   b. Sheet flow is identified by straight contour lines on a topo map.
   c. Maximum length of sheet flow was usually about 300 ft (now 100 ft). In one case in Southeastern Texas, my Tc was through rice paddies. My assumption for sheet-flow length was over 3,400 ft. Elevation at the four corners of a square mile and at its center showed a total difference in elevation of 0.1 ft.

3. Shallow concentrated flow
   a. This type begins when flow is contained in a shallow, parabolic swale or gutter.
   b. A swale is shown on a topographic map as a broad curved contour line.
   c. Gutters are always considered to be SCF.

4. Channel flow
   a. Once SCF passes through a structure, flow becomes and remains CF regardless of what downstream topo indicates.
   b. V-shaped contours on topo maps are considered channels.
   c. Ditches along roads are considered channels. Contours indicating a roadside channel are shown outside a roadway. A cross section is set using field investigation data.
   d. When no data is presented for channel data, use topographic map contour lines to estimate channel bottom widths, side slopes, and depths—or use measured field data.

    e. Lengths are taken to a culvert's inlet at downstream ends of subareas.

    f. With a closed conduit system at a subbasin's outlet, its length is taken to the outlet.

    g. Ignore short culverts between sections of open channel.

    h. If there are two or more pipe sizes, use the dominant size to compute Tc.

    i. When culverts or closed conduit systems are found, use an equivalent pipe diameter that yields a similar hydraulic radius when the culvert is flowing full. This approach can lead to unusual pipe diameters, i.e., 73.5 in. Carry this diameter to the nearest 0.1 ft.

Available topo maps are used to delineate each flow path for each subbasin in its existing condition. Elevations, lengths, and Manning's n values are determined for each flow path. Show these on the topo map. Results for existing conditions should be summarized as part of the database for each subarea. Photographs and tables of collected data for flow paths must be included as part of a design's records.

8.10   What Should I Do?

If required to do so, use a methodology described in a policy statement, criteria, or user's manual of the local jurisdiction. If no method is specified, I recommend you do the following.

Show pathways for the three flow paths on topo maps. Use a different color for each pathway. Also, show points on your map where flow type changes, elevations at beginning and end of each path, and length of each path. If there is more than one slope in any flow path, then include elevations and lengths for each segment. Then use a spreadsheet that estimates Tc incorporating data and calculations for all three pathways using the NRCS method, Eq. (8.3) through Eq. (8.8). These maps and spreadsheets provide documentary evidence to show how you obtained your Tc estimates. Spreadsheets to estimate Tc are contained on McGraw-Hill's website (www.mhprofessional.com/sdsd). Each uses the three-flow paths method.

Some jurisdictions require that Tc times be estimated for both existing and developed conditions. This is because of a local ordinance that requires developed flow rates be less than or equal to undeveloped flow rates. Two sets of topographic maps, flow paths, and calculations must be made for existing and developed conditions, including all BMPs to be included in a design. Using the spreadsheets mentioned above will do this for both sets of conditions.

CHAPTER 9

Sizing BMPs

9.1 Decisions and Calculations Already Made

Several decisions and calculations have been made to this point in the overall development planning and design process. These include:

1. Right-of ways (ROWs) of surrounding and interior existing streets and utilities have been established.

2. Locations of natural creeks and swales have been established. How lots are to be drained while preserving most of the existing terrain features has been decided.

3. Curvilinear streets, building locations, lots, and open space areas have been sited.

4. Best management practice (BMP) types and locations for runoff quantity/quality control to be used have been determined.

5. Methodologies for BMP design meeting local requirements have been decided.

6. Area and time of concentration (Tc) for each BMP have been estimated.

9.2 Steps Needed to Complete the Design Process

Five steps are needed to develop an inflow hydrograph and route it through any BMP. These steps are the same for an existing retrofitted "normal" detention basin, retrofit of an existing culvert installation as a detention basin, and any other BMPs that lead to sustainable projects.

1. Allowable headwater depth

2. Depth or elevation versus storage relationship

3. Inflow hydrographs for various return periods

4. Depth or elevation versus outflow relationship

5. Routing methodology

Each is discussed briefly below and then expanded on in Chaps. 10 through 16 in order for BMPs to fulfill their objectives. Most ordinances require peak flows after development be no > those prior to development for flows between 2- and 100-year rainfalls. Agencies at all levels have rules and ordinances requiring runoff quality enhancement. Developers are encouraged to include recreational opportunities, aesthetics, safety, and sustainability in their developments.

9.3 Allowable Headwater (AHW) Depth

A designer's first task for a site is to determine how deep water can be allowed to pond in a BMP. This is true whether it is a rain garden, a greenroof, a swale, or what I have been terming a "normal" detention basin or retrofit. Allowable depths are discussed in Chap. 10.

Items to be kept in mind are what storm magnitudes are being designed for and what are the elevations of surrounding homes, businesses, streets, and other improvements. Adding to your decision-making process are local terrain features, storage volumes (existing and what can be achieved), regulatory requirements, whether a permanent pond is used, planned recreational opportunities and facilities to be included, and grading requirements for water plants.

Computer programs typically do not ask some of these questions or the questions listed below. You will certainly use one or more of the available software packages or my or other spreadsheets. What I am hoping this book will do is to help you think about these questions when considering your designs.

What happens to upstream/downstream property owners when an outlet plugs or a greater than design storm occurs? How far upstream/downstream do I need to look? In one case, a subdivision a quarter-mile upstream was flooded when a large storm occurred and water rose in an upstream swale to a point where water was two feet deep inside homes. What happens if at a future time half of your storage volume has been lost to sedimentation? What happens if...?

We cannot make our basins bomb proof—because of Murphy's law—but we can try to anticipate problems and guard against them. Working with a team is helpful in posing questions and proposing solutions. Just inputting numbers to computer programs does not design a BMP—people design them. They think about these and more questions, aesthetics, costs, and sustainability, then modify their designs.

9.4 Depth or Elevation versus Storage Relationship

A roof's parapet wall, rain garden or swale depth, berm, or street or highway closes off a BMP so water is stored behind and in it. Three methods for estimating an elevation-storage relationship are discussed in Chap. 11. A berms's, street's, or highway's elevation will be higher than the AHW depth or elevation. Whatever this elevation is, depth-storage calculations should be carried higher than that. Hopefully, some events may not occur during a BMP's useful life, but knowing what happens if they do occur could cause some adjustment to a design.

Use of many BMPs (lawns, rain gardens, greenroofs, swales, porous pavements, and underground and surface storage) can eliminate any need for a "normal" detention basin. The only time I use a "normal" basin is when the terrain is such that storage volume is available for a 100-year, 24-hour storm below the AHW elevation plus freeboard to decrease flow rates to allowable values without moving one shovel of dirt. This saves tons of money for moving dirt, doing rough and final grading, then revegetating it.

Storage volume for a cistern or rain barrel is easy to determine. Calculating storage volumes in a swale, wetland, rain garden, or surface parking lot is easy. Calculating underground storage in a series of pipes under a shopping center's parking is easy. Storage in a layer of gravel on a roof, under a rain garden, under a parking lot, or in an infiltration trench is the volume of voids.

If an entity requires a "normal" detention basin be constructed, then use of other BMPs mentioned above could reduce its size or eliminate it. Your job is to convince a reviewer to allow you to do this. Developers will be on your side if he/she can construct more homes or buildings.

9.5   Inflow Hydrograph Methodology

Jurisdictions use hydrographs to estimate peak flow rates and volumes. They use them to route flows through temporary storage volumes to reduce runoff rates and enhance runoff quality. Inflow hydrograph development using the Natural Resources Conservation Service (NRCS) method is discussed in Chap. 12. Area under a hydrograph is the runoff volume. This is true because of the units used on ordinates and abscissas of hydrographs as described in Chap. 12.

Volumes needed to reduce a peak inflow to a lesser peak outflow is equal to the area between inflow and outflow hydrographs as illustrated in Fig. 9.1. This is the basis for reservoir routing operations, whether done by hand or equations inside computer programs or spreadsheets.

Many programs have been used to develop and route hydrographs. Respondents to a national questionnaire distributed by the American Public Works Association (APWA) in the early 1980s indicated they were using 44 methods to develop hydrographs. Applying these methods to a watershed yields results that are as variable as results from the rational formula. They model all hydrologic cycle portions, but do it differently. The NRCS method uses a runoff curve number (CN) between zero and 100. Another method models infiltration with partial differential equations solved by a finite difference method. Complexity does not guarantee more accurate results.

Designers should use methods required by local reviewing agencies. Two federal agency programs that include the NRCS method and have withstood the test of time are TR-20 (USDA, 1984) and HEC-1 (USCOE, 1981), now HEC-HMS. These programs and user manuals can be downloaded free from the Internet or obtained when attending a short course on their use.

As noted above, there are numerous other hydrograph methodologies in use. However, in the interest of saving time and money, do not use a computer program that must be calibrated with local data. These data are usually not available and require years, if not decades, of collection to acquire the data needed for the calibration process.

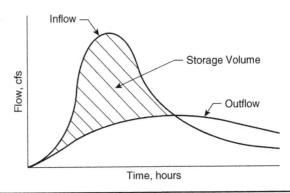

FIGURE 9.1   Volume needed to reduce peak flow.

The method used in the spreadsheets available on the McGraw-Hill website (www.mhprofessional.com/sdsd) is the NRCS method. It includes all aspects of the hydrologic cycle and uses readily obtainable variables. These are drainage area, Tc, land uses, soil types, CNs, and antecedent moisture condition (AMC). A copy of TR-55 (USDA, 1986) is needed for use with these spreadsheets.

9.6   Depth or Elevation versus Outflow Relationship

Outlets for greenroofs are downspouts designed as orifices. Outlets for rain gardens or vegetated swales could be small pipes designed as culverts. The soil itself is the outlet for a number of BMPs: rain gardens, infiltration trenches or basins, vegetated swales and trenches, porous and permeable pavements for streets, driveways, sidewalks, patios, and parking lots.

A depth or elevation versus outflow curve for a retrofitted detention basin or a new one where sloping terrain contains adequate storage is the most difficult step in BMP design to do correctly. An outlet structure could consist of two parts: a conduit under a berm or street and a riser with openings of various sizes, shapes, and elevations to reduce various storm events. An outlet conduit is usually a pipe or box culvert. Basic hydraulic principles are described in Chap. 13. Chapter 14 contains a discussion of culvert design, worthy of a book itself.

The second part is a riser structure. It can be square, rectangular, or circular. Its top can be open or closed. Located on its various sides and elevations, openings of various sizes and shapes allow runoff to flow through them at reduced rates. Sum of these flows become a depth or elevation versus outflow relationship. Openings are used for a water-quality event (from a 6-month to a 2-year, 24-hour rainfall), a 10-year and a 100-year, 24-hour event.

This is difficult for two reasons. First, while weir and orifice equations are simple, many designers and computer programs use them incorrectly, unaware of their subtleties. Second, water depths inside risers can be higher than crests of one or more openings during various portions of rainfall events. When this occurs and an opening acts as an orifice, its definition of head changes. This changes flow rates and the depth/elevation versus outflow curve. After reading about weir and orifice equations and examples of their use in riser design discussed in Chap. 15, you should be able to use them correctly. Whatever your BMP, spreadsheets included on the McGraw-Hill website (www.mhprofessional.com/sdsd) assist you to develop good depth or elevation versus outflow curves.

9.7   Routing Methodology

The routing methodology used in BMP design is the modified Puls method. An inflow hydrograph is combined with a routing curve and routing equation to develop an outflow hydrograph. A routing curve is developed from an elevation-storage and an elevation-outflow relationship. A storage and outflow value at some elevation is plotted. Values of these two at other depths or elevations are plotted and a smooth curve(s) drawn through them.

However, a better curve is outflow versus $2S/\Delta T + O$. This curve used with an inflow hydrograph and routing equation develops the outflow hydrograph. This curve is plotted on log-log graph paper when routing hydrographs manually. For routing hydrographs in a computer program, numerous values are used as a lookup table in the program. Development of the routing equation, routing curve, and routing methodology are described in detail in Chap. 16. Again spreadsheets are available on the McGraw-Hill website (www.mhprofessional.com/sdsd) to correctly route hydrographs.

CHAPTER 10

Allowable Depths

10.1 Definitions

Headwater (HW) depth is defined as the elevation difference between a water surface and invert of a best management practice (BMP) See Fig. 10.1. Allowable headwater (AHW) is defined as the maximum depth of water allowed based on physical conditions and institutional guidelines. AHW is set by a roadway or berm in relation to topographic, stream, and land uses, existing or future. Rules are taken into account and discussed below.

10.2 Roadways for New or Existing Detention Basins

Roadway alignment, grade, and elevation for both existing and proposed conditions must be known. Stream alignment dictates if a structure crosses at a right angle or skew. Its result must be its outlet structure is aligned with a stream or downstream conduit. Difference in a roadway's low point and streambed dictates size and type of the outlet and riser. With a large difference, a single outlet could be used. With little difference, a basin could be infeasible. If there is a larger difference, using storage upstream reduces outlet size, flow rate, and enhances water quality.

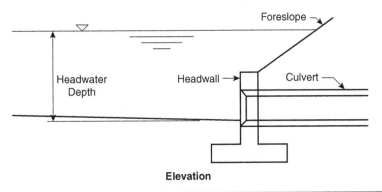

FIGURE 10.1 Detention basin headwater depth.

10.3 Berms for Many BMPs

If a BMP's downstream end is not closed off by a street or highway, then a berm is constructed to confine the storage area. The berm height needed is sum of the following:

1. Depth of permanent pool, if any
2. Maximum depth needed to provide storage volume for the design storm
3. Freeboard

Then route a greater than design flood thru it to determine berm overflow depth. Once determined, check surrounding upstream elevations to ensure that water only exits in a downstream direction through the outlet and berm. If upstream low points are found, they must be filled to provide at least 2-ft of freeboard above a greater-than-design flood's depth. This prevents damage to upstream dwellings or businesses. What effect this greater-than-design flood has on properties and streets must be checked. Storage on greenroofs, rain gardens, gravel under porous concrete, and/or other BMPs could be enough to store runoff from a 100-year, 24-hour storm.

10.4 Site Characteristics

Determine terrain up and downstream; channel/floodplain locations and sizes; road alignment, grades, and elevations; berms; and land uses up and downstream. A topo map for a "normal" basin site reveals much about development and structure types/sizes needed. Contours show existing relief. With little relief, outlet height is restricted, larger conduits are needed, or detention may not be feasible. Channel/floodplain dimensions, roughnesses, and slopes set tailwater (TW) depths for runoff rates, if flow is at, greater than, or less than normal depth (Dn). Tailwater is defined as the depth of water immediately downstream of any type of outlet structure. Normal depth is defined as that depth of water in a closed conduit or open channel in which the friction losses are just overcome by the conduit's or channel's slope. Channel width and depth determines how an outlet fits into it or if widening is needed. Outlets could be placed above streambed. Existing and future buildings must be above water surfaces. Crops and trees cannot be inundated for an extended period of time. If a rural watershed is to be urbanized, should we design for future conditions today or build for existing conditions and rebuild when some urbanization occurs?

10.5 Institutional Guidelines

These guidelines include return periods, freeboards, and others. Most use several storms from 3-months to 100 years. Freeboards vary between jurisdictions. Some use no roadway overtopping during a 25- or 50- or 100-year event. Some allow temporary ponding during storms using freeboards below roadway grade line, i.e., 3-ft. Policies found in design manuals include designing for present or future conditions, types of material allowed, storage to reduce flows and enhance runoff quality, end treatments, allowable velocities, flows, hydrograph methods, and minimum sizes.

10.6   Land Use

Upstream land uses impact AHW. Water must not enter homes, businesses, or factories, either first floors or basements. AHW should be a few feet below first floor or basement window-well elevations. Take into account current and/or future urbanization. Whether this is done now depends on institutional policy. In rural areas, AHW must be below a roadway elevation, first floors of any farmhouse, barn, and other outbuildings, and crop elevations. Some crops can be inundated for a short period of time but any longer and roots die. For corn, this is about a day.

10.7   Return Periods

Return periods for which BMPs are designed is an institutional policy decision and is related to a berm, street, highway, or railroad's importance. Importance is a function of roadway type (interstate, highway, secondary, thoroughfare, collector, or local) or average daily traffic (ADT). For a railroad, it could be whether tracks are a main line, feeder line, or siding. Range is usually from 5 years to 100 years. Table 10.1 was taken from a Department of Transportation (DOT) manual (Commonwealth of Virginia, April, 2002).

Minimum Criteria	
Flood Frequencies for Use in Design	
Type of Road	**Return Period**
Culverts	
Interstate Highway	50-Year Minimum (2%)
Primary Highway	25-Year Minimum (4%)
Secondary Highway	10-Year Minimum (10%)
Storm Sewer Systems	

Inlets
  Principal Arterial
    With and Without Shoulder
    Sag Location
  Minor Arterial, Collector, Local
	
	10-Year (10%)
	50-Year (2%)
	4 in/hr

Storm Drains
  Principal Arterial
  Minor Arterial, Collector, Local
	
	25-Year Minimum (4%)
	10-Year (10%)

Channels	
Roadside and Median Ditches	Capacity—10 Year (10%) Protective Lining—2 Year (50%)
Natural Channels	2-year to 500-year as deemed necessary by the department

TABLE 10.1   Design Return Period Based on Type of Roadway

10.8   Freeboard

Jurisdictions impose limits on AHW in terms of freeboard below top of street, highway, or railroad, and other factors. These amounts can be tied to a return period and/or its importance. Freeboard of 3 ft on a major highway and 1 ft on a rural county road with an average daily traffic of 10 or less is used. Some encourage use of storage upstream of culverts. Then AHW could be tens of feet above an outlet structure. AHW is set based on freeboard below structures or land use. Two AHWs may be set: one for a design flood and one for a larger flood.

Some require 3 ft of freeboard during a design event and zero or 1 ft during a 100-year event. A design event could be a 10- to 50-year event depending on a roadway's importance. Freeboard could be 2 to 3 ft during a design event and zero feet during a rarer event. In some cases, freeboard depends on a building or crop elevation. Many homes and businesses have their first floor or basement window wells below an adjacent street or highway. The same is true of cropland. Care must be taken to become well enough acquainted with your site to understand the governing factors determining AHW elevation.

10.9   Caution

Freeboard is usually measured from a low point in a street, highway, or berm. Since detention basins are usually located in a valley's streambed, a highway could have a sag vertical curve in a basin's vicinity. Freeboard is measured from this lowest elevation in roadway grade. This low point could be directly over an outlet structure or offset from it somewhat depending on magnitudes of the two sag vertical-curve grades.

A case to look out for is that of a continuous grade in a culvert's vicinity with a sag vertical curve some distance down the road (see Fig. 10.2). Many designers look at road elevations directly over culverts and have blinders on for anything else. They use required freeboard at a culvert while a roadway's true low point is several feet lower, hundreds of feet away. A side ditch conveys water to this roadway's low point, and runoff overtops the roadway there.

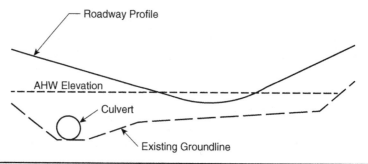

Figure 10.2   AHW in conflict with roadway elevation.

CHAPTER 11

Depth-Storage Relationships

11.1   Definition

A depth-storage relationship is a table or curve of depth or elevation versus total storage volume at each depth or elevation. Units can be feet, meters, or inches on the ordinate and cubic feet, acre-feet or cubic meters on the abscissa. Volume can be existing volume at a site or a volume enhanced by excavation and/or addition of a berm. Storage volumes are available above, on the surface, and below ground. However, in many situations, above and below ground storage is an alternative. A shopping mall is an excellent example for all three locations. Water can be stored temporarily on building roofs, on parking lots, below ground in a series of pipes much larger than needed for simple run-off conveyance, or below ground in a layer of gravel.

11.2   Equations

Storage volume is determined using one of three methods: (1) average-end area, (2) frustum of a cone, and (3) incremental area. An average-end area method is the same as that used for estimating cut and fill quantities for roadway construction and is determined using Eq. (11.1).

$$V = 1/2 \times \text{Depth} \times (\text{Area}_1 + \text{Area}_2) \tag{11.1}$$

A slightly more accurate method is frustum of a cone volume. Determining areas of two successive contour lines and contour interval, then using Eq. (11.2) does this.

$$V = 1/3 \times \text{Depth} \times [\text{Area}_1 + \text{Area}_2 + (\text{Area}_1 \times \text{Area}_2)^{1/2}] \tag{11.2}$$

A third method is the incremental-area method. It is used for estimating volumes of circular containers and other configurations. Areas of a partially filled pipe or structure are determined at each foot of depth. These areas are multiplied by total structure length. Incremental volumes are added together at each depth to obtain total volume of temporary storage.

11.3   Examples

The first examples are solved using the first two methods to show differences between them. All potential storage locations are feasible if a client and jurisdiction allow their use. Depending on type(s) and size(s) of outlets used, they are used for quantity control and/or water quality enhancement. A later example uses the third method.

11.3.1 Example 11.1

An area upstream of a culvert is shown in Fig. 11.1. Its drainage area is 126 acres. Elevation of a 4-lane highway is 856 ft. These 4-ft contours were planimetered and results shown in Table 11.1. Available storage volumes using Eqs. (11.1) and (11.2) are shown in Tables 11.2 and 11.3, respectively. These spreadsheets are contained on the McGraw-Hill website (www.mhprofessional.com/sdsd).

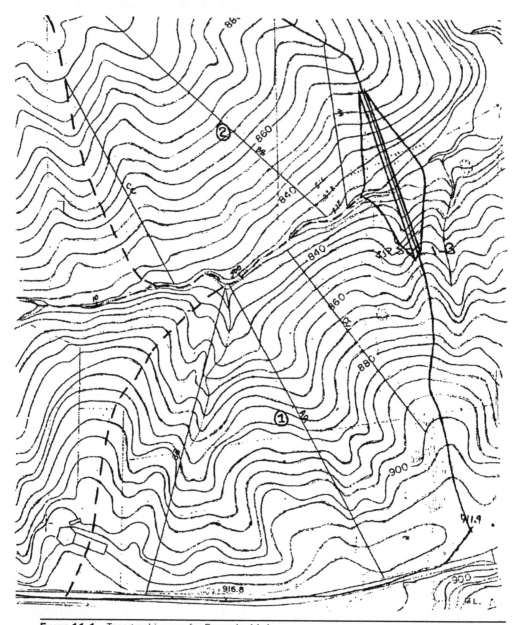

FIGURE 11.1 Topographic map for Example 11.1.

Elevation, Feet	Area, Acres
824	0.00
828	0.24
832	0.84
836	1.90
840	3.54
844	5.28
848	7.56

TABLE 11.1   Elevation-Area Relationship for Example 11.1

Temporary pond areas and storage volumes from Table 11.3 are plotted in Fig. 11.2. Available storage at elevation 848 is 61.5 acre feet (AF). If all rainfall became runoff from a 100-year, 24-hour storm event, runoff would be 126 ac × 6.3 in./12 in./ft = 66.2 AF. Thus, all runoff could be stored below the highway. Any size culvert would be enough.

Elevation/Depth-Storage Calculations							
Average End Area Method							
Perm. Pond Elev. = 824.0			Total Pond Volume = 0.00				
Elev. ft 1	Area ac 2	Aver. Area ac 3	Δ Depth ft 4	Δ Volume ac ft 5	Tot. Vol. ac ft 6	Pond Vol. ac ft 7	Net Vol. ac ft 8
824.0	0.00				0.0	0.0	0.0
		0.12	4.0	0.48			
828.0	0.24				0.5	0.0	0.5
		0.54	4.0	2.16			
832.0	0.84				2.6	0.0	2.6
		1.37	4.0	5.48			
836.0	1.90				8.1	0.0	8.1
		2.72	4.0	10.88			
840.0	3.54				19.0	0.0	19.0
		4.41	4.0	17.64			
844.0	5.28				36.6	0.0	36.6
		6.42	4.0	25.68			
848.0	7.56				62.3	0.0	62.3

TABLE 11.2   Elevation-Storage Calculations for Example 11.1

Elevation/Depth-Storage Calculations							
Frustum of a Cone Method							
Perm. Pond Elev. = 824.0			Total Pond Volume = 0.00				
Elev. ft 1	Area ac 2	(A1*A2)^.5 ac 3	Δ Depth ft 4	Δ Volume ac ft 5	Total Volume ac ft 6	Pond Vol. ac ft 7	Net Volume ac ft 8
824.0	0.00				0.0	0.0	0.0
		0.0	4.0	0.32			
828.0	0.24				0.3	0.0	0.3
		0.4	4.0	2.04			
832.0	0.84				2.4	0.0	2.4
		1.3	4.0	5.34			
836.0	1.90				7.7	0.0	7.7
		2.6	4.0	10.71			
840.0	3.54				18.4	0.0	18.4
		4.3	4.0	17.52			
844.0	5.28				35.9	0.0	35.9
		6.3	4.0	25.54			
848.0	7.56				61.5	0.0	61.5

TABLE 11.3   Elevation-Storage Calculations for Example 11.1

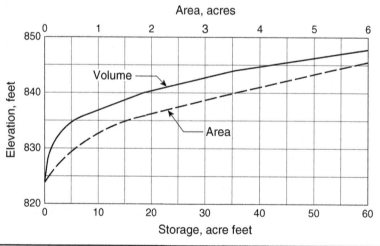

FIGURE 11.2   Elevation-storage curve for Example 11.1.

11.3.2   Example 11.2

An existing culvert is to be modified under Glenkirk Road. A topo map with contours and single-family residential homes is depicted in Fig. 11.3. The street will be left as is so runoff can pond no deeper than elevation 636.7. Contours were planimetered and

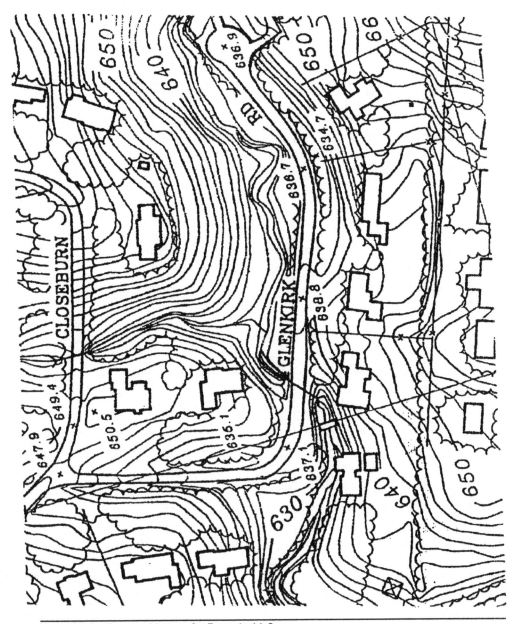

Figure 11.3   Topographic map for Example 11.2.

Elevation, Feet	Area, Acres
627	0.00
628	0.04
630	0.18
632	0.66
634	1.00
636	1.60
636.7	2.00

TABLE 11.4 Elevation-Area Relationship for Example 11.2

results listed in Table 11.4. Determine the elevation-storage relationship using both methods as shown in Tables 11.5 and 11.6. Again, frustum of a cone method yielded slightly smaller values. Table 11.6 is plotted as Fig. 11.4. Total available storage to elevation 636.7 is 6.5 AF.

Elevation/Depth-Storage Calculations							
Average End Area Method							
Perm. Pond Elev. = 627.0			Total Pond Volume = 0.0				
Elev. ft 1	Area ac/sq ft 2	Aver. Area ac/sq ft 3	Δ Depth ft 4	Δ Volume ac ft/ cu ft 5	Tot. Vol. ac ft/ cu ft 6	Pond Vol. ac ft/ cu ft 7	Net Vol. ac ft/ cu ft 8
627.0	0.00				0.0	0.0	0.0
		0.02	1.0	0.02			
628.0	0.04				0.0	0.0	0.0
		0.11	2.0	0.22			
630.0	0.18				0.2	0.0	0.2
		0.42	2.0	0.84			
632.0	0.66				1.1	0.0	1.1
		0.83	2.0	1.66			
634.0	1.00				2.7	0.0	2.7
		1.30	2.0	2.60			
636.0	1.60				5.3	0.0	5.3
		1.80	0.7	1.26			
636.7	2.00				6.6	0.0	6.6

TABLE 11.5 Elevation-Storage Calculations for Example 11.2

Elevation/Depth-Storage Calculations							
Frustum of a Cone Method							
Permanent Pond Elev. = 627.0		Total Pond Volume = 0.0					
Elevation ft 1	Area ac 2	(A1*A2)^.5 ac 3	Δ Depth ft 4	Δ Volume ac ft 5	Total Volume ac ft 6	Pond Vol. ac ft 7	Net Volume ac ft 8
627.0	0.00				0.0	0.0	0.0
		0.0	1.0	0.01			
628.0	0.04				0.0	0.0	0.0
		0.1	2.0	0.20			
630.0	0.18				0.2	0.0	0.2
		0.3	2.0	0.79			
632.0	0.66				1.0	0.0	1.0
		0.8	2.0	1.65			
634.0	1.00				2.7	0.0	2.7
		1.3	2.0	2.58			
636.0	1.60				5.2	0.0	5.2
		1.8	0.7	1.26			
636.7	2.00				6.5	0.0	6.5

TABLE 11.6   Elevation-Storage Calculations for Example 11.2

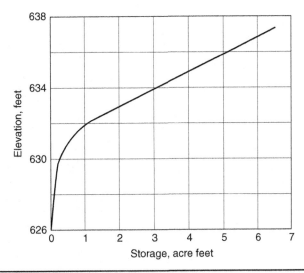

FIGURE 11.4   Elevation-storage curve for Example 11.2.

11.3.3 Example 11.3

A 400-ft-long parking lot slopes away from a building at a 0.4 percent slope with a 7-in.-high earth berm at its downstream end and along its sides. Parking lots should always slope away from buildings. In many cases, parking lots slope towards them resulting in water ponded next to and inside a building. The result is unhappy tenants and possible damage to the building and cars. With a slope of 0.4 percent and a depth of 0.6 ft, water ponds 150 ft wide. Thus, only that half of the parking lot furthest from the building has water ponded on it.

The parking lot and building is shown in Fig. 11.5. Estimate storage on it to a depth of 0.6 ft, using 0.1-ft increments. Areas at various depths are triangles with widths being depth/0.004 and lengths of 400 ft. Use both methods to estimate available storage volumes, shown in Tables 11.7 and 11.8. Frustum of a cone method again yielded slightly smaller values.

Table 11.8 is plotted as Fig. 11.6. While available storage volume is only 17,750 cu ft, the parking lot has a drainage area of only 2.75 ac. Storage is sufficient to reduce the peak-outflow rate to about 18 percent of the peak inflow rate during the 100-year, 24-hour storm.

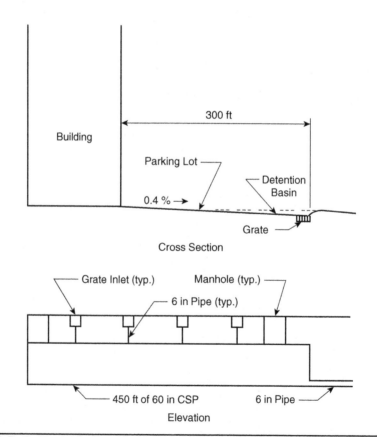

FIGURE 11.5 Parking lot for Example 11.3.

Elevation/Depth-Storage Calculations							
Average End Area Method							
Perm. Pond Elev. = 0.0				Tot. Pond Volume = 0.0			
Depth ft 1	Area sq ft 2	Aver. Area ac/sq ft 3	Δ Depth ft 4	Δ Volume ac ft/cu ft 5	Tot. Vol. ac ft/cu ft 6	Pond Vol. ac ft/cu ft 7	Net Vol. ac ft/cu ft 8
0.0	0				0	0.0	0
		5000	0.1	500			
0.1	10000				500	0.0	500
		15000	0.1	1500			
0.2	20000				2000	0.0	2000
		25000	0.1	2500			
0.3	30000				4500	0.0	4500
		35000	0.1	3500			
0.4	40000				8000	0.0	8000
		45000	0.1	4500			
0.5	50000				12500	0.0	12500
		55000	0.1	5500			
0.6	60000				18000	0.0	18000

TABLE 11.7   Depth-Storage Calculations for Example 11.3

Elevation/Depth-Storage Calculations							
Frustum of a Cone Method							
Permanent Pond Elev. = 0.0				Tot. Pond Volume = 0			
Depth ft 1	Area sq ft 2	(A1*A2)^.5 sq ft 3	Δ Depth ft 4	Δ Volume cu ft 5	Total Vol. cu ft 6	Pond Vol. cu ft 7	Net Volume cu ft 8
0.0	0				0	0.0	0
		0	0.1	333			
0.1	10000				333	0.0	333
		14142	0.1	1471			
0.2	20000				1805	0.0	1805
		24495	0.1	2483			
0.3	30000				4288	0.0	4288
		34641	0.1	3488			
0.4	40000				7776	0.0	7776
		44721	0.1	4491			
0.5	50000				12267	0.0	12267
		54772	0.1	5492			
0.6	60000				17759	0.0	17759

TABLE 11.8   Depth-Storage Calculations for Example 11.3

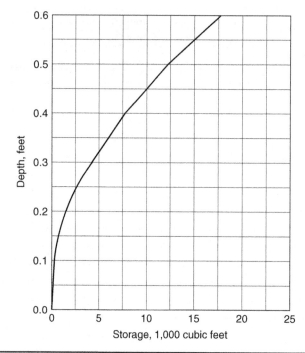

FIGURE 11.6   Depth-storage curve for Example 11.3.

11.3.4   Example 11.4

Outlet structures from the Example 11.3 parking lot are vertical 6-in. diameter pipes at the bottom of 1.5-ft deep grate inlets. These pipes are connected to a horizontal, 450-ft long 60-in. corrugated steel pipe (CSP). This pipe's outlet is a 6-in. diameter orifice that drains to an existing storm sewer.

Values for a partially filled pipe are listed in Table 11.9 (Brater and King, 1976). These values are multiplied by the pipe diameter in feet squared to obtain a partially filled pipe's area. Multiplying these areas by 450-ft yields storage volumes at each depth as a pipe fills. These incremental volumes are then added together to obtain total storage volume at each depth.

Determine a depth-storage relationship for this detention basin, shown in Table 11.10, using an incremental-area method. Its curve is plotted in Fig 11.7. This incremental area method can also be used when an underground-storage container is a box culvert, a gasoline storage tank, an underground vault, a series of arch or oval pipes, or any other shape of underground storage.

11.3.5   Example 11.5

A berm is to be constructed upstream of NW 97th Street in a midwestern state to create a detention basin. Its upstream watershed will be developed with single-family residences on quarter-acre lots. Depths and storage areas upstream of this berm are listed

D/d = Depth/Diameter				Area = Table Value × d²					
D/d	.00	.01	.02	.03	.04	.05	.06	.07	.08	.09
.0	.0000	.0013	.0037	.0069	.0105	.0147	.0192	.0242	.0294	.0350
.1	.0409	.0470	.0534	.0600	.0668	.0739	.0811	.0885	.0961	.1039
.2	.1118	.1199	.1281	.1365	.1449	.1535	.1623	.1711	.1800	.1890
.3	.1982	.2074	.2167	.2260	.2355	.2450	.2546	.2642	.2739	.2836
.4	.2934	.3032	.3130	.3229	.3328	.3428	.3527	.3627	.3727	.2836
.5	.393	.403	.413	.423	.433	.443	.453	.462	.472	.482
.6	.492	.502	.512	.521	.531	.540	.550	.559	.569	.578
.7	.587	.596	.605	.614	.623	.632	.640	.649	.657	.666
.8	.674	.681	.689	.697	.704	.712	.719	.725	.732	.738
.9	.745	.750	.756	.761	.766	.771	.775	.779	.782	.784

TABLE 11.9   Area of a Circular Conduit Flowing Part Full

Elevation/Depth-Storage Calculations					
Incremental Area Method					
Depth ft 1	Area sq ft 2	Incre. Area sq ft 3	Length ft 4	Δ Volume cu ft 5	Total Volume cu ft 6
0.0	0.00				0
		2.79	450	1256	
1.0	2.79				1256
		4.55	450	2048	
2.0	7.34				3303
		4.96	450	2232	
3.0	12.30				5535
		4.55	450	2048	
4.0	16.85				7583
		2.79	450	1256	
5.0	19.64				8838

TABLE 11.10   Depth-Storage Calculations for Example 11.4

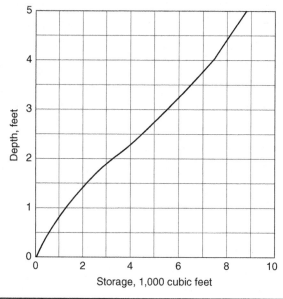

Figure **11.7**   Depth-storage curve for Example 11.4.

in Table 11.11. Calculate available storage volume using Methods 1 and 2, shown in Tables 11.12 and 11.13, respectively. Results for Table 11.13 are plotted in Fig. 11.8.

Total storage volume to a depth of 35 ft is 127.4 ac ft. Depending on the upstream watershed's size, soils, and land uses, this volume will temporarily store a portion of the watershed's runoff volume for some storm magnitude and duration.

Elevation, Feet	Area, Acres
895	0.00
900	0.20
905	0.40
910	1.40
915	3.20
920	5.60
925	10.00
930	17.20

Table **11.11**   Elevation-Area Relationship for Example 11.5.

Elevation/Depth-Storage Calculations							
Average End Area Method							
Perm. Pond Elev. = 915.0			Tot. Pond Volume = 18.0				
Elev. ft 1	Area sq ft 2	Aver. Area ac/sq ft 3	Δ Depth ft 4	Δ Volume ac ft/cu ft 5	Tot. Vol. ac ft/cu ft 6	Pond Vol. ac ft/cu ft 7	Net Vol. ac ft/cu ft 8
895	0.00				0.0	0.0	0.0
		0.10	5.0	0.50			
900	0.20				0.5	0.0	0.0
		0.30	5.0	1.50			
905	0.40				2.0	0.0	0.0
		0.90	5.0	4.50			
910	1.40				6.5	0.0	0.0
		2.30	5.0	11.50			
915	3.20				18.0	18.0	0.0
		4.50	5.0	22.50			
920	5.80				40.5	18.0	22.5
		7.90	5.0	39.50			
925	10.00				80.0	18.0	62.0
		13.60	5.0	68.00			
930	17.20				148.0	18.0	130.0

Table 11.12   Elevation-Storage Calculations for Example 11.5

Elevation/Depth-Storage Calculations							
Frustum of a Cone Method							
Permanent Pond Elev. = 915.0			Tot. Pond Volume = 17.2				
Elev. ft 1	Area sq ft 2	$(A1*A2)^{.5}$ sq ft 3	Δ Depth ft 4	Δ Volume cu ft 5	Total Vol. cu ft 6	Pond Vol. cu ft 7	Net Volume cu ft 8
895	0.00				0.0	0.0	0.0
		0.00	5.0	0.33			
900	0.20				0.3	0.0	0.0
		0.28	5.0	1.47			
905	0.40				1.8	0.0	0.0
		0.75	5.0	4.25			
910	1.40				6.1	0.0	0.0
		2.12	5.0	11.19			
915	3.20				17.2	17.2	0.0
		4.23	5.0	21.72			
920	5.60				39.0	17.2	21.8
		7.48	5.0	38.47			
925	10.00				77.4	17.2	60.2
		13.11	5.0	67.19			
930	17.20				144.6	17.2	127.4

Table 11.13   Elevation-Storage Calculations for Example 11.5

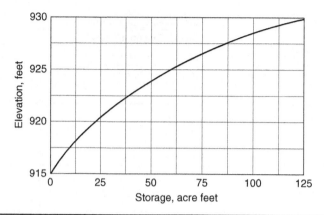

FIGURE 11.8   Elevation-storage curve for Example 11.5.

11.4   Summary

These examples portray the three ways in which depth or elevation versus storage calculations are made. Frustum of a cone method is better than the average end-area method because it yields a slightly more accurate answer. These two methods are used for surface detention. The incremental-area method is used for above- and underground-storage structures of various shapes and sizes. Spreadsheets for these three ways of developing depth or elevation versus storage calculations are contained on the McGraw-Hill website (www.mhprofessional.com/sdsd).

CHAPTER 12

Inflow Hydrographs

12.1 Introduction

In this and other chapters as well as in some appendices, you will need to use other manuals along with this book to follow the use of that step in the design methodology described in that chapter or appendix. In this chapter, it is the United States Department of Agriculture (USDA), Natural Resources Conservation Service (NRCS) Technical Release No. 55 (USDA, 1986) to develop inflow hydrographs. This is my preferred method in this how-to-do-it book to design drainage best management practices (BMPs) as part of the overall design of some type of development.

Chapter 9 contained an introduction to inflow hydrograph development. I prefer using the NRCS method because it includes all hydrologic cycle portions and uses easily obtainable variables. It also develops four hydrographs for design of BMPs such as:

1. Six-month to 2-year, 24-hour rainfall event—the water quality event if storage volume for a permanent pond, water quality event, and water quantity event, plus freeboard is available.

2. Ten-year, 24-hour rainfall event.

3. One-hundred-year, 24-hour rainfall event.

4. Greater than a 100-year, 24-hour rainfall event.

Spreadsheets needed to develop hydrographs are found on the McGraw-Hill website (www.mhprofessional.com/sdsd). Before discussing their use, review the hydrologic cycle's components covered in the next section. When reading them, think about its components from a viewpoint of how to use them to reduce runoff peaks and volumes and ways to reduce pollutants leaving a site.

After the hydrologic cycle, how the spreadsheets are filled in is explained. The chapter's last portion discusses nuances of the NRCS method so that users will have a better understanding of the method and be able to give good answers to the questions, "Why and how did you use that method?" and "Where did you get that number from?" to anyone who asks them.

12.2 Hydrologic Cycle

Authors have described the hydrologic cycle in various ways. However, they simply use different words. As you read them, think about how to use its various portions to design BMPs to reduce rates and volumes of runoff and improve the runoff's quality. Remember,

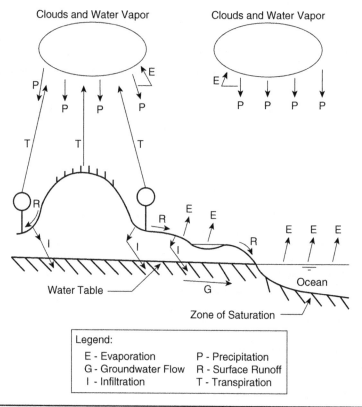

FIGURE 12.1 The hydrologic cycle.

bite off rain in small chunks and keep it close to where it falls. Linsley et al. (1975) wrote:

> This cycle (Fig. 12.1) is visualized as beginning with the evaporation of water from the oceans. The resulting vapor is transported by moving air masses. Under the proper conditions, the vapor is condensed to form clouds, which in turn may result in precipitation. The precipitation which falls upon land is dispersed in several ways. The greater part is temporarily retained in the soil near where it falls and is ultimately returned to the atmosphere by evaporation and transpiration by plants. A portion of the water finds its way over and through the surface soil to stream channels, while other water penetrates farther into the ground to become part of the groundwater. Under the influence of gravity, both surface streamflow and groundwater move toward lower elevations and may eventually discharge into the ocean. However, substantial quantities of surface and underground water are returned to the atmosphere by evaporation and transpiration before reaching the oceans.

Table 12.1 shows it as a three-tier process. The top tier is precipitation (P) in its many forms, but only rain is considered in design. A middle tier has three processes into which rain can be divided: initial abstraction (Ia), infiltration (F), and surface runoff (SRO). A third tier is also composed of three items: evapotranspiration (ET), groundwater (GW), and stream flow (Q).

			P			
	Ia		F		SRO	
ET			GW			Q

TABLE 12.1 Hydrologic Cycle

Ia is made up of interception and depression storage. Interception is rain captured before it falls to earth. Leaves hold large quantities of rain. Studies show in a hardwood forest, if one inch of rain falls, only a quarter inch is caught in a gage. If wind blows, rain can blow off and become infiltration or surface runoff. Depression storage is rain captured in surface depressions in fields, parking lots, and gutters. Ia is subtracted from rain since it is not part of runoff.

Infiltration is entrance of water through the earth's surface and is subtracted from rain since it is not part of runoff. Hydrograph methods include infiltration in many ways: some are simple, and some are complex. Surface runoff is rain's remainder after subtracting Ia and F that flows into drainage ways, such as a swale, creek, stream, river, lake, and/or ocean.

ET is evapotranspiration in inches: evaporation from surfaces and upper layers of soil plus transpiration from leaves. Groundwater is a portion of infiltration that moves through root zones and penetrates a soil mass to reach groundwater. Stream flow is surface runoff that reaches drainage networks plus groundwater, when a water table is above a channel's invert.

A goal of hydrograph methods is to model the hydrologic cycle and estimate the amounts of rain that become streamflow, both as flow rates and volumes. Some are simple models such as the rational formula (Kuichling, 1889, with permission from ASCE) and some are complex such as the Environmental Protection Agency (EPA) Stormwater Management Model (SWMM) computer program (Metcalf and Eddy, 1971). More complex models do not guarantee better results. My preference is the NRCS model (USDA, 1986).

Unfortunately, either simple or complex, these models yield results that for a given set of storm and watershed factors are hardly ever in agreement. Their range at times is an order of magnitude apart. It is left to you to decide which method to use. My preference is never to use the rational formula or other peak flow methods for BMP design. BMPs need a hydrograph to estimate both peak runoff rates and volumes.

12.3 Hydrograph Methods

A hydrograph plot includes both a peak flow and runoff volume (area under a hydrograph). This is true due to units used on ordinates (cfs) and abscissas (hours) as shown in Fig. 12.2. An area of a small square under it has a unit of cfs times hours or cfs-hours. An acre's area is 43,560 ft². If water ponds a foot deep over an acre, there is a volume of 43,560 times one foot or 43,560 ft³ or one acre-foot (AF). If we pond water an inch deep over an acre, there is a volume of 43,560 times 1/12 ft or 3,630 ft³ or one acre-inch.

If we run water into a tank at a rate of 1 cfs for 1 hour, we have a volume of water equal to 1 cfs times 60 seconds per minute times 60 minutes per hour, which is 3,600 cubic feet or one cfs-hour. One cfs-hour equals one acre-inch, 3,630/3,600 = 1.008, within 0.8 of 1.0 percent. Since 3,630 and 3,600 are volumes, an acre-inch and a cfs-hour are volumes. Thus, area under a hydrograph *is* the runoff volume. Volume needed to reduce a peak inflow to a lesser peak outflow is the area between the two hydrographs. See Fig. 12.3.

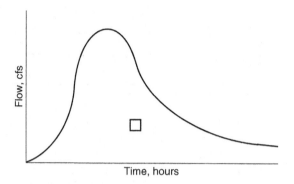

FIGURE 12.2   A typical runoff hydrograph.

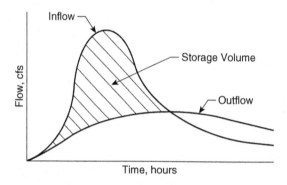

FIGURE 12.3   Storage volume needed.

Respondents to a questionnaire distributed by the American Public Works Association (APWA) in the 1980s said they used 44 methods to develop hydrographs. Applying these methods yield highly variable results. Each models the hydrologic cycle but do it differently. NRCS method uses a runoff curve number (CN), a number between zero and 100. Another method models the infiltration portion with partial differential equations solved by a finite difference method.

Federal agency methods that have withstood the test of time are TR-20 (SCS, 1984) and HEC-1 (USCOE, 1981), now Hydrologic Engineering Center-Hydrologic Modeling System (HEC-HMS). These programs can be downloaded free from the Internet. The NRCS method is used in both TR-20 and HEC-HMS. These two programs just develop and rout hydrographs. Depth-storage and depth-outflow relationships are also needed. For the TR-20 and HEC-HMS programs, these are inputs needing other programs or manual methods to estimate them. Then, the results are input to either program to route hydrographs. The McGraw-Hill website (www.mhprofessional.com/sdsd) has spreadsheets needed to develop hydrographs, depth-storage curves, depth-outflow curves, and to route them. The TR-55 (USDA, 1986) manual is needed for use with the spreadsheets I developed.

Just because you have a program or spreadsheets that do this, does not mean you can design BMPs. To be an effective designer requires that you be in control of the process. First, you must understand all equations involved, their nuances, and values used for variables in certain situations. Second, you must be confident the programs or spreadsheets are correct. Third, you must be aware of many other items, take them into consideration, make decisions about them, and incorporate them into your design before you truly have designed a BMP. The following chapters and Apps. A through J discuss and illustrate these.

12.4 Hydrograph Development

Four spreadsheets are used to develop inflow hydrographs using the NRCS method. These are time of concentration (Tc), CN, hydrograph variables, and hydrograph ordinates.

12.4.1 Time of Concentration

There are three Tc spreadsheets that look different but contain the same data. Table 12.2 or Table 12.3 is used for a few subareas or a single subarea in which one or more flow paths has more than one slope. Table 12.4 is for multiple subareas or for a watershed in which one or more flow paths has more than one slope. These use NRCS' method that estimates velocities in three paths: sheet flow (SF), shallow concentrated flow (SCF), and channel flow (CF). Developing Tc estimates was covered in Chapter 8. Explanations of each spreadsheet line and column are included on the McGraw-Hill website (www.mhprofessional.com/sdsd).

If one or more flow paths is not needed, fill in values for those variables to eliminate all divide-by-zero symbols. If you have one, two, or three subareas with a single slope in each flow path, then read total Tc for each subarea at the form's bottom. If you have more than one slope in any flow path in a single subarea, then read Tc in the form's lower right-hand corner. If you have more than three subareas or more than one slope in more than one path, then use Table 12.4.

For Table 12.4, begin by filling in a table number and complete its title at the form's top. Five values must be filled in near this form's top even if one or more flow paths is not present in your site. Values are filled in, but you may need to change some or them for your site. The first is Manning's n for sheet flow. Table 3-1 in TR-55 (USDA, 1986) lists several values for various covers.

The second is rainfall amount for a 2-year, 24-hour event. Appendix B in TR-55 includes a map for the eastern two-thirds of the United States (USDA, 1986). If your site is in the northern portion of the Western States, use Atlas 2 (NWS, 1983). If your site is in the southern portion of the Western States, use Atlas 14 (NWS, 2003). The third is Manning's n for open channels (See App. J). The fourth is street cross slope. Input whatever value is used in your community. The last is Manning's n in a gutter. Street roughness values range from 0.012 to 0.016 depending on surface type and finish.

Now add numbers to the form. Values in rows and columns must be input only if different for your site. Change numbers as needed for different flow paths and subareas. Each line is a different subarea or, if some flow paths have more than one slope, use a different line for each slope. In these cases, use the same subarea value in Col. 1. Tc is total time for all rows with the same value in Col. 1. Remember, Tc values are only estimates.

Overland (Sheet)	1	2	3	Total
Pathway Length, ft*	0.0	0.0	0.0	
Upstream Elevation, ft*	0.1	0.1	0.1	
Downstream Elevation, ft*	0	0	0	
Pathway Slope, ft/ft	10.00000	10.00000	10.00000	
Manning's n*				
2-yr, 24-hr Rainfall, in.*				
Flow Velocity, fps	#DIV/0!	#DIV/0!	#DIV/0!	
Travel Time, min	#DIV/0!	#DIV/0!	#DIV/0!	#DIV/0!
Shallow Concentrated	**1**	**2**	**3**	
Pathway Length, ft*	0.0	0.0	0.0	
Upstream Elevation, ft*	0.1	0.1	0.1	
Downstream Elevation, ft*	0	0	0	
Pathway Slope, ft/ft	10	10	10	
Equation Coefficient*				
Flow Velocity, fps	0.00	0.00	0.00	
Travel Time, min	#DIV/0!	#DIV/0!	#DIV/0!	#DIV/0!
Channel	**1**	**2**	**3**	
Pathway Length, ft*	0.0	0.0	0.0	
Upstream Elevation, ft*	0.1	0.1	0.1	
Downstream Elevation, ft*	0	0	0	
Pathway Slope, ft/ft	10	10	10.00000	
Bottom Width, ft*	1	1	1	
Flow Depth, ft*	1	1	1	
Side Slope, H:V*	1	1	1	
Area, sq ft	2.0	2.0	2.0	
Wetted Perimeter, ft	3.8	3.8	3.8	
Manning's n*				
Flow Velocity, fps	#DIV/0!	#DIV/0!	#DIV/0!	
Travel Time, min	#DIV/0!	#DIV/0!	#DIV/0!	#DIV/0!
Gutter	**1**	**2**	**3**	
Pathway Length, ft*	0.0	0.0	0.0	
Upstream Elevation, ft*	0.1	0.1	0.1	
Downstream Elevation, ft*	0	0	0	
Pathway Slope, ft/ft	10.00000	10.00000	10.00000	
Street Cross Slope, ft/ft*	0.02	0.02	0.02	
Flow Depth, ft*	0.25	0.25	0.25	
Flow Top Width, ft	12.5	12.5	12.5	
Manning's n*	0.016	0.016	0.016	
Flow Velocity, fps	87.84	87.84	87.84	
Travel Time, min	0.0	0.0	0.0	0.0
Total Travel Time, Min	#DIV/0!	#DIV/0!	#DIV/0!	#DIV/0!

*Insert a number in those columns that are currently blank or change the numbers in those columns for which the values are different for your site.

TABLE 12.2   Time of Concentration Spreadsheet

Time of Concentration Calculations

Nsh = 0.011
0.240

P2 = 3.2 in.

	Sheet Flow					Shallow Concentrated Flow						
Subarea	Length	El. Up	El. Dn.	Slope	Travel Time	Length	El. Up	El. Dn.	Slope	Travel Time	Length	El. Up
	ft	ft	ft	ft/ft	min	ft	ft	ft	ft/ft	min	ft	ft
1	2	3	4	5	6	7	8	9	10	11	12	13
Parking	1	0.00		0.00000	#DIV/0!	1	0.0		0.01000	0.0	1	0.0
Spreader	1	0.00		0.00000	#DIV/0!	1	0.0		0.01000	0.0	1	0.0
Swale	1	0.0		0.00000	#DIV/0!	1	0.0		0.01000	0.0	1	0.0
	1	0.0		0.00000	#DIV/0!	1	0.0		0.01000	0.0	1	0.0
	1	0.0		0.00000	#DIV/0!	1	0.0		0.01000	0.0	1	0.0
	1	0.0		0.00000	#DIV/0!	1	0.0		0.01000	0.0	1	0.0
	1	0.0		0.00000	#DIV/0!	1	0.0		0.01000	0.0	1	0.0
	1	0.0		0.00000	#DIV/0!	1	0.0		0.01000	0.0	1	0.0

Nch = 0.05   Sx = 0.0208 ft/ft   Ngu = 0.016

Channel Flow						Gutter Flow						Tc
El. Dn.	Slope	Area	Wet. Per.	R	Travel Time	Length	El. Up	El. Dn.	Slope	Top Width	Travel Time	
ft	ft/ft	sq ft	ft	ft	min	ft	ft	ft	ft/ft	ft	min	min
14	15	16	17	18	19	20	21	22	23	24	25	26
	0.01000	1.0	1.0	1.000	0.0	1	0.0		0.01000	1.0	0.0	#DIV/0!
	0.01000	1.0	1.0	1.000	0.0	1	0.0		0.01000	1.0	0.0	#DIV/0!
	0.01000	1.0	1.0	1.000	0.0	1	0.0		0.01000	1.0	0.0	#DIV/0!
	0.01000	1.0	1.0	1.000	0.0	1	0.0		0.01000	1.0	0.0	#DIV/0!
	0.01000	1.0	1.0	1.000	0.0	1	0.0		0.01000	1.0	0.0	#DIV/0!
	0.01000	1.0	1.0	1.000	0.0	1	0.0		0.01000	1.0	0.0	#DIV/0!
	0.01000	1.0	1.0	1.000	0.0	1	0.0		0.01000	1.0	0.0	#DIV/0!
	0.01000	1.0	1.0	1.000	0.0	1	0.0		0.01000	1.0	0.0	#DIV/0!

TABLE 12.3   Time of Concentration Spreadsheet

Nsh = 0.24					P2 = 3.5 in.						
	Sheet Flow					**Shallow Concentrated Flow**						
Subarea	Length ft	El. Up ft	El. Dn. ft	Slope ft/ft	Travel Time min	Length ft	El. Up ft	El. Dn. ft	Slope ft/ft	Travel Time min	Length ft	El. Up ft
1	2 *	3 *	4 *	5	6	7 *	8 *	9 *	10	11	12 *	13 *
	0.1	0.0		0.30000	0.0	1	0.0		0.01000	0.0	0	0.0
	0.1	0.0		0.30000	0.0	1	0.0		0.01000	0.0	1	0.0
	0.1	0.0		0.30000	0.0	1	0.0		0.01000	0.0	1	0.0
	0.1	0.0		0.30000	0.0	1	0.0		0.01000	0.0	1	0.0
	0.1	0.0		0.30000	0.0	1	0.0		0.01000	0.0	1	0.0
	0.1	0.0		0.30000	0.0	1	0.0		0.01000	0.0	1	0.0
	0.1	0.0		0.30000	0.0	1	0.0		0.01000	0.0	1	0.0
	0.1	0.0		0.30000	0.0	1	0.0		0.01000	0.0	1	0.0
	0.1	0.0		0.30000	0.0	1	0.0		0.01000	0.0	1	0.0
	0.1	0.0		0.30000	0.0	1	0.0		0.01000	0.0	1	0.0
	0.1	0.0		0.30000	0.0	1	0.0		0.01000	0.0	1	0.0
	0.1	0.0		0.30000	0.0	1	0.0		0.01000	0.0	1	0.0
	0.1	0.0		0.30000	0.0	1	0.0		0.01000	0.0	1	0.0
	0.1	0.0		0.30000	0.0	1	0.0		0.01000	0.0	1	0.0
	0.1	0.0		0.30000	0.0	1	0.0		0.01000	0.0	1	0.0
	0.1	0.0		0.30000	0.0	1	0.0		0.01000	0.0	1	0.0
	0.1	0.0		0.30000	0.0	1	0.0		0.01000	0.0	1	0.0
	0.1	0.0		0.30000	0.0	1	0.0		0.01000	0.0	1	0.0
	0.1	0.0		0.30000	0.0	1	0.0		0.01000	0.0	1	0.0
	0.1	0.0		0.30000	0.0	1	0.0		0.01000	0.0	1	0.0
	0.1	0.0		0.30000	0.0	1	0.0		0.01000	0.0	1	0.0
	0.1	0.0		0.30000	0.0	1	0.0		0.01000	0.0	1	0.0
	0.1	0.0		0.30000	0.0	1	0.0		0.01000	0.0	1	0.0
	0.1	0.0		0.30000	0.0	1	0.0		0.01000	0.0	1	0.0

TABLE 12.4   Time of Concentration Calculations

Nch = 0.05						Sx = 0.0208 ft/ft						Ngu = 0.016	
Channel Flow						**Gutter Flow**							**Tc**
El. Dn. ft 14 *	Slope ft/ft 15	Area sq ft 16 *	Wet. Per. ft 17 *	R ft 18	Trav. Time min 19	Length ft 20 *	El. Up ft 21 *	El. Dn. ft 22 *	Slope ft/ft 23	Top Wid. ft 24 *	Trav. Time min 25	min 26	
	1.00000	1.0	1.0	1.000	0.00	0	0.0		0.10000	1.0	0.001	0.0
	0.01000	1.0	1.0	1.000	0.0	0	0.0		0.10000	1.0	0.00	0.0
	0.01000	1.0	1.0	1.000	0.0	0	0.0		0.10000	1.0	0.0	0.0
	0.01000	1.0	1.0	1.000	0.0	0	0.0		0.10000	1.0	0.0	0.0
	0.01000	1.0	1.0	1.000	0.0	0	0.0		0.10000	1.0	0.0	0.0
	0.01000	1.0	1.0	1.000	0.0	0	0.0		0.10000	1.0	0.0	0.0
	0.01000	1.0	1.0	1.000	0.0	0	0.0		0.10000	1.0	0.0	0.0
	0.01000	1.0	1.0	1.000	0.0	0	0.0		0.10000	1.0	0.0	0.0
	0.01000	1.0	1.0	1.000	0.0	0	0.0		0.10000	1.0	0.0	0.0
	0.01000	1.0	1.0	1.000	0.0	0	0.0		0.10000	1.0	0.0	0.0
	0.01000	1.0	1.0	1.000	0.0	0	0.0		0.10000	1.0	0.0	0.0
	0.01000	1.0	1.0	1.000	0.0	0	0.0		0.10000	1.0	0.0	0.0
	0.01000	1.0	1.0	1.000	0.0	0	0.0		0.10000	1.0	0.0	0.0
	0.01000	1.0	1.0	1.000	0.0	0	0.0		0.10000	1.0	0.0	0.0
	0.01000	1.0	1.0	1.000	0.0	0	0.0		0.10000	1.0	0.0	0.0
	0.01000	1.0	1.0	1.000	0.0	0	0.0		0.10000	1.0	0.0	0.0
	0.01000	1.0	1.0	1.000	0.0	0	0.0		0.10000	1.0	0.0	0.0
	0.01000	1.0	1.0	1.000	0.0	0	0.0		0.10000	1.0	0.0	0.0
	0.01000	1.0	1.0	1.000	0.0	0	0.0		0.10000	1.0	0.0	0.0
	0.01000	1.0	1.0	1.000	0.0	0	0.0		0.10000	1.0	0.0	0.0
	0.01000	1.0	1.0	1.000	0.0	0	0.0		0.10000	1.0	0.0	0.0
	0.01000	1.0	1.0	1.000	0.0	0	0.0		0.10000	1.0	0.0	0.0
	0.01000	1.0	1.0	1.000	0.0	0	0.0		0.10000	1.0	0.0	0.0

*Insert a number in those columns that are currently blank or change the numbers in those columns for which the values are different for your site.

Table 12.4 Time of Concentration Calculations (*Continued*)

12.4.2   Runoff Curve Number

For Table 12.5, begin by filling in a table number and complete its title at the form's top. The site could have just a single subarea, a few, or several subareas. Each subarea contains only a single land use and soil type. Use a separate line for each subarea. Hydrologic soil group (HSG) A, B, C, or D for over 5,000 soil names are listed in TR-55, App. A (USDA, 1986). Each HSG for some land use has a different CN, listed in Chap. 2 of TR-55 (USDA, 1986). Acres or percent of each CN in a subarea is listed in Col. 5.

Column 6 is calculated and uses totals of Col. 5 and Col. 6 to determine an average CN. Then, on the line "Use CN=", enter CN to the nearest whole number from that on the line above. Near the bottom, enter years and rainfall amounts for storm return periods you will be using. The spreadsheet calculates runoff amounts for these rainfall amounts and CNs.

Worksheet 2						
Project:		By:			Date:	
Location:		Ckd:			Date:	
Outline one: Present Developed						

1. Runoff Curve Number (CN)

Subarea Name 1	Soil Name and Hydrologic Group (App. A) 2	Cover Description (cover type, treatment, and hydrologic condition: percent imperviousness; unconnected/ connected impervious area ratio) 3	Curve Number 4		Area acres or percent 5		Product of CN × Area 6
							0
							0
							0
							0
							0
							0
							0
							0
							0

Total =   [0]   [0]

Weighted Curve Number =   product/area = 0   over 0 = #DIV/0!

Use CN =

2. Runoff

	Storm #1	Storm #2	Storm #3
Frequency . . . . yr			
Rainfall, (P) 24-hour . . . . in			
Runoff, Q . . . . in	#DIV/0!	#DIV/0!	#DIV/0!

TABLE 12.5   Curve Number and Runoff

12.4.3 Subareas

Each subarea contains a land use and a soil type. There is just one subarea for most BMPs. Dividing a watershed or retrofit site into a number of subareas is an art. In some cases, subareas are needed for a good hydrograph representation of a basin or retrofit site for these reasons.

1. Change in land use
2. Change in soil type
3. Confluence of two channels
4. Street crossings where a hydrograph is need to determine size of culvert or bridge
5. Locations of an existing or a proposed detention basin or BMP site
6. In order to maintain a maximum subarea size, i.e., 40 acres

Never include a park and a mall in a single subarea. Runoff from them are too different. Its hydrograph is a poor representation of either. Use two subareas, then add the two hydrographs.

12.4.4 Hydrograph Variables

For Table 12.6, begin by filling in a table number and complete its title at the form's top. Enter a project name, location, return period, and underline "undeveloped" or "developed." A single or several return periods can be entered on the form. Simply change rain amount in Col. 7. Use a separate line for each subarea. You must enter values in Cols. 1 through 8 and 11.

Columns 4 through 6 need some explanation. Hydrographs are developed at a subarea's downstream end. Figure 12.4 portrays a site with seven subareas with a basin proposed at subarea 7's downstream end. Thus, hydrographs for subareas 1 and 2 are at their confluence at subarea 3's upstream end. With a detention basin at subarea 7's downstream end, both hydrographs must be transposed through stream reaches in subareas 3, 5, and 7. Hydrographs from subareas 3 through 6 must be transposed to subarea 7's downstream end. Subarea 7's hydrograph is already there.

Travel time in a reach for a subarea listed in Col. 1 is placed in Col. 4. Estimate velocities using Fig. 3-1 in TR-55 (USDA, 1986), then use a reach's length to get travel time in hours. Names of subareas through which a hydrograph flows are listed in Col. 5. Place total travel time in Col. 6. For areas 1 and 2, this would be total travel times in subareas 3, 5, and 7. See TR-55 for examples.

Values of precipitation (P) and CNs from the CN table are placed in Cols. 7 and 8. Cols. 9 and 10 have equations buried in them. Col. 10 values are the product of Cols. 2 and 9; use five decimal places. Ia in Col. 11 is obtained from Table 5-1 in TR-55 (USDA, 1986). Column 12 values are Col. 11/Col. 7.

12.4.5 Hydrograph Ordinates

The upper portion of Table 12.7 is filled in similar to that for Table 12.6. However, hydrograph ordinates for only a single return period can be developed in Table 12.7. If you are developing hydrographs for five storm magnitudes, you must fill in five tables. The only difference between these tables will be the value from Col. 10 in Table 12.6.

Worksheet 5a

Project: _____ Location: _____ By: _____ Date: _____

Outline one: Present   Developed   Frequency (yr): _____ Checked by: _____ Date: _____

Subarea Name	Drainage Area sq. mi.	Time of Concen. hr.	Travel Time thru Subarea hr.	Downstream Subarea Names	Travel Time Summation to Outlet	24-hr Rainfall in.	Runoff Curve Number	Runoff in.	AmQ sq. mi.-in.	Initial Abstraction in.	Ia/P
1	2	3	4	5	6	7	8	9	10	11	12
								#DIV/0!	#DIV/0!		#DIV/0!
									#DIV/0!		#DIV/0!
								#DIV/0!	#DIV/0!		#DIV/0!
								#DIV/0!	#DIV/0!		#DIV/0!
								#DIV/0!	#DIV/0!		#DIV/0!
								#DIV/0!	#DIV/0!		#DIV/0!
								#DIV/0!	#DIV/0!		#DIV/0!
								#DIV/0!	#DIV/0!		#DIV/0!

TABLE 12.6 Hydrograph Variables

118

		#DIV/0!								#DIV/0!
		#DIV/0!								#DIV/0!
		#DIV/0!								#DIV/0!
		#DIV/0!								#DIV/0!
		#DIV/0!								#DIV/0!
		#DIV/0!								#DIV/0!
		#DIV/0!								#DIV/0!

TABLE **12.6** Hydrograph Variables (Continued)

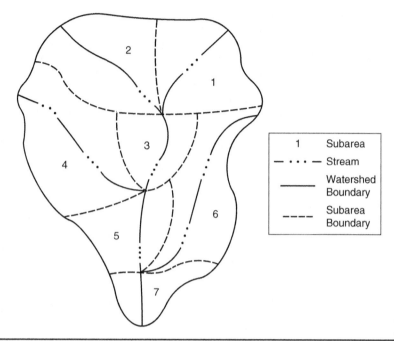

Figure 12.4 Watershed with seven subareas.

Chapter 5 in TR-55 (USDA, 1986) has numerous tables for hydrograph ordinates. Each page is for a specific storm type and Tc. Figure B-2 in App. B of TR-55 (UDSA, 1986) is a U.S. map showing those areas in which these storm types are used.

First five columns are listed in lines 20 and 38 of Table 12.7 for a single subarea, list more subareas on even lines. List names in Col. 1. In Table 12.6, Col. 3 and Col. 6 were entered to two decimals. In Table 12.7, Cols. 2 and 3 are entered to one decimal place since the tables in Chap. 5 of TR-55 tables list them that way. This excerpt from TR-55 (USDA, 1986) converts two places to one place:

> Since the timing of peak discharges changes with Tc and Tt, interpolation of peak discharge for Tc and Tt values for use in exhibit 5 is not recommended. Interpolation may result in an estimate of peak discharge that would be invalid because it would be lower than either of the hydrographs. Therefore, round the actual values of Tc and Tt to values presented in exhibit 5. Perform this rounding so that the sum of the selected table values is close to the sum of actual Tc and Tt. An acceptable procedure is to use the result of one of three rounding operations:
>
> 1. Round Tc and Tt separately to the nearest table value and sum;
> 2. Round Tc down and Tt up to nearest table value and sum; and
> 3. Round Tc up and Tt down to nearest table value and sum.

From these three alternatives, choose the pair of rounded Tc and Tt values whose sum is closest to the sum of the actual Tc and Tt. If two rounding methods produce sums equally close to the actual sum, use the combination in which the rounded Tc is closest to actual Tc.

An illustration of this rounding procedure is shown in Table 12.8.

Worksheet 5b

Project: Location: By: Date:

Outline one: Present   Developed   Frequency (yr):   Checked by: Date:

Subarea Name 1	Subarea Tc hr 2	ΣTt to Outlet hr 3	Ia/P 4	AmQ sq. mi.-in. 5	11.0 / 6	11.3 / 7	11.6 / 8	11.9 / 9	12.0 / 10	12.2 / 11	12.4 / 12	12.6 / 13	12.8 / 14	13.0 / 15	13.2 / 16	13.4 / 17
					0.0	0.0	0.0	0.0	0.0	0.0	0.0	0.0	0.0	0.0	0.0	0.0
					0.0	0.0	0.0	0.0	0.0	0.0	0.0	0.0	0.0	0.0	0.0	0.0
					0.0	0.0	0.0	0.0	0.0	0.0	0.0	0.0	0.0	0.0	0.0	0.0
					0.0	0.0	0.0	0.0	0.0	0.0	0.0	0.0	0.0	0.0	0.0	0.0
					0.0	0.0	0.0	0.0	0.0	0.0	0.0	0.0	0.0	0.0	0.0	0.0
					0.0	0.0	0.0	0.0	0.0	0.0	0.0	0.0	0.0	0.0	0.0	0.0
Total					0.0	0.0	0.0	0.0	0.0	0.0	0.0	0.0	0.0	0.0	0.0	0.0

Basic Watershed Data Used — Select and enter hydrograph times in hours from exhibit 5- — Discharge at selected hydrograph times, cfs

TABLE 12.7 Hydrograph Ordinates

Worksheet 5b

Project:			Location:		By:		Date:
Outline one: Present Developed			Frequency (yr):		Checked by:		Date:

	Basic Watershed Data Used				Select and enter hydrograph times in hours from exhibit 5-											
Subarea Name	Subarea Tc hr	ΣTt to Outlet hr	Ia/P	AmQ sq. mi.–in.	13.6	13.8	14.0	14.3	14.6	15.0	16.0	17.0	18.0	20.0	22.0	26.0
1	2	3	4	5						Discharge at selected hydrograph times cfs						
					6	7	8	9	10	11	12	13	14	15	16	17
					Select and enter hydrograph times in hours from exhibit 5-											
					13.6	13.8	14.0	14.3	14.6	15.0	16.0	17.0	18.0	20.0	22.0	26.0
					0.0	0.0	0.0	0.0	0.0	0.0	0.0	0.0	0.0	0.0	0.0	0.0
					0.0	0.0	0.0	0.0	0.0	0.0	0.0	0.0	0.0	0.0	0.0	0.0
					0.0	0.0	0.0	0.0	0.0	0.0	0.0	0.0	0.0	0.0	0.0	0.0
					0.0	0.0	0.0	0.0	0.0	0.0	0.0	0.0	0.0	0.0	0.0	0.0
					0.0	0.0	0.0	0.0	0.0	0.0	0.0	0.0	0.0	0.0	0.0	0.0
Total					0.0	0.0	0.0	0.0	0.0	0.0	0.0	0.0	0.0	0.0	0.0	0.0

TABLE **12.7** Hydrograph Ordinates (*Continued*)

122

Variable	Actual Values	Table Values by Rounding Method		
		1	**2**	**3**
Tc	1.1	1.0	1.0	1.25
Tt	1.7	1.5	2.0	1.50
Sum	2.8	2.5	3.0	2.75

TABLE **12.8**   TR-55 Methods for Rounding

Place rounding results for each subarea in Cols. 2 and 3 of Table 12.7 for Tc and Tt. Round values of Ia/P in Col. 12 of Table 12.6 for each subarea to 0.1, 0.3, or 0.5 and place them in Col. 4 of Table 12.7. Place values of AmQ in Col. 10 of Table 12.6 in Col. 5 of Table 12.7 to five places. Remember to place your values in Cols. 1 through 5 of Table 12.7 on lines 20 and 38 for a single subarea and even numbered lines for more subareas.

Values in Cols. 6 through 17 at various times are selected as follows. Select a time for ΔT. As a general rule, for subareas less than 25 to 50 acres, use ΔT as 0.1 hour; 50 to 200 acres, use ΔT as 0.2 hour; for larger subareas, use ΔT as 0.5 or 1.0 hour. Times on the form range from 11.0 to 26.0. Other copies of this spreadsheet begin at hour 9 for other storm types. This form uses 0.2 hour between 11.0 and 26.0. Between hours 11 and 12 and after hour 13, use the values listed. Between the hours of 12.0 and 13.0, remember to use only those values for 0.2 hour, i.e., 12.2, 12.4, 12.6, and 12.8. Other copies of this spreadsheet have times every 0.1 hour between hours 11 and 14.

Open TR-55 to page 5-31. Each page is for a storm type and Tc. Its title is "Exhibit 5-II," a Type II storm. Tc is listed as 0.3 hours in the page's middle. The page is divided vertically into three parts. Near this page's left-hand side, IA/P is listed as 0.1, 0.3, and 0.5. If a subarea has Ia/P of 0.1 in Col. 4 of Table 12.7, then use the page's top third. Finally, if total travel time is 0.2 hour in Col. 3 of Table 12.7, then use the third line in the top third of this page. Values along this line are placed on even numbered lines of Table 12.7 for times listed on Table 12.7.

Lastly, page 5-31's title in TR-55 includes units of "csm/in," cfs per square mile of drainage area per inch of runoff. If our watershed was one square mile in area and our runoff was one inch, then values listed on each line of page 5-31 would be hydrograph ordinates for the values listed in Col. 2, 3, and 4 in Table 12.7. This is the reason for Col. 5 on Table 12.7. If we multiply cfs/sq. mi. in. by sq. mi. in., we get cfs. This is done automatically in the spreadsheet. To set up Cols. 6 thru 17 in Table 12.7, enter values in Cols. 1 through 5 from Table 12.6. Then use the above procedure. Values in Cols. 6 through 17 on odd numbered lines are added to obtain our final hydrograph values for each time listed above on the two lines labeled "total."

12.5   Potential Attenuation of Inflow Hydrographs

As hydrographs enter a BMP, a portion of its volume is stored forming a temporary pond. This storage and a reduced outlet size changes an inflow hydrograph's shape by reducing its peak and extending its time base. Peak time is delayed from a half to three or more hours for a water quantity storm event.

NRCS (USDA, 1986) developed Fig. 12.5 to estimate peak flow reduction by equating it to a ratio of a BMP's storage volume to inflow volume. A peak is reduced by 90 percent for Type II storms if half its runoff volume is stored. I developed a spreadsheet (Table 12.9) to estimate peak flow reduction for various storm magnitudes.

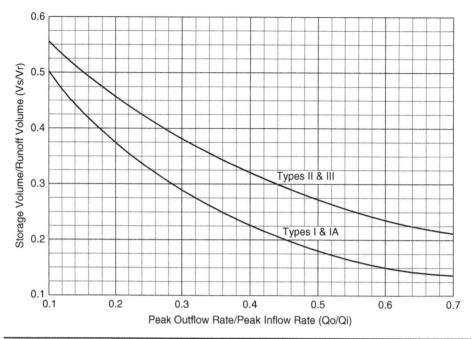

FIGURE 12.5   Approximate detention basin routing for NRCS rainfall types.

Flow Rate Reduction Calculations				
Drainage Area = * **sq. mi. =** *			**Rainfall Distri. =** *	
Frequency	**2-Year**		**10-Year**	**100-Year**
Peak Inflow∗				
Peak Outflow∗				
Qin/Qout	#DIV/0!		#DIV/0!	#DIV/0!
Vs/Vr∗				
Runoff, Q∗				
Runoff, Vol.	#Value!		#Value!	#Value!
Storage Vol.	#Value!		#Value!	#Value!
Max. Elev.∗				

TABLE 12.9   Flow-Rate Reductions

Enter numbers into those rows and locations that have an asterisk in them. The other rows have equations buried in them. Obtain values for the "Max. Elev." row from your elevation-storage curve. This spreadsheet is available on the McGraw-Hill website (www.mhprofessional.com/sdsd).

12.6   Example 12.1

This example illustrates use of the NRCS forms to generate inflow hydrographs. Figure 12.6 is a topographic map of a site in a midwestern state divided into four subareas.

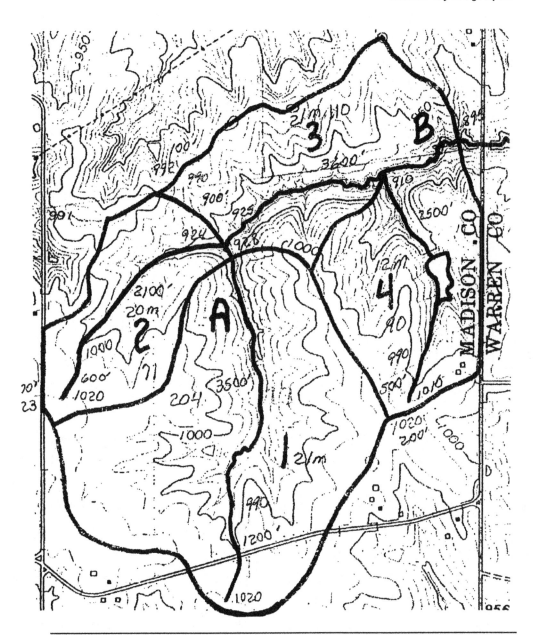

FIGURE 12.6   Topographic map for Example 12.1.

12.6.1   Times of Concentration

Tc subarea values are estimated in Table 12.10. Locations, elevations, and lengths for the three flow path types were estimated based on examining the topo's contours. You might select other locations, elevations, and lengths. Tc varied from 32.4 to 52.9 minutes, 0.54 to 0.88 hours.

Nsh = 0.24   P2 = 3.2 in.

Subarea	Sheet Flow					Shallow Concentrated Flow						
	Length	El. Up	El. Dn.	Slope	Travel Time	Length	El. Up	El. Dn.	Slope	Travel Time	Length	El. Up
	ft	ft	ft	ft/ft	min	ft	ft	ft	ft/ft	min	ft	ft
1	2	3	4	5	6	7	8	9	10	11	12	13
	*	*	*			*	*	*			*	*
1	100.0	1022.0	1020.0	0.02000	14.27	1200	1020.0	990.0	0.02500	7.84	3500	990.0
	0.1	0.0		0.30000	0.02	1	0.0		0.01000	0.01	1	0.0
2	300.0	1023.0	1020.0	0.01000	45.35	600	1020.0	1000.0	0.03333	3.39	2100	1000.0
	0.1	0.0		0.30000	0.02	1	0.0		0.01000	0.01	1	0.0
3	100.0	992.0	990.0	0.02000	14.27	900	990.0	925.0	0.07222	3.46	3600	925.0
	0.1	0.0		0.30000	0.02	1	0.0		0.01000	0.01	1	0.0
4	200.0	1020.0	1010.0	0.05000	17.22	500	1010.0	990.0	0.04000	2.58	3600	990.0
	0.1	0.0		0.30000	0.02	1	0.0		0.01000	0.01	1	0.0
Nch = 0.05				Sx = 0.0208		ft/ft					Ngu = 0.015	

El. Dn.	Channel Flow					Gutter Flow						Tc
ft	Slope	Area	Wet. Per.	R	Trav. Time	Length	El. Up	El. Dn.	Slope	Top Wid.	Trav. Time	min
	ft/ft	sq ft	ft	ft	min	ft	ft	ft	ft/ft	ft	min	
14	15	16	17	18	19	20	21	22	23	24	25	26
*		*	*			*	*	*		*		
928.0	0.01771	16.0	16.4	0.976	14.95	0	0.0		0.10000	1.0	0.001	37.06
	0.01000	1.0	1.0	1.000	0.01	0	0.0		0.10000	1.0	0.00	0.04
824.0	0.08381	16.0	16.4	0.976	4.12	0	0.0		0.10000	1.0	0.0	52.87
	0.01000	1.0	1.0	1.000	0.01	0	0.0		0.10000	1.0	0.0	0.04
895.0	0.00833	16.0	16.4	0.976	22.42	0	0.0		0.10000	1.0	0.0	40.15
	0.01000	1.0	1.0	1.000	0.01	0	0.0		0.10000	1.0	0.0	0.04
895.0	0.02639	16.0	16.4	0.976	12.60	0	0.0		0.10000	1.0	0.0	32.41
	0.01000	1.0	1.0	1.000	0.01	0	0.0		0.10000	1.0	0.0	0.04

TABLE 12.10 Times of Concentration Calculations for Example 12.1

			Worksheet 2			

Project: NW 97th Street   By: Date:
Location: Midwestern United States   Ckd: Date:
Outline one: Present   ☐ Developed

1. Runoff Curve Number (CN)

Subarea Name 1	Soil Name and Hydrologic Group (App. A) 2	Cover Description (cover type, treatment, and hydrologic condition: percent imperviousness; unconnected/connected impervious area ratio) 3		Curve Number 4	Area acres or percent 5	Product of CN × Area 6
all	C	Single-Family Residential		83	100	8300
						0
						0
						0
						0
						0
						0
						0
						0

Total = 100   8300

Weighted Curve Number = product/area = 8300   over 100 = 83.0

Use CN = 83

2. Runoff

	Storm #1	Storm #2	Storm #3
Frequency . . . . yr *	2	10	100
Rainfall, (P) 24-hour . . . . in *	3.2	4.6	6.6
Runoff, Q . . . . in	1.6	2.8	4.7

TABLE 12.11   Curve Number and Runoff for Example 12.1

12.6.2   Curve Number

Each of these four subareas were developed with single-family homes on quarter-acre lots. Soil type for the watershed is HSG C. CNs and runoff amounts for this site are estimated in Table 12.11. CN for each subarea are obtained from Chap. 2 in TR-55.

12.6.3   Hydrograph Variables

Variables are listed in Table 12.12. Note each subarea was repeated three times. First four lines are a 2-year; second four lines are a 10-year; third four lines are a 100-year, 24-hour storm event. The only column changed was Col. 7 with values taken from Table 12.11. Values in Col. 2 were obtained by planimetering each subarea on the topo map. Column 3 values were taken from Table 12.10. Values in Col. 4 were obtained

Worksheet 5a

Project: NW 97th Street
Outline one: Present | Developed
Location: Midwestern United States
Frequency (yr):
By:
Checked by:
Date:
Date:

Subarea Name 1	Drainage Area sq. mi. 2	Time of Concen. hr. 3	Travel Time thru Subarea hr. 4	Downstream Subarea Names 5	Travel Time Summation to Outlet 6	24-hr Rainfall in. 7	Runoff Curve Number 8	Runoff in. 9	AmQ sq. mi.–in. 10	Initial Abstraction in. 11	Ia/P 12
1	0.319	0.62		3	0.37	3.2	83	1.6	0.5133	0.410	0.128
2	0.111	0.88		3	0.37	3.2	83	1.6	0.1786	0.410	0.128
3	0.172	0.67	0.37	—	0.00	3.2	83	1.6	0.2768	0.410	0.128
4	0.141	0.54		—	0.12	3.2	83	1.6	0.2269	0.410	0.128
1	0.319	0.62		3	0.37	4.6	83	2.8	0.8979	0.410	0.089
2	0.111	0.88		3	0.37	4.6	83	2.8	0.3124	0.410	0.089
3	0.172	0.67	0.37	—	0.00	4.6	83	2.8	0.4841	0.410	0.089
4	0.141	0.54		—	0.12	4.6	83	2.8	0.3969	0.410	0.089
1	0.319	0.62		3	0.37	6.6	83	4.7	1.4838	0.410	0.062
2	0.111	0.88		3	0.37	6.6	83	4.7	0.5163	0.410	0.062
3	0.172	0.67	0.37	—	0.00	6.6	83	4.7	0.8000	0.410	0.062
4	0.141	0.54		—	0.12	6.6	83	4.7	0.6558	0.410	0.062

TABLE 12.12 Hydrograph Variables for Example 12.1

from Fig. 3-1 in TR-55 by lengths of the appropriate subareas. Column 5 numbers were taken from the topo map. Values in Col. 6 were taken from Col. 4. Columns 7 and 8 are values from Table 12.11. Values in Cols. 9 and 10 are calculated from equations buried in these columns. Ia in Col. 11 is obtained from Table 5-1 in TR-55 (USDA, 1986). Values in Col. 12 are calculated from an equation buried in the column.

12.6.4   Hydrograph Ordinates

Fill in the top of Table 12.13 as you did for Table 12.12. Use one return period on a form. If you want to develop three hydrographs, fill out three forms. However, just change the value Am times Q in Col. 5. Enter subarea names in Col. 1. Section 12.8 explained how to obtain values for Cols. 2, 3, and 4. Copy Col. 5 from Col. 10 of Table 12.12 to five places. Review Section 12.9 for values in Cols. 6 thru 17. Table 12.13 has hydrograph ordinates for a 2-year, Table 12.14 for a 10-year, and Table 12.15 for a 100-year, 24-hour storm. Each hydrograph should have at least five time periods before its peak flow and several periods after it.

12.6.5   Attenuation of Inflow Hydrographs

Assume a pond elevation is set at 915. Tables 12.13 through 12.15 estimated inflow peaks of 394, 689, and 1,138 cfs for return periods of 2, 10, and 100 years. Figure 11.8 is a graph of the site's storage volume. Fill in Table 12.16 as follows. Take peak inflows from Tables 12.13 to 12.15. Then select peak outflows but no less than 10 percent of peak inflows. Qin/Qout values are calculated for you. Select values of Vs/Vr from Fig. 12.5. Enter runoff values from Table 12.11. Runoff and storage volume are calculated for you. Then use Fig. 11.8 to obtain values of the pond's maximum water surface elevation. Results show the site can reduce inflows by about 90 percent. Maximum water surface during a 100-year storm is 928.5, 1.5 ft below the top of berm.

12.7   Example 12.2: Catfish Creek Tributary

12.7.1   Drainage Areas

Subarea sizes were estimated using a planimeter. See Figs. 12.7 and 11.1. They are plan views of a Catfish Creek tributary located in Northeast Iowa and a portion of its topo map, respectively. The berm on the topo map is currently a 4-lane divided highway. Sizes of the four subareas and total tributary area are listed in Table 12.17.

12.7.2   Soil Types

Soils in this 126-acre subwatershed are HSG B with infiltration rates of 0.30 in./h.

12.7.3   Antecedent Moisture Condition

There are three antecedent moisture conditions (AMCs) used in the NRCS methodology: I, II, and III. AMC I is used when plants are at the wilting point. AMC II is used when soils are at field capacity. AMC III is used when soils are saturated. Curve numbers and initial abstractions listed in TR-55 (USDA, 1986) are for AMC II, the usual moisture content used for design calculations.

Worksheet 5b

Project: NW 97th Street   [Developed]   Location: Midwestern Untied States   By: Date:

Outline one: Present   Frequency (yr): 2   Checked by: Date:

Basic Watershed Data Used · Select and enter hydrograph times in hours from exhibit 5- · Discharge at selected hydrograph times (cfs)

Subarea Name	Subarea Tc hr	ΣTt to Outlet hr	Ia/P	AmQ sq. mi.-in.	11.0	11.3	11.6	11.9	12.0	12.2	12.4	12.6	12.8	13.0	13.2	13.4
1	2	3	4	5	6	7	8	9	10	11	12	13	14	15	16	17
1	0.75	0.30	0.10	0.5133	11	14	19	26	30	47	113	256	379	360	277	196
					5.6	7.2	9.8	13.3	15.4	24.1	58.0	131.4	194.5	184.8	142.2	100.6
2	1.00	0.30	0.10	0.1786	9	12	16	22	24	35	70	152	256	323	310	254
					1.6	2.1	2.9	3.9	4.3	6.3	12.5	27.1	45.7	57.7	55.4	45.4
3	0.75	0.00	0.10	0.2768	13	18	24	36	46	115	294	424	369	252	172	123
					3.6	5.0	6.6	10.0	12.7	31.8	81.4	117.4	102.1	69.8	47.6	34.0
4	0.50	0.00	0.10	0.2269	17	23	32	57	94	308	529	402	226	140	96	74
					3.9	5.2	7.3	12.9	21.3	69.9	120.0	91.2	51.3	31.8	21.8	16.8
					0.0	0.0	0.0	0.0	0.0	0.0	0.0	0.0	0.0	0.0	0.0	0.0
					0.0	0.0	0.0	0.0	0.0	0.0	0.0	0.0	0.0	0.0	0.0	0.0
Total					14.7	19.5	26.5	40.2	53.7	132.1	271.9	367.1	393.7	344.0	266.9	196.8

Table 12.13 Hydrograph Ordinates for Example 12.1 at B—2-Year

| | | | | | Select and enter hydrograph times in hours from exhibit 5- | | | | | | | | | | | |
					13.6	13.8	14.0	14.3	14.6	15.0	15.5	16.0	16.5	17.0	17.5	18.0
1	0.75	0.30	0.10	0.5133	140	103	80	60	48	38	33	29	26	23	21	20
					71.9	52.9	41.1	30.8	24.6	19.5	16.9	14.9	13.3	11.8	10.8	10.3
2	1.00	0.30	0.10	0.1786	193	146	113	81	61	46	36	36	31	27	24	22
					34.5	26.1	20.2	14.5	10.9	8.2	6.4	6.4	5.5	4.8	4.3	3.9
3	0.75	0.00	0.10	0.2768	93	74	61	49	41	35	31	27	24	22	20	19
					25.7	20.5	16.9	13.6	11.3	9.7	8.6	7.5	6.6	6.1	5.5	5.3
4	0.50	0.00	0.10	0.2269	61	53	47	41	36	32	29	26	23	21	20	19
					13.8	12.0	10.7	9.3	8.2	7.3	6.6	5.9	5.2	4.8	4.5	4.3
					0.0	0.0	0.0	0.0	0.0	0.0	0.0	0.0	0.0	0.0	0.0	0.0
					0.0	0.0	0.0	0.0	0.0	0.0	0.0	0.0	0.0	0.0	0.0	0.0
Total					145.9	111.5	88.8	68.1	55.1	44.7	38.5	34.7	30.7	27.5	25.1	23.8

TABLE 12.13 Hydrograph Ordinates for Example 12.1 at B—2-Year (*Continued*)

Worksheet 5b

Project: NW 97th Street Location: Midwestern Untied States By: Date:

Outline one: Present [Developed] Frequency (yr): 10 Checked by: Date:

Basic Watershed Data Used					Select and enter hydrograph times in hours from exhibit 5-											
Subarea Name	Subarea Tc hr	ΣTt to Outlet hr	Ia/P	AmQ sq. mi.-in.	Discharge at selected hydrograph times cfs											
					11.0	11.3	11.6	11.9	12.0	12.2	12.4	12.6	12.8	13.0	13.2	13.4
1	2	3	4	5	6	7	8	9	10	11	12	13	14	15	16	17
1	0.75	0.30	0.10	0.8979	11	14	19	26	30	47	113	256	379	360	277	196
					9.9	12.6	17.1	23.3	26.9	42.2	101.5	229.9	340.3	323.2	248.7	176.0
2	1.00	0.30	0.10	0.3124	9	12	16	22	24	35	70	152	256	323	310	254
					2.8	3.7	5.0	6.9	7.5	10.9	21.9	47.5	80.0	100.9	96.8	79.3
3	0.75	0.00	0.10	0.4841	13	18	24	36	46	115	294	424	369	252	172	123
					6.3	8.7	11.6	17.4	22.3	55.7	142.3	205.3	178.6	122.0	83.3	59.5
4	0.50	0.00	0.10	0.3969	17	23	32	57	94	308	529	402	226	140	96	74
					6.7	9.1	12.7	22.6	37.3	122.2	210.0	159.6	89.7	55.6	38.1	29.4
					0.0	0.0	0.0	0.0	0.0	0.0	0.0	0.0	0.0	0.0	0.0	0.0
					0.0	0.0	0.0	0.0	0.0	0.0	0.0	0.0	0.0	0.0	0.0	0.0
Total					25.7	34.2	46.4	70.3	94.0	231.1	475.6	642.2	688.6	601.7	466.9	344.3

TABLE 12.14 Hydrograph Ordinates for Example 12.1 at B—10-Year

					13.6	13.8	14.0	14.3	14.6	15.0	15.5	16.0	16.5	17.0	17.5	18.0
										Select and enter hydrograph times in hours from exhibit 5-							
1	0.75	0.30	0.10	0.8979	140	103	80	60	48	38	33	29	26	23	21	20
					125.7	92.5	71.8	53.9	43.1	34.1	29.6	26.0	23.3	20.7	18.9	18.0
2	1.00	0.30	0.10	0.3124	193	146	113	81	61	46	36	36	31	27	24	22
					60.3	45.6	35.3	25.3	19.1	14.4	11.2	11.2	9.7	8.4	7.5	6.9
3	0.75	0.00	0.10	0.4841	93	74	61	49	41	35	31	27	24	22	20	19
					45.0	35.8	29.5	23.7	19.8	16.9	15.0	13.1	11.6	10.7	9.7	9.2
4	0.50	0.00	0.10	0.3969	61	53	47	41	36	32	29	26	23	21	19	19
					24.2	21.0	18.7	16.3	14.3	12.7	11.5	10.3	9.1	8.3	7.9	7.5
					0.0	0.0	0.0	0.0	0.0	0.0	0.0	0.0	0.0	0.0	0.0	0.0
					0.0	0.0	0.0	0.0	0.0	0.0	0.0	0.0	0.0	0.0	0.0	0.0
Total					255.2	195.0	155.3	119.2	96.3	78.1	67.4	60.7	53.8	48.1	44.0	41.6

TABLE 12.14 Hydrograph Ordinates for Example 12.1 at B—10-Year (Continued)

133

Worksheet 5b

Project: NW 97th Street		Location: Midwestern Untied States	By:		Date:
Outline one: Present [Developed]		Frequency (yr): 100	Checked by:		Date:

Basic Watershed Data Used

Select and enter hydrograph times in hours from exhibit 5-
Discharge at selected hydrograph times
cfs

Subarea Name	Subarea Tc hr	ΣTt to Outlet hr	Ia/P	AmQ sq. mi.-in.	11.0	11.3	11.6	11.9	12.0	12.2	12.4	12.6	12.8	13.0	13.2	13.4
1	2	3	4	5	6	7	8	9	10	11	12	13	14	15	16	17
1	0.75	0.30	0.10	1.4838	11	14	19	26	30	47	113	256	379	360	277	196
					16.3	20.8	28.2	38.6	44.5	69.7	167.7	379.9	562.4	534.2	411.0	290.8
2	1.00	0.30	0.10	0.5163	9	12	16	22	24	35	70	152	256	323	310	254
					4.6	6.2	8.3	11.4	12.4	18.1	36.1	78.5	132.2	166.8	160.1	131.1
3	0.75	0.00	0.10	0.8000	13	18	24	36	46	115	294	424	369	252	172	123
					10.4	14.4	19.2	28.8	36.8	92.0	235.2	339.2	295.2	201.6	137.6	98.4
4	0.50	0.00	0.10	0.6558	17	23	32	57	94	308	529	402	226	140	96	74
					11.1	15.1	21.0	37.4	61.6	202.0	346.9	263.6	148.2	91.8	63.0	48.5
					0.0	0.0	0.0	0.0	0.0	0.0	0.0	0.0	0.0	0.0	0.0	0.0
					0.0	0.0	0.0	0.0	0.0	0.0	0.0	0.0	0.0	0.0	0.0	0.0
Total					42.5	56.5	76.6	116.1	155.4	381.8	785.9	1061.2	1137.9	994.3	771.6	568.9

TABLE 12.15 Hydrograph Ordinates for Example 12.1 at B—100-Year

134

					13.6	13.8	14.0	14.3	14.6	15.0	15.5	16.0	16.5	17.0	17.5	18.0
1	0.75	0.30	0.10	1.4838	140	103	80	60	48	38	33	29	26	23	21	20
					207.7	152.8	118.7	89.0	71.2	56.4	49.0	43.0	38.6	34.1	31.2	29.7
2	1.00	0.30	0.10	0.5163	193	146	113	81	61	46	36	36	31	27	24	22
					99.6	75.4	58.3	41.8	31.5	23.7	18.6	18.6	16.0	13.9	12.4	11.4
3	0.75	0.00	0.10	0.8000	93	74	61	49	41	35	31	27	24	22	20	19
					74.4	59.2	48.8	39.2	32.8	28.0	24.8	21.6	19.2	17.6	16.0	15.2
4	0.50	0.00	0.10	0.6558	61	53	47	41	36	32	29	26	23	21	20	19
					40.0	34.8	30.8	26.9	23.6	21.0	19.0	17.1	15.1	13.8	13.1	12.5
					0.0	0.0	0.0	0.0	0.0	0.0	0.0	0.0	0.0	0.0	0.0	0.0
					0.0	0.0	0.0	0.0	0.0	0.0	0.0	0.0	0.0	0.0	0.0	0.0
Total					421.8	322.2	256.7	196.9	159.1	129.1	111.4	100.3	88.9	79.4	72.7	68.7

TABLE **12.15** Hydrograph Ordinates for Example 12.1 at B—100-Year (Continued)

Flow Rate Reduction Calculations			
Drainage Area = 0.743 sq. mi.		**Rainfall Distri. = II**	
Frequency	**2-Year**	**10-Year**	**100-Year**
Peak Inflow*	394	689	1138
Peak Outflow*	40	70	120
Qin/Qout	0.10	0.10	0.11
Vs/Vr*	0.55	0.55	0.54
Runoff, Q*	1.60	2.80	4.70
Runoff, Vol.	63.4	110.9	186.2
Storage Vol.	34.9	61.0	100.6
Max. Elev.*	922.0	925.0	928.5

TABLE 12.16   Flow Reductions for Example 12.1

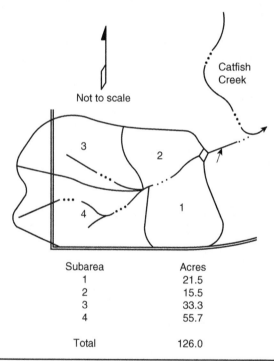

Subarea	Acres
1	21.5
2	15.5
3	33.3
4	55.7
Total	126.0

FIGURE 12.7   Watershed Map and Areas for Example 12.2.

12.7.4   Land Uses

Watershed land uses, now almost totally developed, consists of single-family residences, apartments, and a proposed neighborhood shopping center upstream of the road in subarea 4.

Subarea	Size, acres	Size, sq. mi.
1	21.5	0.024
2	15.5	0.024
3	33.3	0.052
4	55.7	0.087
Total	126.0	0.187

TABLE 12.17   Catfish Creek Tributary Drainage Areas

12.7.5   Curve Numbers

Curve numbers (CNs) for numerous land uses, both urban and rural, are listed in Chap. 2 of TR-55 (USDA, 1986). These were derived from rainfall and runoff records in rural areas. Curve numbers for urban areas were derived assuming that impervious areas have a CN of 98, and the remainder had a CN of pasture (grass) in good condition. CNs for the four subareas and total subwatershed were obtained from a study conducted by the Soil Conservation Service (SCS) for the developed condition over 40 years ago. These CNs and calculations are shown in Table 12.18. Subareas are necessary when soil type or land use changes. Subareas are also needed when a hydrograph is desired at particular locations. Subareas should be added when CNs for a certain land use and soil type differ by 10 or more.

Subarea	Land Use	Percent	CN	Product	CN
1	Commercial	5	92	4.6	
	Residential—1/4 ac.	65	75	48.8	
	Streets w/curbs	5	98	4.9	
	Open space—fair	25	69	17.3	76
2	Residential—1/4 ac.	80	75	60.0	
	Streets w/curbs	5	98	4.9	
	Open space—fair	15	69	10.4	75
3	Residential—1/4 ac.	75	75	56.3	
	Cult. w/o cons.	25	81	20.3	77
4	Commercial	65	92	59.8	
	Cult. w/o cons.	35	81	28.4	88
Total	Commercial	30	92	27.6	
	Residential—1/4 ac.	41	75	30.8	
	Streets w/curbs	1	98	1.0	
	Cult. w/o cons.	22	81	17.8	
	Open space—fair	6	69	4.1	81

TABLE 12.18   Curve Numbers for a Developed Subwatershed of Catfish Creek

Subarea	Tc	Tt
1	0.14	0.00
2	0.16	0.00
3	0.19	0.23
4	0.17	0.17
Total	0.42	0.00

TABLE 12.19   Tc and Tt for the Catfish Creek Tributary

12.7.6   Times of Concentration and Travel Times

Developed Tc and Tt values are listed in Table 12.19 as determined by the SCS. A hydrograph is needed at the Catfish Creek tributary's downstream end shown in Fig. 12.7, just upstream of the four-lane divided highway. Tc is estimated at the downstream ends of each of the subareas and the total area. Tc for subareas 1 and 2 and total area have their downstream ends at the highway. Therefore, time of travel for these three areas at the detention basin location is zero. But, subareas 3 and 4 have their downstream ends at points upstream of the highway by 1,550 ft and 1,150 ft, respectively. The channel's slope is 1.9 percent. Their hydrographs need to be translated downstream. This is done by using Fig. 3-1 in TR-55 (USDA, 1986).

12.7.7   Rainfall

Rainfall amounts for the 2- and 100-year, 24-hour storms are 3.1 and 6.3 in, respectively, and were obtained from App. B of TR-55 (USDA, 1986).

12.7.8   Inflow Hydrographs

Values for CN, Tc, and Tt had already been estimated by the SCS. Worksheet 5a (Table 12.20) contains the variables needed to determine an inflow hydrograph. Values for both the 2- and 100-year, 24-hour storm events are listed. The only difference in the numbers are the rainfall amounts. Separate forms for the hydrograph ordinates must be used for each storm magnitude. These are developed in Tables 12.21 and 12.22 for the 2- and 100-year, 24-hour storms, respectively. Peak-flow rate for the 2-year storm is 190 cfs at hour 12.4. Peak-flow rate for the 100-year storm is 539 cfs at hour 12.4.

12.8   Nuances of the Method

It is easy to fill out forms. The hard part is to truly understand the methodology and be able to input good numbers for the particular situation at hand. Hopefully, the following comments will help to guide you.

12.8.1   Rainfall

Rainfall amounts for various storm durations and return periods have been developed and published by various federal agencies over the past several decades. These are based on statistical analyses of rainfall records. At this point in time, these are the best data that most jurisdictions have. Some states have developed more comprehensive data and published them. Use the best you have available.

Hydrograph Development—Worksheet 5a

Project: Catfish Creek
Outline one: Present   [Developed]

Location: Dubuque, Iowa
Frequency (yr): 2 & 100

By:
Checked by:

Date:
Date:

Subarea Name 1	Drainage Area sq. mi. 2	Time of Concen. hr. 3	Travel Time thru Subarea hr. 4	Downstream Subarea Names 5	Travel Time Summation to Outlet 6	24-hr Rainfall in. 7	Runoff Curve Number 8	Runoff in. 9	AmQ sq. mi.-in. 10	Initial Abstraction in. 11	Ia/P 12
1	0.034	0.14	0.23	—	0.00	3.1	76	1.1	0.0368	0.632	0.204
2	0.024	0.16	0.17	—	0.00	3.1	75	1.0	0.0246	0.667	0.215
3	0.052	0.19	0.00	1,2	0.23	3.1	77	1.1	0.0593	0.597	0.193
4	0.087	0.17	0.00	1,2	0.17	3.1	88	1.9	0.1659	0.273	0.088
								#DIV/0!	#DIV/0!		#DIV/0!
Total	0.197	0.42	0.00	—	0.00	3.1	81	1.4	0.2740	0.469	0.151
								#DIV/0!	#DIV/0!		#DIV/0!
1	0.034	0.14	0.23	—	0.00	6.3	76	3.6	0.1238	0.632	0.100

TABLE 12.20 Two-Year and 100-Year Hydrograph Variables for Catfish Creek Tributary

Hydrograph Development—Worksheet 5a

Project: Catfish Creek
Outline one: Present

Location: Dubuque, Iowa
Frequency (yr): 2 & 100

By:
Checked by:

Date:
Date:

Developed

Subarea Name 1	Drainage Area sq. mi. 2	Time of Concen. hr. 3	Travel Time thru Subarea hr. 4	Downstream Subarea Names 5	Travel Time Summation to Outlet 6	24-hr Rainfall in. 7	Runoff Curve Number 8	Runoff in. 9	AmQ sq. mi.–in. 10	Initial Abstraction in. 11	Ia/P 12
2	0.024	0.16	0.17	—	0.00	6.3	75	3.5	0.0849	0.667	0.106
3	0.052	0.19	0.00	1,2	0.23	6.3	77	3.7	0.1946	0.597	0.095
4	0.087	0.17	0.00	1,2	0.17	6.3	88	4.9	0.4276	0.273	0.043
								#DIV/0!	#DIV/0!		#DIV/0!
Total	0.197	0.42	0.00	—	0.00	6.3	81	4.2	0.8191	0.469	0.074
								#DIV/0!	#DIV/0!		#DIV/0!
								#DIV/0!	#DIV/0!		#DIV/0!

TABLE 12.20 Two-Year and 100-Year Hydrograph Variables for Catfish Creek Tributary (*Continued*)

Hydrograph Development—Worksheet 5b

Project: Catfish Creek
Outline one: Present [Developed]

Location: Dubuque, Iowa
Frequency (yr): 2

By: ___ Date: ___
Checked by: ___ Date: ___

Select and enter hydrograph times in hours from exhibit 5-
Discharge at selected hydrograph times cfs

Subarea Name	Subarea Tc hr	ΣTt to Outlet hr	Ia/P	AmQ sq. mi.–in.	11.0	11.3	11.6	11.9	12.0	12.2	12.4	12.6	12.8	13.0	13.2	13.4
1	2	3	4	5	6	7	8	9	10	11	12	13	14	15	16	17
1	0.10	0.00	0.10	0.0368	24	34	53	334	647	1010	628	147	104	76	57	51
					0.9	1.3	2.0	12.3	23.8	37.2	23.1	5.4	3.8	2.8	2.1	1.9
2	0.20	0.00	0.10	0.0246	23	41	47	209	403	800	250	128	86	70	61	54
					0.6	1.0	1.2	5.1	9.9	19.7	6.2	3.1	2.1	1.7	1.5	1.3
3	0.20	0.20	0.10	0.0593	17	23	32	49	76	282	652	436	207	115	81	67
					1.0	1.4	1.9	2.9	4.5	16.7	38.7	25.9	12.3	6.8	4.8	4.0
4	0.20	0.10	0.10	0.1659	19	26	39	86	168	601	733	564	229	122	83	69
					3.2	4.3	6.5	14.3	27.9	99.7	121.6	93.6	38.0	20.2	13.8	11.4
					0.0	0.0	0.0	0.0	0.0	0.0	0.0	0.0	0.0	0.0	0.0	0.0
					0.0	0.0	0.0	0.0	0.0	0.0	0.0	0.0	0.0	0.0	0.0	0.0
Total					5.6	7.9	11.5	34.6	66.1	173.3	189.5	128.0	56.2	31.6	22.2	18.6

TABLE 12.21 Two-Year Hydrograph Ordinates for Catfish Creek Tributary

Hydrograph Development—Worksheet 5b

Project: Catfish Creek Location: Dubuque, Iowa By: Date:

Outline one: Present [Developed] Frequency (yr): 2 Checked by: Date:

Note: Discharge at selected hydrograph times (cfs). "Select and enter hydrograph times in hours from exhibit 5-." Each subarea lists the exhibit hydrograph ordinate (integer) and the computed discharge in cfs (decimal).

Basic Watershed Data Used

Subarea Name (1)	Subarea Tc, hr (2)	ΣTt to Outlet, hr (3)	Ia/P (4)	AmQ sq. mi.–in. (5)
1	0.10	0.00	0.10	0.0368
2	0.20	0.00	0.10	0.0246
3	0.20	0.20	0.10	0.0593
4	0.20	0.10	0.10	0.1659

Hydrograph ordinates

Hydrograph time (hr)	11.0	11.3	11.6	11.9	12.0	12.2	12.4	12.6	12.8	13.0	13.2	13.4
Column no.	6	7	8	9	10	11	12	13	14	15	16	17
(second band times)	13.6	13.8	14.0	14.3	14.6	15.0	15.5	16.0	16.5	17.0	17.5	18.0
Subarea 1 ordinate	46	42	38	34	32	29	26	23	21	20	19	18
Subarea 1 discharge	1.7	1.5	1.4	1.3	1.2	1.1	1.0	0.8	0.8	0.7	0.7	0.7
Subarea 2 ordinate	49	44	40	35	33	30	27	24	21	20	19	18
Subarea 2 discharge	1.2	1.1	1.0	0.9	0.8	0.7	0.7	0.6	0.5	0.5	0.5	0.4
Subarea 3 ordinate	58	51	46	40	35	32	29	26	23	21	20	19
Subarea 3 discharge	3.4	3.0	2.7	2.4	2.1	1.9	1.7	1.5	1.4	1.2	1.2	1.1
Subarea 4 ordinate	53	47	43	37	34	31	28	25	23	21	19	18
Subarea 4 discharge	8.8	7.8	7.1	6.1	5.6	5.1	4.6	4.1	3.8	3.5	3.2	3.0
(blank)	0.0	0.0	0.0	0.0	0.0	0.0	0.0	0.0	0.0	0.0	0.0	0.0
(blank)	0.0	0.0	0.0	0.0	0.0	0.0	0.0	0.0	0.0	0.0	0.0	0.0
Total	15.1	13.4	12.2	10.6	9.7	8.8	8.0	7.1	6.5	6.0	5.5	5.2

TABLE 12.21 Two-Year Hydrograph Ordinates for Catfish Creek Tributary (*Continued*)

Hydrograph Development—Worksheet 5b

Project: Catfish Creek Outline one: Present [Developed] Location: Dubuque, Iowa Frequency (yr): 100 By: Checked by: Date: Date:

Select and enter hydrograph times in hours from exhibit 5-

Discharge at selected hydrograph times, cfs

| | Basic Watershed Data Used | | | | 11.0 | 11.3 | 11.6 | 11.9 | 12.0 | 12.2 | 12.4 | 12.6 | 12.8 | 13.0 | 13.2 | 13.4 |
| Subarea Name | Subarea Tc hr | ΣTt to Outlet hr | Ia/P | AmQ sq. mi.–in. | 6 | 7 | 8 | 9 | 10 | 11 | 12 | 13 | 14 | 15 | 16 | 17 |
1	2	3	4	5												
1	0.10	0.00	0.10	0.1238	24	34	53	334	647	1010	628	147	104	76	57	51
					3.0	4.2	6.6	41.3	80.1	125.0	77.7	18.2	12.9	9.4	7.1	6.3
2	0.20	0.00	0.10	0.0849	23	31	47	209	403	800	250	128	86	70	61	54
					2.0	2.6	4.0	17.7	34.2	67.9	21.2	10.9	7.3	5.9	5.2	4.6
3	0.20	0.20	0.10	0.1948	17	23	32	49	76	262	652	435	207	115	81	67
					3.3	4.5	6.2	9.5	14.8	51.0	127.0	84.7	40.3	22.4	15.8	13.1
4	0.20	0.10	0.10	0.4276	19	26	39	86	168	601	733	564	229	122	83	69
					8.1	11.1	16.7	36.8	71.8	257.0	313.4	241.2	97.9	52.2	35.5	29.5
					0.0	0.0	0.0	0.0	0.0	0.0	0.0	0.0	0.0	0.0	0.0	0.0
					0.0	0.0	0.0	0.0	0.0	0.0	0.0	0.0	0.0	0.0	0.0	0.0
Total					16.4	22.4	33.5	105.4	201.0	501.0	539.4	355.0	158.4	89.9	63.5	53.5

TABLE 12.22 One-hundred-Year Hydrograph Ordinates for Catfish Creek Tributary

Hydrograph Development—Worksheet 5b

Project: Catfish Creek Location: Dubuque, Iowa By: Date:
Outline one: Present [Developed] Frequency (yr): 100 Checked by: Date:

Basic Watershed Data Used

Select and enter hydrograph times in hours from exhibit 5-
Discharge at selected hydrograph times, cfs
Select and enter hydrograph times in hours from exhibit 5-

Subarea Name (1)	Subarea Tc hr (2)	ΣTt to Outlet hr (3)	Ia/P (4)	AmQ sq. mi.-in. (5)	11.0 (6)	11.3 (7)	11.6 (8)	11.9 (9)	12.0 (10)	12.2 (11)	12.4 (12)	12.6 (13)	12.8 (14)	13.0 (15)	13.2 (16)	13.4 (17)
(times, second set)					13.6	13.8	14.0	14.3	14.6	15.0	15.5	16.0	16.5	17.0	17.5	18.0
1	0.10	0.00	0.10	0.1238	46	42	38	34	32	29	26	23	21	20	19	18
					5.7	5.2	4.7	4.2	4.0	3.6	3.2	2.8	2.6	2.5	2.4	2.2
2	0.20	0.00	0.10	0.0849	49	44	40	35	33	30	27	24	21	20	19	18
					4.2	3.7	3.4	3.0	2.8	2.5	2.3	2.0	1.8	1.7	1.6	1.5
3	0.20	0.20	0.10	0.1948	58	51	46	40	35	32	29	26	23	21	20	19
					11.3	9.9	9.0	7.8	6.8	6.2	5.6	5.1	4.5	4.1	3.9	3.7
4	0.20	0.10	0.10	0.4276	53	47	43	37	34	31	28	25	23	21	19	18
					22.7	20.1	18.4	15.8	14.5	13.3	12.0	10.7	9.8	9.0	8.1	7.7
					0.0	0.0	0.0	0.0	0.0	0.0	0.0	0.0	0.0	0.0	0.0	0.0
					0.0	0.0	0.0	0.0	0.0	0.0	0.0	0.0	0.0	0.0	0.0	0.0
Total					43.8	39.0	35.4	30.8	28.1	25.6	23.1	20.6	18.7	17.2	16.0	15.2

TABLE **12.22** One-hundred-Year Hydrograph Ordinates for Catfish Creek Tributary (*Continued*)

A few communities keep track of local data and use their results. I forget the name of the city, but some decades ago, it had its own rain gages and kept track of the ten-largest storm events. When a large storm occurred, it was placed in their top ten and dropped the lowest rain storm (amount, duration, and time distribution) from the ten. Then, all designs used each of the top ten to size drainage facilities. In this way, they hoped to prevent future flooding.

As noted in Chap. 6, the NRCS uses four storm types based on their analyses of rainfall records. See Fig. B.2 in App. B of TR-55 (USDA, 1986). Storm duration used is 24 hours. Its time distribution of rain during the 24 hours is known as a nested distribution. Table 6.1 for Madison, Wisconsin listed the federal data for both duration and return period.

Nesting is done in the following manner. A rain amount in a short time such as 5 or 10 minutes with the highest intensity is placed sometime before hour twelve. Then the next highest rain amount of the same duration is placed in the time period before the first one, Then the third highest rain amount of the same duration is placed in the time period after the first one. This same procedure, before and after, is used until a 24-hour storm time distribution is built up.

In this way, each storm type distribution includes all depths for each community for each storm duration such as is listed in Table 6.1 for durations shorter than 24 hours for all return periods. Some designers complain this is too severe a distribution. No storm could have exactly the correct amount of rain for each portion of a storm shorter than 24 hours. However, since rain is a random event it could happen this way. I like it for two reasons. First, our rainfall records are our best estimate based on the problems I discussed in Chap. 6. Using a conservative estimate of time distribution of rain leads to a conservative design.

Second, Tc for urban BMPs is always less than 24 hours, ranging from 6 minutes to a few hours. A nested distribution gives us a good rainfall amount no matter what Tc is. Also a 24-hour duration yields more depth of rain. Volume considerations in detention basin design (whether it is a rain garden or includes a multi-acre pond) is just as important as peak flow rate.

12.8.2 Time of Concentration

If we want to develop, retrofit, and redevelop land using low impact development (LID) and the triple bottom line (TBL) of people, planet, and profit, we need to look no further than how we allow rain to run off land. Water on a concrete or asphalt street or parking lot on a 2 percent slope will run off at a velocity of about 1.5 fps. Water flowing through a lawn on a 2 percent slope will run off at a velocity of about 0.1 fps. That is about 15 times faster. The longer Tc is, the lower the peak runoff rate will be.

It rains 6 in in 24 hours. Being impervious, about 98 percent of water on a street or parking lot (5.8 in) will run off to somewheres. Being pervious, based on underlying soil type and degree of saturation, about 7 to 63 percent of water falling on it (0.5 to 3.8 in) will run off to somewheres. That is 1.5 to 11.6 times more volume of runoff, if it is a sandy or clayey soil.

Therefore, slow water down and allow some of it to infiltrate. Run roof runoff through downspouts onto grass not a driveway. Have parking lot runoff flow onto a grassy area rather than into storm sewer inlets. Have street runoff flow through a porous concrete curb and gutter into a swale connected to a rain garden. These reduce runoff rates and volumes plus traps some pollutants. Many other BMPs can be used to reduce rates, volumes, and pollutants even more.

12.8.3 Runoff Curve Number

When and why do we input a number 98 or 75 when a program asks for a CN? A CN of 100 means 100 percent of rain runs off. The higher a CN, the more runoff there is. From page 2–7 of TR-55 (USDA, 1986), pasture in good condition with a C soil has a CN of 74. From page 2–5 of TR-55, open space in good condition with a C soil has a CN of 74. The other soil types have similar CNs.

Impervious areas have a CN of 98. Why not 100? Some rain wets pavements and infiltrates through cracks and joints. A ¼-acre lot on a C soil has a CN of 83 on page 2–5. Look at footnote 2 at its bottom. A ¼-acre lot is 38% impervious. CN = 0.38 × 98 + 0.62 × 74 = 83.1 or 83.

Assume instead the undeveloped land is a pasture in poor condition with a C soil. Its CN is 86 from page 2–7. But if we create ¼-acre lots, its CN is 83. The site becomes 38 percent impervious but the CN is reduced from 86 to 83. Less runoff after development? Why? Because 62 percent has become lush lawns rather than poor meadow.

Just because we urbanize does not mean runoff rates and volumes increase. The other factor is Tc. If we make runoff wander around in swales we can increase Tc from an existing condition Tc. We can have an existing meadow and develop a shopping mall with 95 percent imperviousness but reduce runoff from the site to almost zero by using greenroofs, some rain gardens, and porous pavement parking lots underlain with a layer of gravel. Tc values to these BMPs are short but the water simply infiltrates rather than running off to storm sewers or to a large detention basin. The developer is able to use 100 percent of his/her land for profitable construction.

Look at the existing terrain on your site and think about the proposed land use. Are there existing drainageways onsite, how can lots and street circulation fit, what types of BMPs will work, how will rates, volumes, and pollutants be reduced, change your mind several times, then produce a result so everything fits as one harmonious whole, without spending an arm and a leg.

12.8.4 Hydrograph Variables and Ordinates

Going from the hydrograph variables form to the hydrograph ordinates spreadsheet, requires changing Tc, Tt, Ia/P to one decimal place. Rainfall is the driving force for all drainage facility designs. Since we are not sure our rainfall numbers are good, I tend to lean to the conservative side. Therefore, when my values for Tc, Tt, and Ia/P are between Chap. 5 values in TR-55 (USDA, 1986), I tend to select the smaller value of Tc and Tt and the smaller value of 0.1, 0.3, or 0.5 for Ia/P since they result in higher values of hydrograph ordinates.

12.9 Summary

This chapter presented details on using spreadsheets with the NRCS methodology to develop inflow hydrographs using TR-55 (USDA, 1986). The above nuances are good if you use a computer program with the exception of Sec. 12.8.4 above. When using TR-20, HEC-HMS, or some other computer program, it can utilize any values of Tc, Tt, and Ia/P if those variables are used in the program.

Whatever method you use to develop hydrographs, continue to ask the two questions.

1. Why and how am I using this particular methodology to develop an inflow hydrograph?

2. How and where did I get those numbers from?

CHAPTER 13

Basic Hydraulics

13.1 Introduction

This chapter discusses the basic hydraulic principles that all hydraulic structures must obey. Whatever best management practices (BMPs) are used in developments, whatever their outlets look like, whatever their shapes, sizes, and configurations are, their outlets must obey these principles. Weirs and orifices are discussed in Chap. 16. Use of these principles and related equations can be found as part of the calculations for every example used in the several appendices of this book.

BMPs' outlet structures utilize equations such as conservation of mass, total energy, Bernouilli's equation, specific energy, Froude number, critical depth, normal depth, Manning's equation, hydraulic jump, and friction loss. The following explanation of these principles is part of the notes used in a short course I used to teach once or twice a year (Rossmiller, 2008).

13.2 Conservation of Mass

Conservation of mass is usually expressed as Eq. 13.1.

$$Q_1 = Q_2 = A_1 V_1 = A_2 V_2 \tag{13.1}$$

where Q is flow rate in cfs, A is flow area in sq ft, and V is velocity in fps. Subscripts 1 and 2 are locations of a reach's upstream and downstream ends, respectively. Flow entering a conduit or outlet equals flow exiting it. If its cross section changes in size or shape without a change in flow rate, flow area multiplied by velocity in a first cross section is equal to flow area multiplied by velocity in a second section.

$$A = Q/V$$

If conditions remain the same, area remains constant. Conditions include conduit shape, size, slope, roughness, and Q. If any of them change, then area must change. Area has a width and depth; so if some condition changes, then area constraints cannot be exceeded.

13.3 Total Energy

Total energy is expressed as Eq. (13.2) and as shown in Fig. 13.1.

$$H = Z + y + p/\partial + V^2/2g \tag{13.2}$$

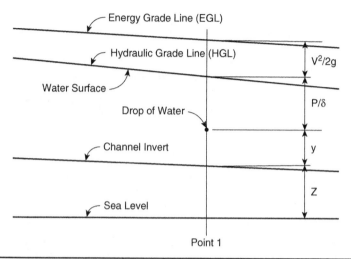

FIGURE 13.1 Total energy at a point.

where H is total head in ft, Z is head between an invert and a datum (usually sea level) in ft, y is streamline height above an invert in ft, p is streamline pressure in lb/sq ft, ∂ is water's weight (62.4 lb/cu ft), V is as above, and g is acceleration due to gravity (32.16 fps/sec).

Total energy is what drives runoff to a point of lesser energy, usually downhill. It is made up of two parts: (1) static energy which is energy of position (elevation), and (2) kinetic energy, which is energy of movement (velocity). We keep track of both because as long as a drop of water has some type of energy, it is capable of doing something, such as erosion or flooding or drowning someone. If water is knee deep and flowing at four fps, it can knock a person down.

13.4 Bernouilli's Equation

The distance between parallel lines is equal at all points. Bernoulli said total energy at two points some distance apart is equal (Fig. 13.2). The bottom line is a datum line, sea level. A second line is drawn through a horizontal line at an upstream cross section.

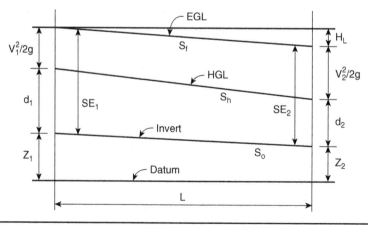

FIGURE 13.2 Bernouilli's equation.

His equation is:

$$Z_1 + d_1 + V_1^2/2g = Z_2 + d_2 + V_2^2/2g + \text{head losses} \qquad (13.3)$$

where d is defined as flow depth in ft. All other terms are as defined previously.

In drainage, we use inverts and depths, not streamlines and pressures. Head losses are friction plus minor losses, such as entrance, exit, junction, transition, bend, and angle point. An energy grade line (EGL) is the friction slope (Sf) and is one velocity head, $V^2/2g$, above the hydraulic grade line (HGL), the water surface (Fig. 13.2). This equation allows us to develop water-surface profiles, which let us know the depth, elevation, and velocity at each point in a drainage system, such as a storm sewer, swale, or creek.

13.5   Specific Energy

Specific energy (SE) is total energy at a point when the datum is assumed to be the channel invert and is:

$$SE = d + V^2/2g \qquad (13.4)$$

where SE is specific energy in feet (Fig. 13.3), asymptotic to the abscissa and a 45-degree line. For some channel size and shape, there is a different specific energy curve for each flow rate (Q).

A vertical line cuts the curve at two depths, A and B, known as alternate depths. Both have the same SE but one has a greater depth and lower velocity, the other has a lower depth and higher velocity. Depth water flows at is dependent on channel roughness and slope. A vertical line cuts the curve at only one point. This point is minimum SE, labeled as point C, and depth is known as critical depth, Dc. Water wants to flow at normal depth, Dn. Whatever depth water is, depth moves on the curve until Dn is reached. If Dn is on its upper leg and a beginning depth is on its lower leg, water can reach Dn only through a hydraulic jump.

Swale or channel linings must be strong to withstand water's velocity flowing along it. If not, a channel's bottom and sides erode. Sediment and debris is conveyed and deposited in unwanted locations, such as plugging a channel or culvert causing flooding. A hydraulic jump causes a sudden loss of energy in a short distance that creates turbulence. This definitely causes channel erosion and more sediment and debris.

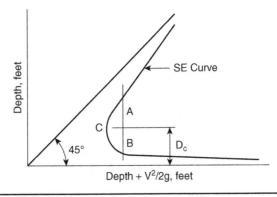

Figure 13.3   Specific energy curve.

13.6   Froude Number

Froude number is expressed as:

$$\mathbf{F} = V/(g\,Dm)^{0.5} \tag{13.5}$$

where $\mathbf{F}$ is Froude number, ratio of inertial forces to gravitational effects, and Dm is hydraulic mean depth in feet, equal to flow area divided by top width of flow. All other terms are as defined above. When $\mathbf{F}$ is 1.00, flow is critical; when $\mathbf{F}$ is less than 1.00, flow is subcritical, mild, or tranquil; when $\mathbf{F}$ is greater than 1.00, flow is supercritical, fast, or shooting.

Knowing $\mathbf{F}$ yields ideas about conduit flow conditions. If $\mathbf{F} < 1.0$, depths are greater and flow is very slowly moving along. If $\mathbf{F} > 1.0$, depths are shallow and water roars along our conduit. Pebbles and cobbles erode concrete pipes down to reinforcing steel and tear out metal pipe linings. Open channels' bottoms and sides are protected with concrete, riprap, or natural or other manmade materials. Toe-of-slope protection is important to prevent undermining it.

13.7   Manning's Equation

Manning's equation (Manning, 1891) is written as:

$$V = 1.49\,R^{0.667}\,Sf^{0.5}/n \tag{13.6}$$

and
$$R = A/WP \tag{13.7}$$

where R is hydraulic radius in feet, Sf is friction slope (EGL) in ft/ft, n is roughness coefficient, WP is wetted perimeter in feet, and other variables are as defined previously. WP is a cross section's perimeter wet by water. See Fig. 13.4. Manning's equation is used to estimate friction loss along a conduit. We know Q and $Q = AV$; so we multiply both sides of the equation by A, then solve it for Sf. Sf multiplied by length is friction loss. Manning's

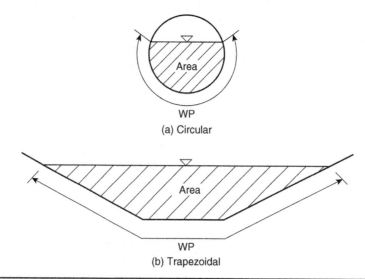

(a) Circular

(b) Trapezoidal

FIGURE 13.4   Culvert and channel area and wetted perimeter.

and Bernouilli's equations are used to develop all types of water-surface profiles. We need to know flow depths to prevent flooding.

13.8   Critical Depth

Critical depth (Dc) is depth at which **F** is 1, and specific energy is a minimum. It is also that depth where inertial forces equal gravitational forces. *Dc* is a function of just *Q*, size, and shape. See Eq. (13.8).

$$Q^2/g = A^3/T \tag{13.8}$$

where *T* is top width of flow in feet, and all other terms are as defined previously. An equation for area is substituted for *A*, and Eq. (13.8) is solved for *Dc*. *Dc* can be determined in three ways:

1. By using Eq. (13.8).
2. From tables contained in *Handbook of Hydraulics*, (Brater and King, 1976).
3. From nomographs contained in Hydraulic Design Series (HDS) No. 3 (FHWA, 1980).

13.9   Normal Depth

Dn is defined as that depth where friction loss is just overcome by a conduit's slope. Conduit slope, water surface (HGL), and EGL (*Sf*) are all parallel. Thus, in Manning's equation, bed slope is used as *Sf* when flow is at *Dn*. See Fig. 13.5. *Dn* is a function of flow rate; cross-section size, shape, and length; slope; and conduit's roughness. It is determined in three ways:

1. Using Manning's equation by setting *Sf* equal to *So*, then using a trial-and-error procedure.
2. From tables in *Handbook of Hydraulics*, (Brater and King, 1976) for various shapes.
3. From nomographs contained in HDS No. 3 (FHWA, 1980).

When *Dn* < *Dc*, a conduit is said to be on a hydraulically steep slope. When *Dn* > *Dc*, a conduit is said to be on a hydraulically mild slope.

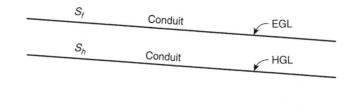

Figure 13.5   Slopes at normal depth.

13.10   Example 13.1

A 10-ft wide by 8-ft high reinforced concrete box (RCB) culvert on a 1.4 percent slope is conveying 1,000 *cfs*. Compute the following:

    a. Critical depth—three different ways

    b. Normal depth—three different ways

    c. Froude number

    d. Is this culvert on a hydraulically mild or a hydraulically steep slope?

In order to solve this problem you need to have a copy of the following book. If you or your employer does not have a copy, buy one. The information contained in the book is needed to give you a good understanding of the hydraulics of flow and the methods and equations needed to solve hydraulics problems. The book is: Brater and King (1976) *Handbook of Hydraulics*, published by the McGraw-Hill Book Company.

13.10.1   Critical Depth

Critical depth is determined using each of the three above methods as shown by the calculations below:

    1. Using Eq. (13.8)

$$Q^2/g = A^3/T$$
$$(1{,}000)^2/32.16 = (10\,Dc)^3/10$$
$$1{,}000{,}000/32.16 = 100\,Dc^3$$
$$Dc = 6.77 \text{ ft}$$

    2. From Table 8-5 in *Brater and King, (1976)*

$$Kc' = Q/B^{5/2}$$
$$Kc' = 1{,}000/(10)^{5/2}$$
$$Kc' = 1{,}000/316.2$$
$$Kc' = 3.162$$

From Table 8-5, for $Kc' = 3.162$, the factor is between $Dc/B = 0.67$ and 0.68

Interpolating	0.68	3.180	0.68	3.180	0.018/0.070 = 0.26
	0.67	<u>3.162</u>	0.67	<u>3.110</u>	
		0.018		0.070	0.68 − 0.0026 = 0.6774

$$Dc = 0.6774 \times 10 = 6.77 \text{ ft}$$

    3. From Chart 9 in HDS No. 3

$$Dc = 6.8 \text{ ft}$$

13.10.2   Normal Depth

Normal depth is determined using each of the three above methods as shown by the calculations on the next page.

1. Use Manning's equation and solve for $AR^{2/3}$. Do a trial-and-error solution for Dn.

$$Q = 1.49\, AR^{2/3}\, S^{1/2}/n$$
$$1{,}000 = 1.49\, AR^{2/3}\, (0.014)^{1/2}/0.015$$
or $\;AR^{2/3} = 85.1$

Depth	Area	Wetted P	Hyd. Rad.	$R^{2/3}$	$AR^{2/3}$
2.50	25.0	15.00	1.667	1.406	35.1
4.70	47.0	19.40	2.423	1.804	84.8
4.71	47.1	19.42	2.425	1.805	85.0

2. From Table 7-11 in Brater and King, (1976)

$$K' = Qn/B^{8/3}\, S^{1/2}$$
$$K' = (1{,}000 \times 0.015)/((10)^{8/3}\, (0.014)^{1/2})$$
$$K' = 15/(464.2 \times 0.1183)$$
$$K' = 0.273$$

From Table 7-11, for $K' = 0.273$, the factor is between $Dn/B = 0.47$ and 0.48

Interpolating   0.48   0.279   0.48   0.279 $0.006/0.008 = 0.75$

$$\frac{0.273}{0.006} \quad 0.47 \quad \frac{0.271}{0.008} \qquad 0.48 - 0.0075 = 0.4725$$

$$Dn = 0.4725 \times 10 = 4.72\ \text{ft}$$

3. From Chart 9 in HDS No. 3

$$Dn = 4.7\ \text{ft}$$

13.10.3 Froude Number

The Froude number is determined by first calculating the flow velocity and then solving the Froude number equation.

$$Vn = Q/A \qquad\qquad F = V/(g\, Dm)^{0.5}$$
$$Vn = 1{,}000/(4.71 \times 10) \qquad F = 21.2\,(32.16 \times 4.71)^{0.5}$$
$$Vn = 21.2\ \text{fps} \qquad\qquad F = 1.72$$

13.10.4 Is This a Hydraulically Mild or Hydraulically Steep Slope?

This is a steep slope because Dn is less than Dc and the Froude number is greater than 1.

13.11 Spreadsheets

Spreadsheets were developed to calculate Dc and Dn and are located on McGraw-Hill's website (www.mhprofessional.com/sdsd) using the first solution for Dc and Dn with results shown on Tables 13.1 and 13.2.

Critical Depth for Rectangular Cross Sections						
1 Depth D ft	2 Area A sq ft	3 Top Width T ft	4 A^3/T	5 Mean Depth Dm ft	6 Velocity V fps	7 Froude Number F
7.0000	70.00	10.00	34300.00	7.00	14.29	0.9521
6.8000	68.00	10.00	31443.20	6.80	14.71	0.9944
6.7500	67.50	10.00	30754.69	6.75	14.81	1.0055
6.7700	67.70	10.00	31028.87	6.77	14.77	1.0011
6.7750	67.75	10.00	31097.67	6.78	14.76	0.9999
6.7748	67.75	10.00	31094.92	6.77	14.76	1.0000
6.7747	67.75	10.00	31093.54	6.77	14.76	1.0000
6.7748	67.75	10.00	31094.51	6.77	14.76	1.0000
6.7748	67.75	10.00	31094.53	6.77	14.76	1.0000

Q = 1000 cfs; Bottom Width = 10 ft; A^3/T = 31094.53; Height = 8; Enter height equals 50 if section is not a box.

TABLE 13.1 Critical Depth in Example 13.1

Normal Depth for Rectangular Cross Sections								
1 Depth ft.	2 Top Width ft	3 Area sq ft	4 Wet. Per. ft	5 Hyd. Rad. ft	6 $R^{2/3}$	7 $A*R^{2/3}$	8 Velocity fps	9 Froude Number
5.0000	10.00	50.00	20.00	2.50	1.842	92.1008	20.00	1.5772
4.7000	10.00	47.00	19.40	2.42	1.804	84.7804	21.28	1.7306
4.7500	10.00	47.50	19.50	2.44	1.810	85.9937	21.05	1.7033
4.7100	10.00	47.10	19.42	2.43	1.805	85.0228	21.23	1.7251
4.7120	10.00	47.12	19.42	2.43	1.805	85.0713	21.22	1.7240
4.7125	10.00	47.13	19.43	2.43	1.805	85.0834	21.22	1.7237
4.7125	10.00	47.12	19.42	2.43	1.805	85.0826	21.22	1.7237

Q = 1000 cfs; Bottom Width = 10.0 ft; n = 0.015; Slope = 0.01400 ft/ft; Height = 8.00 ft; $Q = 1.49*A*R^{2/3}*S^{1/2}/n$; $A*R^{2/3} = Q*n/1.49*S^{1/2} = 85.0826$

TABLE 13.2 Normal Depth in Example 13.1

13.12 Conduit Slope

A conduit's slope tells us much about its flow type. If a concrete conduit's slope is 0.8 percent or less, it is hydraulically mild, subcritical, and tranquil. It has greater depths and lower velocities ($Dn > Dc$). F should be less than 0.8. If it is higher, flatten the slope. Between 0.8 and 1.0, its water surface is unstable since it is too close to a *SE* curve's nose. A small change in energy causes water to yo-yo between depths 3-ft apart.

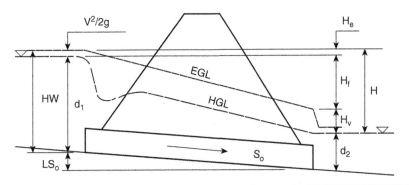

FIGURE 13.6   Slopes, depths, and heads within a culvert.

If its slope is >0.8 percent, it is hydraulically steep, supercritical flow. Depths are less with higher velocities. **F** must be greater than 1.2. If lower, steepen the slope. Between 1.0 and 1.2, its water surface is unstable. Any change in energy causes water to yo-yo between depths 3-ft apart.

HGL and EGL in a culvert are shown in Fig. 13.6, taken from HEC-5 (USDOT, Reprinted 1977). Water is assumed to be ponded at its entrance so velocity is small or zero. Designers assume it is zero so water surface and EGL are the same just upstream of the culvert's entrance. When flowing full, HGL and EGL are parallel to each other. When it is not flowing full, i.e., open-channel flow, HGL and EGL are not parallel. Depth of flow moves towards Dn as water flows through the culvert.

13.13   Hydraulic Jump

As water changes from depths less than Dc to greater than Dc, it goes thru a jump. Its start is estimated by equating pressure plus momentum at flow depths before and after it using Eq. (13.9).

$$Q^2/(A_1 g) + A_1 \hat{y}_1 = Q^2/(A_2 g) + A_2 \hat{y}_2 \tag{13.9}$$

$\hat{y}$ is depth in feet between the water surface and centroid of a flow's prism. Hydraulic jump length is a function of **F** and depth after a jump and is determined from Fig. 13.7 (Chow, 1959). Containing a hydraulic jump within a culvert helps prevent potential erosion in its downstream channel. This is because with deeper depths, exit velocities are less.

13.14   Friction Loss

The major energy loss in open channels and closed conduits is due to friction. Friction loss is estimated by multiplying length by average friction slope as shown in Eq. (13.10). Average S_f at a culvert and a channel reach is determined by solving Manning's equation.

$$H_f = L^*(S_{f_1} + S_{f_2})/2 \tag{13.10}$$

where H_f is friction loss in feet, L is culvert or channel length in feet, and Sf_1 and Sf_2 are friction slopes at a culvert's ends or ends of a channel's reach in feet per foot.

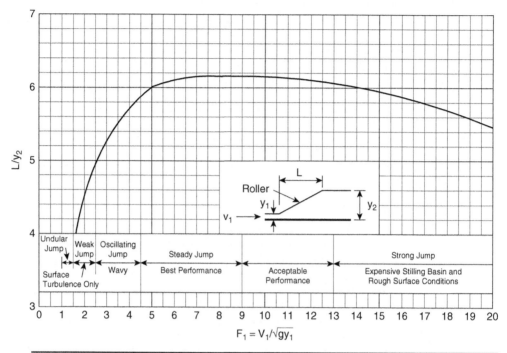

Figure 13.7 Lengths of jumps in terms of sequent depth (Y2) in horizontal rectangular channels.

13.15 Summary

All BMPs must obey basic hydraulic principles. For basins and retrofits, these structures are outlets and risers. Outlet conduits for many BMPs are culverts of some size, shape, and material. Other outlets are the earth itself. These are covered in Chap. 14. Risers are vertical structures of some size, shape, and material. Their openings are either weirs or orifices and are covered in Chap. 15. All BMP outlets must obey these basic hydraulic principles as illustrated in the appendices' examples.

CHAPTER 14

Culvert Hydraulics

14.1   Introduction

Culverts are outlets for many best management practices (BMPs). However, the main reason for including them and the next chapter in this book is because of the numerous low-cost retrofits available. Some serve existing detention basins that were constructed decades ago for water quantity purposes with extra storage capacity. They can be retrofitted to serve both runoff quantity and quality requirements. Also, millions of culverts exist throughout the United States. Some portion of them with certain terrains and land uses could also be retrofitted to serve both requirements.

Hydraulic design of culverts as outlets for various BMPs use nomographs in a publication titled HEC No. 5 (USDOT, Reprinted 1977). You will need a copy of this for your designs using the spreadsheets available on the McGraw-Hill website (**www .mhprofessional.com/sdsd**). Flow rates were covered in Chap. 12 and hydraulic structure design principles in Chap. 13.

14.2   Design Sequence

Culvert hydraulic design involves four steps:

1. Select an allowable headwater (AHW) depth or elevation based on site characteristics and institutional guidelines.

2. Estimate a design flow rate based on an approved method and institutional guidelines.

3. Size a culvert based on the first two steps. This is a trial and error process with these steps:

   a. Select a culvert size, shape, slope, material, and some entrance condition.
   b. Estimate tailwater (TW) depth using a depth-discharge curve or water-surface profile.
   c. Calculate headwater (HW) depth using both inlet and outlet control.
   d. Actual HW depth is greater of these two depths calculated in (c.) above.
   e. Compare actual HW depth with AHW depth.
   f. If actual HW depth is >AHW depth, try a larger culvert and repeat steps (c.) thru (e.).
   g. If actual HW depth is <AHW depth, try a smaller culvert and repeat steps (c.) thru (e.).
   h. If actual HW depth is just below AHW depth, assume you have a correctly sized culvert.

      i. Determine critical depth (Dc) and normal depth (Dn) to determine a culvert's hydraulic slope.

      j. Develop a depth- or elevation-discharge curve for the culvert.

    4. Add appurtenant structures to ensure that the culvert remains in place.

14.3   Determine AHW

AHW is based on two data sets: site characteristics and institutional guidelines. A designer is free to select an AHW consistent with these two data sets. A BMP's outlet could be a culvert whether for a rain garden, vegetative swale, or a retrofitted detention basin. For a rain garden or a swale this would be its berm elevation minus freeboard. A culvert's AHW could be its depth inside the basin's riser. A first opening is an orifice whose crest is a pond's elevation. This and other openings convey flows from a water quality up to and >100-year storm. AHW inside a riser should be a foot or more below its first opening's elevation. This ensures all riser openings function with a free discharge through a 100-year storm.

14.3.1   Site Characteristics

These include developed lot and site characteristics, or for new or retrofitted detention basin sites: terrain both up and downstream; location/dimensions of channels and floodplains; roadway alignment, grades, and elevations; and land uses down and upstream. Channel and floodplain dimensions, roughness, and slope determine TW depths for flow rates, if at Dn. If not then water surface profiles must be run to estimate TWs for design return periods.

14.3.2   Institutional Guidelines

Institutional guidelines include design return periods and freeboard requirements for all BMPs. Most entities require one or two return periods for design ranging from 5- to 100-years. Some design for a 25-year and then check that size for a 100-year to ensure no unwanted damage occurs or if a roadway is overtopped. Others use the 10- and 50-year. Use whatever is required in your area.

    Other policies adopted by local jurisdictions are contained in design standards. These include designing for present or future conditions, types(s) of material used, and using available storage to reduce culvert size and/or enhance water quality, end treatments, velocities, peak flows, hydrograph methods, minimum culvert size, and maximum size before using a bridge.

14.4   Estimate Design Flow Rates

Design flow rate(s) are estimated based on a governing jurisdiction's guidelines. For detention basins, retrofits and other BMPs, runoff hydrographs are needed. A method used by some local and state agencies was detailed in Chap. 12. Flows for outlet structures are determined for the 2-, 10-, and 100-year events.

14.5   Select Culvert Characteristics

Culvert characteristics consist of its size, shape, slope, length, material, and type(s) of entrance and exit. Minimum culvert size is set by a governing jurisdiction. A culvert's shape is set by designers and is dictated by site characteristics, i.e., a small elevation difference between culvert invert and roadway or berm crest elevations.

Reinforced concrete pipe (RCP) and corrugated steel pipe (CSP) or square and rectangular reinforced concrete box (RCB) or steel-box culverts are used. Ovals and arches are also used. RCPs are manufactured in sizes from 6 in to 12 ft. Most contractors have form sizes available for on-site RCB construction. Suppliers also have precast RCB sections. Precast sections allow for shorter construction times since no on-site curing is needed. Plastic pipes are also used for culverts. A culvert's slope and length are dictated by on-site slopes and alignment and elevations of roadway and channel as well as its size and shape.

A life cycle costing study for the Denver Urban Drainage and Flood Control District (UDFCD) under my direction found that RCP and plain galvanized CSP were about equal in cost over a 50-year period. RCPs had a useful life of 50 years whereas CSP had a 30-year life. Maintenance costs were assumed equal since little data were available. CSP cost included replacing pipe and street once. CSP needed to be one gage thicker than specified by manufacturers to increase time for pitting and/or abrasion to eat through a pipe. CSPs coated with bituminous material only added to its cost. Linings were eaten away in a few years because of lining abrasion by sand, gravel, and debris conveyed in the runoff.

14.6 Location of Control Section

Flow through a culvert is controlled at either its inlet or outlet. A simple method is to calculate both and determine which has a greater HW depth. This is the governing control type. Most culverts on hydraulically mild slopes act in inlet control at lower flows and change to outlet control at higher flow rates. The same is sometimes true for hydraulically steep slopes.

14.6.1 Inlet Control Factors

Factors used in determining inlet control HW depth are:

1. Flow rate
2. Culvert size
3. Culvert shape
4. Entrance type

These factors deal with a culvert's entrance conditions and barrel flow is controlled by them. It can convey more flow than an inlet can deliver to it so there is open-channel flow through it. This condition exists with a steep barrel slope, lower flow rate, and/or a short culvert length.

Figure 14.1 shows four flow profiles for inlet control (USDOT, reprinted 1977). Sketches A and B have a projecting inlet for submerged conditions at an inlet and both unsubmerged and submerged conditions at an outlet. Sketches C and D also have submerged entrance conditions. Note that tailwater again is low and high in Sketches C and D. The barrel flows as an open channel in all four cases. A hydraulic jump occurs inside a culvert's barrel in Sketches B and D.

14.6.2 Outlet Control Factors

Factors used in determining outlet-control HW depth are:

1. All inlet control factors
2. Barrel length, roughness, and slope
3. TW depth

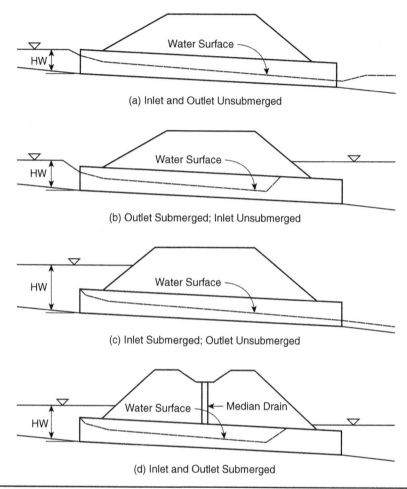

(a) Inlet and Outlet Unsubmerged

(b) Outlet Submerged; Inlet Unsubmerged

(c) Inlet Submerged; Outlet Unsubmerged

(d) Inlet and Outlet Submerged

Figure 14.1 Types of inlet control.

An inlet can deliver more water to a barrel than a barrel is capable of conveying, so a culvert's barrel usually flows full. This occurs when a barrel is on a hydraulically mild slope, larger flow rates, and/or a high TW depth.

Figure 14.2 indicates flow profiles for outlet control (USDOT, Reprinted 1977). Sketches A and B show a projecting inlet for inlet submerged conditions with a full barrel for its entire length. However, TW is greater than culvert height in Sketch A and less than culvert height in Sketch B. Sketches C and D are a projecting entrance with submerged entrance conditions but the barrel does not flow full through its length. TW is less than culvert height in both cases.

Note that H, total head loss, is measured differently. Sketches A and B have H measured as difference between upstream and downstream water surfaces. In B, Dc is equal to barrel height. Sketches C and D have H measured as difference between upstream water surface and an equivalent hydraulic grade line at its exit. This difference is explained later in this chapter.

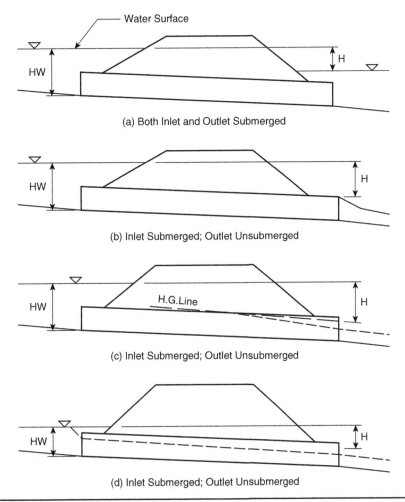

FIGURE 14.2   Types of outlet control.

14.7   Hydraulic Design

Hydraulic culvert sizing is done by using design charts in either HEC-5 (USDOT, Reprinted 1977) or HDS-5 (USDOT, 1985). At least one of these should be on every designer's shelf. They estimate HW depths for inlet and outlet control and Dc for various shapes. Pipe charts are based on tests by French (unknown) and Bossy (1961) plus data for box culverts with headwalls and wingwalls obtained from an unpublished United States Geological Survey (USGS) report. You need one of these manuals to design outlets with spreadsheets available on McGraw-Hill's website (www.mhprofessional.com/sdsd).

A form was developed by the Federal Highway Administration (FHWA) (USDOT, Reprinted 1977) for pipe and arch culverts. See Fig. 14.3. It has a site sketch, inlet and outlet control calculations, and space for a summary and recommendations. Its description includes shape, number of barrels, entrance type (projecting, mitered, end section, headwall, wingwall), and if its entrance is rounded or beveled. Q is flow rate in cfs. If it

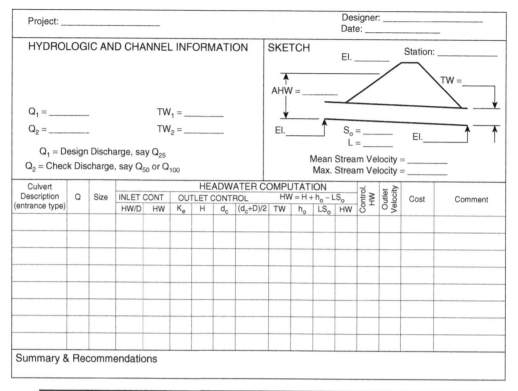

FIGURE **14.3**   Culvert design form for pipes and arches.

has more than one barrel, flow is assumed to be equally divided between barrels, assuming same-sized barrels and the same invert elevations. Flow to be listed is flow per barrel.

Size is listed as "48"" for a pipe or "58" × 36"" for an arch. If an RCB is used, most jurisdictions list width first and height second. An RCB design form has a column for HW/B, cfs/ft of width. Inlet control charts for RCBs use cfs/ft of width. See Fig. 14. 4.

Next two columns are needed to estimate HW depth for inlet control. First of these two columns is HW/D, HW depth divided by culvert height. This value is obtained from an inlet control chart in HEC-5 (USDOT, Reprinted 1977). Read to two decimal places whenever possible. These charts' use is explained later in this chapter. Its second column is HW, water depth at a culvert's entrance, assuming inlet control. This value is obtained by multiplying HW/D obtained from an inlet control chart by culvert height and rounding it to tenths of a foot.

The next eight columns are needed to determine HW depth in outlet control. First column is Ke, entrance-loss coefficient. This value is obtained from Table 14.1 based on data listed in column 1.

Next column is H, head loss, beginning just upstream of the entrance to just inside the exit. Obtain this value from outlet control charts in HEC-5 (USDOT, Reprinted 1977) and read to one decimal place. These charts are explained later in this chapter. H includes all head losses including one velocity head at an entrance to get the water started, an entrance loss, and friction loss through a culvert. This is expressed in equation form as Eq. (14.1).

$$H = V^2/2g + KeV^2/2g + Hf \tag{14.1}$$

Type of Structure and Design of Entrance	Coefficient, Ke
Pipe, Concrete	
Projecting from fill, groove-end	0.2
Projecting from fill, square-cut end	0.5
Headwall or headwall and wingwalls	
Socket end of pipe, groove-end	0.2
Square-edge	0.5
Rounded to radius = 1/12 diameter	0.2
Mitered to conform to fill slope	0.7
End section conforming to fill slope*	0.5
Beveled edges, 33.7° or 45° bevels	0.2
Side- or slope-tapered inlets	0.2
Pipe or Pipe Arch, Corrugated Steel	
Projecting from fill, no headwall	0.9
Headwall or headwall and wingwalls, square edge	0.5
Mitered to conform to fill slope, paved or unpaved slope	0.7
End section conforming to fill slope*	0.5
Beveled edges, 33.7° or 45° bevels	0.2
Side- or slope-tapered inlets	0.2
Box, reinforced concrete	
Headwall parallel to embankment, no wingwalls	0.5
Rounded on three edges to radius of 1/12 barrel height dimension, or beveled edges on three sides	0.2
Wingwalls at 30° to 75° to barrel	
Square-edged at crown	0.4
Rounded on three edges to radius of 1/12 barrel height dimension, or beveled edges on three sides	0.2
Wingwalls at 10° to 25° to barrel	
Square-edged at crown	0.5
Wingwalls parallel, extension of sides	0.7
Side- or slope-tapered inlets	0.2

Note: "End section conforming to fill slope," made of either metal or concrete are the sections comonly available from manufacturers. From limited hydraulic tests, they are equivalent in operation to a headwall in both inlet and outlet control. Some end sections, incorporating a closed taper in their design, have a superior hydraulic performance. These later sections can be designed using the information given for the beveled inlet.

TABLE 14.1 Culvert Entrance Loss Coefficients

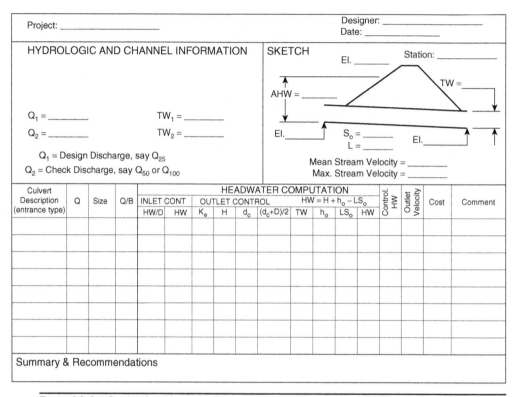

FIGURE 14.4   Design form for box culverts.

Hf is determined from Eq. (14.2). *Sf* is evaluated from Manning's equation as follows:

$$V = 1.49\,R^{2/3}\,Sf^{1/2}/n$$

$$Sf^{1/2} = Vn/1.49\,R^{2/3}$$

$$Sf = V^2 n^2/2.22\,R^{4/3}$$

$$Sf = (V^2 n^2/2.22\,R^{4/3}) \times 64.32/2g$$

$$Hf = Sf \times L = 29\,n^2\,V^2/2gL/R^{4/3} \tag{14.2}$$

The next column is Dc. Dc values are evaluated from Eq. (13.8) in Chap. 13 or by using charts contained in HEC-5 (USDOT, Reprinted 1977). A note on these charts "Dc cannot exceed top of culvert" is explained as follows by using specific energy (SE) curves. Dc occurs at minimum SE as shown in Fig. 13.3. A SE curve for a 4-ft wide channel conveying 300 cfs is calculated, shown in Table 14.2 and plotted in Fig. 14.5. In like manner, a specific energy curve is developed for a 4-ft by 4-ft RCB, shown in Table 14.3 and plotted in Fig. 14.5. Instead of this curve for a RCB being asymptotic to a 45° line, it is parallel to it. Dc is depth at minimum SE. Thus, Dc for a 4-ft-wide open channel is about 5.5 ft while Dc for a 4-ft by 4-ft RCB is 4.0 ft. Thus, a note

Depth	Area	Velocity	Velocity Head, Hv	Depth + Hv
0.0	0	0.00	0.00	0.00
1.0	4	75.00	87.45	88.45
2.0	8	37.50	21.86	23.86
3.0	12	25.00	9.72	12.72
4.0	16	18.75	5.46	9.46
5.0	20	15.00	3.50	8.50
5.5	22	13.64	2.89	8.39
6.0	24	12.50	2.43	8.43
7.0	28	10.71	1.78	8.78

TABLE 14.2   Specific Energy Calculations for a 4-ft Wide Open Channel

on Dc charts in HEC-5 (USDOT, Reprinted 1977) states that Dc cannot exceed a culvert's top.

The next column, (Dc + D)/2, is the average of Dc and culvert height, but the reason for this column requires explanation. To do this, refer to Figs. 14.6 and 14.7. These two figures are taken from HEC-5 (USDOT, Reprinted 1977) and their explanation is taken from it.

Headwater depth, HW, can be expressed by a common equation for all outlet-control conditions, including all depths of tailwater. This is accomplished by designating the vertical dimension from the culvert invert at the outlet to the elevation from which H is measured as h_o. The headwater depth equation is:

$$HW = H + h_o - LS_o \tag{14.3}$$

All the terms in this equation are in feet. H is found from the full-flow nomographs. L is the length of the culvert in feet, and S_o is the barrel slope in feet per foot. The distance h_o is discussed in the following paragraphs for the various conditions of outlet-control flow. Headwater, HW, is the distance in feet from the invert of the culvert at the inlet to the water surface of the pool.

Depth	Area	Velocity	Velocity Head, Hv	Depth + Hv
0.0	0	0.00	0.00	0.00
1.0	4	75.00	87.45	88.45
2.0	8	37.50	21.86	23.86
3.0	12	25.00	9.72	12.72
4.0	16	18.75	5.46	9.46
5.0	16	18.75	5.46	10.46
5.5	16	18.75	5.46	10.96
6.0	16	18.75	5.46	11.46
7.0	16	18.75	5.46	12.46

TABLE 14.3   Specific Energy Calculations for a 4' × 4' Box Culvert

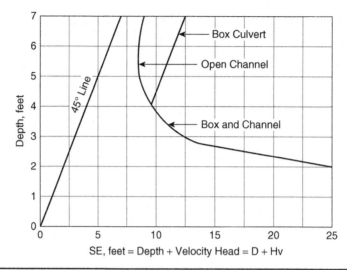

FIGURE 14.5 Specific energy curves for a 4-ft-wide-open channel and a 4-ft by 4-ft box culvert.

When the elevation of the water surface in the outlet channel is equal to or above the elevation of the top of the culvert opening at the outlet (Fig. 14.6), h_o is equal to the tailwater depth. Tailwater depth, TW, is the distance in feet from the culvert invert at the outlet to the water surface in the outlet channel. The relationship of HW to the other terms in Eq. (14.3) is illustrated in Fig. 14.6.

If TW elevation is below the culvert's top at the outlet (Fig. 14.7), h_o is difficult to determine. Discharge, culvert size, and the TW must be considered. In these cases, h_o is the greater of two values (1) TW depth as defined above or (2) $(Dc + D)/2$. The latter dimension is distance to an *equivalent* hydraulic grade line.

In this fraction, Dc is critical depth and D is the culvert height. The value of Dc can not exceed D, making the upper limit of the fraction equal to D. Where TW is the greater of the two, Dc is submerged sufficiently to make TW effective in increasing the headwater. Figure 14.7 shows terms of Eq. (14.3) for this low tailwater condition. Figure 14.7 is drawn similarly to Fig. 14.2c, but a change in discharge can change the water surface profile to that of Fig. 14.2b or Fig. 14.2d.

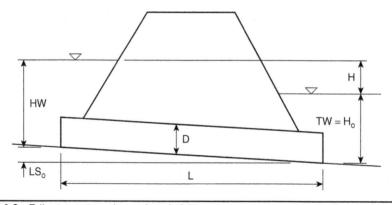

FIGURE 14.6 Tailwater greater than culvert height.

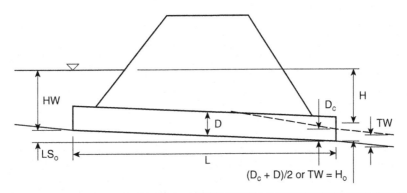

FIGURE **14.7**   Tailwater less than culvert height.

Equation 14.3 in this quote results from writing Bernouilli's equation from a culvert's inlet to its outlet. Use of an equivalent hydraulic grade line results from experiments ran at the St. Anthony Falls laboratory on the campus of the University of Minnesota in Minneapolis to determine flow through culverts under various conditions. When TW was greater than or equal to a culvert's height, Fig. 14.6, the terms in Bernoulli's equation (LS_o plus HW at its inlet and TW plus H at its outlet) were equal to each other. However, when *TW* was less than a culvert's height, as seen in Fig. 14.7, terms in Bernoulli's equation were not equal to each other. It was only when researchers developed the concept of using an equivalent hydraulic grade line at an outlet that this equation balanced. This explanation is the reason that the column, $(Dc + D)/2$, is in this table.

The next column, *TW*, is tailwater depth at its outlet. The next column is h_o and is the greater of the previous two columns, $(Dc + D)/2$ or *TW*. Next column is LS_o. The last column is *HW*, assuming a culvert is in outlet control. It uses Bernouilli's equation written just above, $HW = H + h_o - L \times S_o$, and is written between just outside a culvert's inlet and just inside its outlet.

The next column is controlling *HW*. This is the greater of *HW* in inlet control or outlet control. This controlling *HW* is okay if it is equal to or slightly less than *AHW* for a site. The next column is outlet velocity. Depth to be used is usually *Dc* or *Dn* in a culvert to calculate area. The next column is cost. Culvert cost includes barrel, type of ends used, excavation and backfill, and material used for energy dissipation. Some type of energy dissipation is usually needed because outlet velocity is greater than an existing channel can withstand without damage. Materials used consist of riprap, concrete, or some manufactured hard material. These days, natural materials such as logs, rocks, brush, or other vegetation are being used. A final column is comments and include okay, no good, >AHW, < AHW, or some other comment.

14.8   Determine Critical and Normal Depths

Determine *Dc* and *Dn* for your final culvert size. Three methods for estimating each of these depths were discussed in Chap. 13. Charts in HEC-5 (USDOT, Reprinted 1977) can also be used to determine *Dc*. These depths are calculated once a culvert has been sized, then used to determine whether a culvert is on a hydraulically mild or a hydraulically steep slope.

14.9 Inlet Control Charts

Inlet control charts for culvert types and shapes are contained in HEC No. 5 (USDOT, Reprinted 1977). A nomograph schematic is shown in Fig. 14.8. The left-hand line is culvert size or height. The next line is flow rate in cfs or cfs/ft of width. Lines on the right are entrance types. There can be two or three lines, and each is described or labeled on these inlet control charts.

HW is estimated as follows. Draw a line through its size and flow rate and extend it to line 1, a turning line. Depending on entrance type, use a HW/D value on line 1 or extend it horizontally to line 2 or 3 and read a HW/D value from some line. Enter this value in the design form's HW/D column. Next column's HW value is determined by multiplying a value of HW/D by D. Compare this HW with AHW. If it is ≤ AHW, then determine HW in outlet control. If it is > AHW, then insert "no good" in the comment column and select a somewhat larger culvert size.

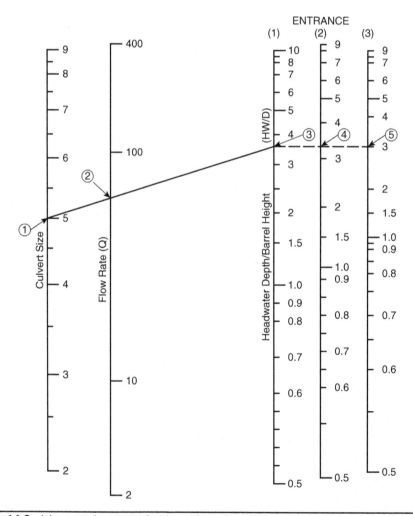

Figure 14.8 Inlet control nomograph schematic.

The hydraulic efficiency of various inlet types is shown on these nomographs. Using an example shown on Fig. 14.8 for a 5-ft RCB (Chart 1 in HEC-5), values for lines 1, 2, and 3 for HW/D are 3.6, 3.2, and 3.0, respectively. For a 5-ft height, these HW depths are 18.0, 16.0, and 15.0 ft, respectively. Thus, using a headwall instead of a projecting condition reduces HW by 3.0 ft. Differences in HW depths are more dramatic for larger culverts. Improving an entrance type enhances a culvert's performance and is important for limited AHW depths.

14.10   Outlet Control Charts

Outlet control charts for culvert types and shapes are also contained in HEC-5 (USDOT, Reprinted 1977). A nomograph schematic is shown in Fig. 14.9. *HW* in outlet control is estimated as follows. Use an entrance loss coefficient, Ke, from Table 14.1

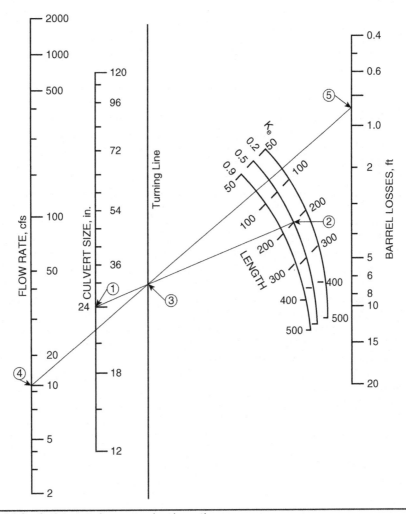

Figure 14.9   Outlet control nomograph schematic.

and put it in the Ke column. Select an outlet control nomograph from HEC-5. Draw a line between culvert size and its length on an appropriate entrance loss coefficient curve to determine a turning-line point. Draw another line from flow rate through this turning line point to the head loss line. Note its value and write it in the column labeled *H* to one decimal place.

An important comment needs to be made here. Manning's roughness coefficient used in outlet-control nomographs for concrete box culverts, pipes, and arches is 0.012. CSPs and structural plate CSPs have other values listed in HEC-5. If a different value is required by a jurisdiction in which your culvert is being designed, then Eq. (14.4) must be used to determine an equivalent culvert length. Use this length in an appropriate outlet control chart to determine head loss. If your equivalent length is greater than 500 ft, use Eq. (2) in HEC-5 to determine *H*.

$$L' = L \ (n1/n)^2 \tag{14.4}$$

where L' = equivalent culvert length, feet
        L = actual culvert length, feet
        $n1$ = actual Manning's n in culvert
        n = Manning's n value listed in an outlet control chart

A culvert must serve its purposes for several decades. Some culverts installed over a hundred years ago are still in use today. Over time, joints settle or pull apart slightly. Debris conveyed in the flow cause culverts to be somewhat rougher. Joint movement and debris increase its roughness. I recommend using $n = 0.015$ instead of 0.012 for RCP and RCB culverts to account for this change in roughness. Applying Eq. (14.4) increases a culvert's length by 56 percent.

Determine *Dc* from a nomograph in HEC-5 and place it in its column to one decimal. Calculate an average of *Dc* and culvert height and place it in the $(Dc + D)/2$ column. Estimate *TW* depth for your design flow rate and write it in the *TW* column.

In the column labeled h_o, enter the larger of values listed in columns $(Dc + D)/2$ or TW. List elevation differences between inlet and outlet inverts in the LS_o column. Calculate HW using $HW = H + h_o - LS_o$, shown above this column, and write it in the HW column.

Finish the line by comparing HWs, inlet or outlet control, and enter it in the controlling HW column. If the result is an unacceptable design, change the culvert size, larger if controlling HW is high and smaller if controlling HW is low, and complete another line on the form. Enter flow velocity at its outlet in its column. Estimate its cost, and enter it in its column. Add any remarks.

The largest head loss listed on outlet control nomographs is 20 ft. In some cases, a line through flow rate and turning line point results in a line below this 20-ft mark. Rather than guessing, do as follows. Extend the head loss line downward. Make tic marks on a piece of paper opposite the 10-ft, 20-ft point, and point where a line through the turning point intersects the head loss line. Move these tic marks up along the head loss line until your first tic mark is opposite a 1.0 ft head loss. Then read head loss opposite the third tick mark and multiply it by 10. This results in a correct head loss. This is because this line is a log scale that repeats itself.

14.11   Performance Curves

Performance curves consist of a family of curves for a culvert size, shape, length, slope, and roughness that include an inlet control curve and curves for different outlet control slopes. Ordinate is depth and abscissa is flow rate. From this figure, designers can determine flow depth for a flow rate and whether a culvert is operating under inlet or outlet control at that flow rate.

Curves can be constructed for culvert shapes and sizes used by a firm. They are useful during preliminary design to get a quick size needed. Calculations for performance curve construction for a 200-ft long, 48-in. CSP culvert with projecting inlet are contained in Table 14.4 for outlet control and Table 14.5 for inlet control. The result is depicted in Fig. 14.10.

Assume Q	H	Dc	(Dc + D)/2	Headwater for Various Percent So					
cfs	ft	ft	ft	0.0	0.5	1.0	1.5	2.0	2.5
20	0.2	1.3	2.6	2.8	-	-	-	-	-
40	0.8	1.9	3.0	3.8	2.8	1.8	0.8	-	-
60	1.9	2.3	3.2	5.1	4.1	3.1	2.1	1.1	0.1
80	3.3	2.7	3.4	6.7	5.7	4.7	3.7	2.7	1.7
100	5.2	3.1	3.6	8.8	7.8	6.8	5.8	4.8	3.8
120	7.5	3.3	3.6	11.1	10.1	9.1	8.1	7.1	6.1
140	10.2	3.5	3.8	14.0	13.0	12.0	11.0	10.0	9.0
160	13.6	3.7	3.8	17.4	16.4	15.4	14.4	13.4	12.4

TABLE 14.4   Data for a 48" CSP's Outlet Control Curves

Assume HW/D	Read Q	HW = HW/D × 4
0.5	21	2.0
0.6	29	2.4
0.7	37	2.8
0.8	46	3.2
0.9	56	3.6
1.0	65	4.0
1.1	74	4.4
1.3	90	5.2
1.5	102	6.0
1.7	112	6.8
2.0	126	8.0
2.5	145	10.0
3.0	165	12.0

TABLE 14.5   Data for a 48" CSP's Inlet Control Curve

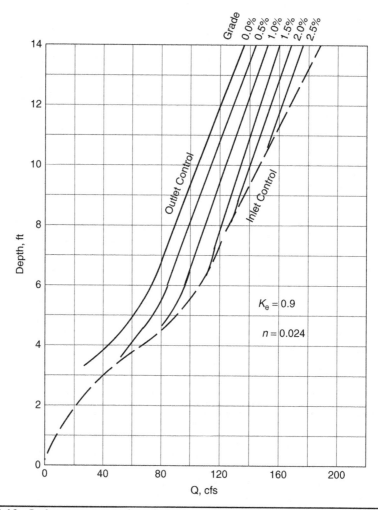

Figure 14.10   Performance curves for a 200-ft long, 48-in. CSP.

14.12   Outlet Transitions

Some jurisdictions use outlet transitions developed by FHWA (1975) to reduce culvert outlet velocities before runoff exits a culvert. They include barrel enlargement and/or placing steel angles before the culvert's downstream end. These outlet transitions are shown in Figs. 14.11 and 14.12. There is an extra cost to purchase and install materials for these alternative outlets.

The purpose of outlet transitions is to create turbulence by changing flow direction to upward and downward in addition to downstream. Turbulence is a visual sign of energy loss, resulting in lower velocities. Outlet transitions can sometimes eliminate a need for bed and bank protection downstream or reduce a need for a more expensive solution. If they are used, downstream protection is a good idea and a cheap insurance policy to protect a culvert's outlet.

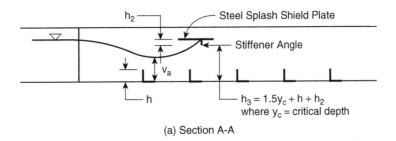

(a) Section A-A

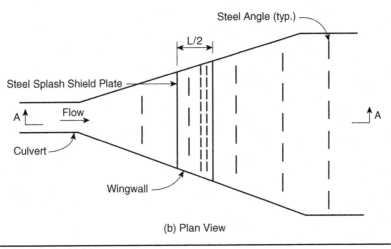

(b) Plan View

Figure 4.11 Splash shield.

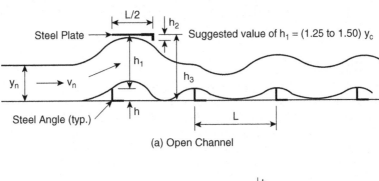

(a) Open Channel

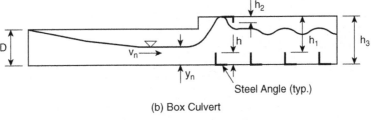

(b) Box Culvert

Figure 14.12 Splash shield definition sketch for tumbling flow.

14.13   Pipe Protective Measures Affecting Culvert Hydraulics

Pipe protection measures are added to parent material to reduce corrosion and abrasion effects. These coatings are added to CSP's insides and/or outsides to extend their useful lives. They reduce Manning's roughness coefficient by making pipe interiors smoother. This increases a pipe's capacity by increasing flow velocity and reducing head losses due to friction.

Some coatings are useful for only a few years if used in poor locations. Sand and gravel in runoff abrade these linings to parent material. High-velocity flows convey cobbles and rocks. These abrade RCP and RCB culverts to reinforcing steel. In a case in Denver, a solution for a larger RCP culvert was to fill in lost material with epoxy and then cover the lower third of its diameter with a ¼-in. thick curved steel plate, held in place by shooting bolts into the RCP.

Asphalt, concrete, fiberglass, epoxy, and steel have been used to provide this protection, reducing Manning's coefficient. Manning's *n* values for these linings are listed in Table 14.6.

Material	Manning's *n*
Full Perimeter Asphalt (thin coat)	0.024
Asphalt Smooth Flow (60° to 180°)	0.021
Asphalt Smooth Flow (360°)	0.012
Concrete	0.015
Fiberglass	0.009
Epoxy	0.009
Steel	0.009

TABLE 14.6   Manning's *n* Values for Various CSP Linings

14.14   Summary

This chapter on culvert design provides only a framework for the many items that should be considered in their design. Culvert design is itself worthy of an entire book. Hundreds of millions of dollars are spent each year for constructing new culverts, repairing and maintaining existing culverts, and reconstructing washed out culverts. Much of new and replacement culverts' construction costs could be eliminated by designers understanding the many ways in which culverts can be designed.

Understanding how to design, construct, and maintain culverts can save large amounts of money, enhance sustainability efforts, prevent accidents and loss of life from washed-out culverts, extend their useful lives, and reduce long-term operation and maintenance costs.

CHAPTER 15

Riser Structure Design

15.1 Introduction

Risers at upstream ends of detention basin outlets are used by jurisdictions whose ordinances require they be designed for several return periods: one for a water-quality event and storm events from a 2- or 5-year through a 100-year. They can be square, round, or rectangular. Tops are open, closed, or partially open, acting as weirs at shallow depths and as orifices at deeper depths. Other openings at various sides and elevations of the riser are either weirs or orifices, sometimes both, depending on a detention basin's pond depth and water depth inside the riser. Riser depth-outflow relationships is one element needed to estimate flow attenuation and how long runoff is detained within a facility. Outlet sizes, shapes, and elevations affect their outflows.

This is true unless a temporary pond's depth is limited due to existing terrain. In this case, some or all riser outlets may be submerged. Then, weir and orifice equation use becomes more complex. Unfortunately, some designers and program developers are not fully aware of the nuances of these equations. As a result, their elevation-outflow characteristics are incorrect.

15.2 A Caution

One word of caution. The orifice equation is usually a well-behaved equation, but it has one quirk that causes development of a poor elevation-outflow relationship. The weir equation is even simpler than the orifice equation but you cannot trust anything about it. Unless you truly understand these two equations, you should not design best management practices (BMPs) or retrofits.

15.3 Orifices

The basic equation for an orifice is:

$$Q = C A (2g)^{1/2} (H)^{1/2} \tag{15.1}$$

where Q = peak discharge rate, cfs
         C = coefficient of discharge, dimensionless
         A = cross-sectional area of orifice, ft^2
         G = acceleration due to gravity: 32.16 fps/sec
         H = head on an orifice, ft.

15.3.1   Area and Gravity

An orifice equation is a well-behaved equation. It has a prismatic shape: round, oval, square, rectangular, or triangular. Simple equations are available to determine its area. "A" in the orifice equation is an orifice's entire cross sectional area. The square root of 2g is 8.02.

15.3.2   Coefficient of Discharge

Tables of discharge coefficients are listed in Brater and King (1976). It is a function of an orifice's size and shape, head on the orifice, sharpness of an orifice's edge, roughness of its inner surface, and degree to which the jet's contraction is suppressed. Experiments on submerged orifices indicate this does not affect the coefficient.

An orifice's coefficient of discharge includes two items. The first is caused by an entrance loss. Water moves from a pond of some size to and through a comparatively small orifice opening. This entrance loss coefficient is akin to those for a culvert's entrance. The second is due to presence of a vena contracta caused by a jet's degree of contraction (see Fig. 15.1a). As noted above, area used in an orifice equation is its full area. However, an area of a vena contracta is usually smaller. This reduction in area is also accounted for in the coefficient of discharge.

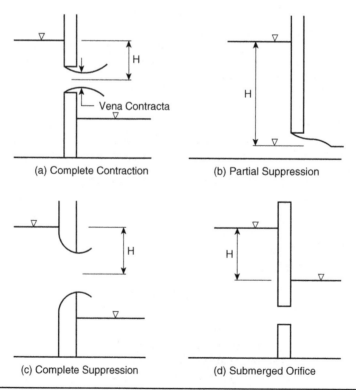

(a) Complete Contraction

(b) Partial Suppression

(c) Complete Suppression

(d) Submerged Orifice

Figure **15.1**   Contraction of a jet through orifices.

A coefficient for a sharp-edged orifice with complete contraction of a jet varies from 0.59 to 0.66 (Fig. 15.1a). A nominal value of 0.60 is used for types of orifices and range of heads used for riser structures. For orifices with partially suppressed contractions (Fig. 15.1b), a coefficient ranges from 0.62 to 0.71. A nominal value of 0.67 can be used. For fully rounded edges, i.e., a jet's contraction is fully suppressed (Fig. 15.1c), its coefficient ranges from 0.94 to 0.95.

If holes are burned or drilled through a steel plate or concrete and its surface and edges are left rough, its coefficient of discharge can be as low as 0.40. The degree of the jet's contraction for structures depends on how much reduction in peak-flow rate you want to achieve. The lower this coefficient is, the smaller the discharge that flows through an orifice at some head.

15.3.3 Heads on an Orifice

There are two definitions of head for orifice flow:

1. In free discharge, water surface downstream of an orifice is below an orifice's centroid (center of gravity). In this case, head is measured from an upstream water surface to an orifice's centroid (Fig. 15.1a through c).

2. For a submerged orifice, water surface downstream of an orifice is above its centroid. In this case, head equals difference between upstream and downstream water surfaces (Fig. 15.1d). Submergence has no effect on an orifice's coefficient of discharge.

An orifice's head is a variable that causes problems for designers and writers of computer programs. At times, both groups just use head's first definition. They neglect the possibility that depth just downstream of openings can be deeper than their centroids, changing flow rates. Pond and riser depths must be checked every time step of hydrograph routing through a BMP.

Structures downstream of orifices are channels, culverts, or storm sewers—or a basin's outlet conduit. These depths are just downstream of orifices for all flows. For risers with a permanent pond, depths inside risers should be lower than openings for water quality and rarer storms _if_ outlet conduits are large enough. In this case, risers' openings always act with a free discharge.

This is difficult to achieve when facilities drain completely after rainfall events. Many facilities are designed with no pond because water quality enhancement was not a facility objective or not enough depth to create a pond. In these cases, ensure outflow rates are correct as water rises in temporary ponds, since definition of head could change from free to submerged for one or more riser openings. I have performed experiments that show as depths increase during runoff events, flows through openings rise to a peak then decrease after head definition shifts from free discharge to submerged.

15.4 Weirs

A weir is defined as a horizontal crested, vertical sided hydraulic structure. Its basic equation is:

$$Q = CLH^{3/2}$$

(15.2)

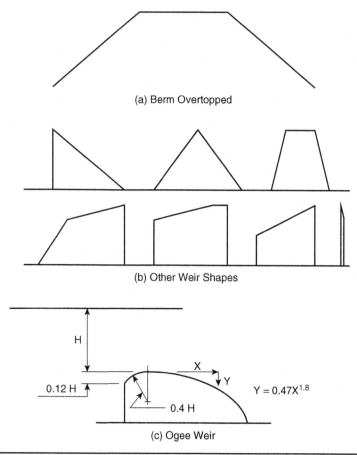

(a) Berm Overtopped

(b) Other Weir Shapes

(c) Ogee Weir

$$Y = 0.47X^{1.8}$$

FIGURE 15.2   Various shapes of weirs.

where Q = flow rate, cfs
   C = coefficient of discharge, dimensionless
   L = length of weir, ft
   H = head on weir, ft

However, this equation is not well-behaved. C is not a constant as some designers and program developers think and varies from 2.6 to over 4.0. It depends on a weir's shape, breadth of crest, depth of flow above its crest, and crest heighth above its invert. With a 10-ft length, length used in its equation is not 10.0 ft. See 15.4.2 for an explanation. Head is measured well upstream of the crest and can take on one of two values. Where and how do you get the numbers you use?

There are two weir types: sharp-crested weirs and weirs that are not sharp-crested. Weirs not sharp-crested are known as broad-crested weirs. See Fig 15.2. Water touches sharp-crested weirs' crests at only a single point. Depending on its configuation, it can be broad-crested at low heads, sharp-crested at higher heads, and an orifice at even higher heads. See Fig. 15.3.

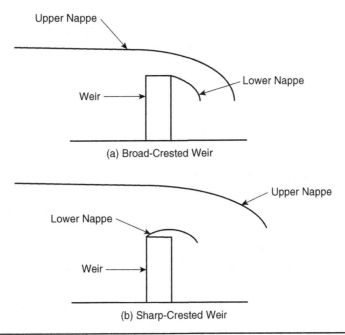

FIGURE 15.3 Types of flow over a weir.

Variable values change depending on how the situation changes during a flood routing.

15.4.1 Coefficient of Discharge

It varies based on the variables listed above. A broad-crested weir is defined as one with a nearly horizontal crest long enough in the flow direction so its lower nappe is supported and hydrostatic pressures are developed for a short length. If its upstream corner is so rounded and its crest's slope is so great as entirely to prevent contraction, flow is at critical depth (Dc), and its flow is given by Eq. (15.3). A value of 3.087 is the largest coefficient value attained by broad-crested weirs.

$$Q = 3.087\, LH^{3/2} \qquad\qquad (15.3)$$

At greater depths, the coefficient increases through a transition stage until it is 3.32, a sharp-crested weir's coefficient. Coefficients for a range of depths from 0.2 through 5.5 ft for a weir shown in Fig. 15.3 are listed in Table 15.1 (Brater & King, 1976). Assume a low wall has a 9-in thickness. It is a broad-crested weir until flow depth is about 0.9 ft. It then transitions into a sharp-crested weir at a depth of 1.8 ft. This is true because sharp-crested weirs have water touching a weir's crest at only a single point as illustrated in Fig. 15.3 (Fig. 14-1, Chow, 1959). The lower nappe springs clear of and above the crest of a sharp-crested weir.

Figure 15.3a has a wall with a low depth of flow. Its lower nappe clings to the crest for its width of 9 in. In Fig. 15.3b, depth is deeper and its lower nappe springs clear of its crest. Water below a weir's crest hits the wall and is diverted upwards. Water above the crest moves horizontally, and their combination of velocities lifts water above the

Measured	Breadth of Weir Crest, Feet										
Head, Feet	0.50	0.75	1.00	1.50	2.00	2.50	3.00	4.00	5.00	10.00	15.00
0.2	2.80	2.75	2.69	2.62	2.54	2.48	2.40	2.38	2.34	2.49	2.68
0.4	2.92	2.80	2.72	2.64	2.61	2.60	2.58	2.54	2.50	2.56	2.70
0.6	3.08	2.89	2.75	2.64	2.61	2.60	2.68	2.69	2.70	2.70	2.70
0.8	3.30	3.04	2.85	2.68	2.63	2.60	2.67	2.68	2.68	2.69	2.64
1.0	3.32	3.14	2.98	2.75	2.65	2.64	2.65	2.67	2.68	2.69	2.63
1.2	3.32	3.20	3.08	2.86	2.70	2.65	2.64	2.67.	2.66	2.67	2.64
1.4	3.32	3.26	3.20	2.92	2.77	2.68	2.64	2.65	2.65	2.64	2.64
1.6	3.32	3.29	3.28	3.04	2.84	2.71	2.68	2.66	2.63	2.64	2.63
1.8	3.32	3.32	3.30	3.07	2.88	2.74	2.68	2.66	2.65	2.64	2.63
2.0	3.32	3.32	3.31	3.14	2.95	2.76	2.72	2.68	2.65	2.64	2.63
2.5	3.32	3.32	3.32	3.28	3.07	2.69	2.81	2.72	2.66	2.64	2.63
3.0	3.32	3.32	3.32	3.32	3.20	3.05	2.92	2.73	2.67	2.64	2.63
3.5	3.32	3.32	3.32	3.32	3.32	3.19	2.97	2.76	2.68	2.64	2.63
4.0	3.32	3.32	3.32	3.32	3.32	3.32	3.07	2.79	2.70	2.64	2.63
4.5	3.32	3.32	3.32	3.32	3.32	3.32	3.32	2.88	2.74	2.64	2.63
5.0	3.32	3.32	3.32	3.32	3.32	3.32	3.32	3.07	2.79	2.64	2.63
5.5	3.32	3.32	3.32	3.32	3.32	3.32	3.32	3.32	2.88	2.64	2.63

TABLE 15.1   Values of C in the Broad-Crested Weir Equation

crest. If the crest's breadth is short enough, gravity does not bring the lower nappe below the weir's crest until downstream of the crest's end. Thus, water touches it at only a single point, and it is now a sharp-crested weir.

A coefficient is a function of flow depth and breadth and its variation is shown in Table 15.1. Narrow breadths transition quickly to sharp-crested weirs. Weir breadths of 5 to 10 ft and greater never do change from broad- to sharp-crested. Weir shape also affects the coefficient. One is a 15-ft wide berm being overtopped in Fig. 15.2a. Table 15.1 lists its discharge coefficients as ranging from 2.63 to 2.70, a broad-crested weir due to friction loss. Other shapes are shown in Fig. 15.2b with coefficients listed in Tables 5-4 to 5-13 in Brater and King (1976).

Figure 15.2c is an ogee weir. Its shape is the under nappe of a sharp-crested weir. Dams use it because its coefficient ranges to over 4.0 (USDI, Revised Reprint 1977). These higher values of C allows lengths to be much shorter, saving millions in construction costs. C values are shown in Fig. 15.4. P is elevation difference between its crest and channel invert. Ha is velocity head of approach. He is actual head being considered. He/Ho is equal to one when an ogee has an ideal nappe shape.

The United States Geological Survey (USGS) has conducted flow measurements over roadways during storms (USDI, Reprinted 1976). They were used to develop C values for these broad-crested weirs. Values are depicted in Figs. 15.5 and 15.6 and can also be used for overtopped berms. Values for a submergence factor, K, are shown in

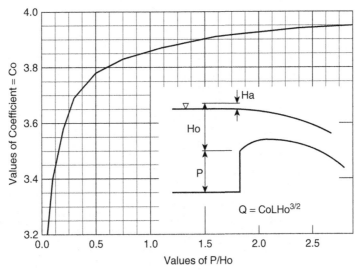

(a) Discharge Coefficients for Vertical-Faced Ogee Weirs

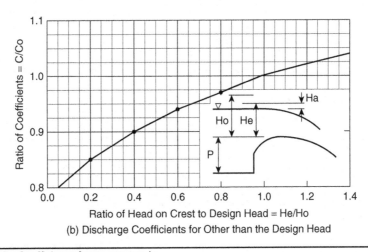

(b) Discharge Coefficients for Other than the Design Head

FIGURE 15.4   Coefficients for an ogee weir.

Fig. 15.6a. Variable sketches in Figs. 15.5 and 15.6 are shown in Fig. 15.6b. This K value is applied to the equation as shown in Eq. (15.4). Another set of K factors for other weir types are shown in Fig. 15.7 (Brater and King, 1976).

$$Q = KCLH^{3/2} \tag{15.4}$$

Figure 15.8 is a V-notch weir, used as an outlet for smaller storage depths. Two are shown in Fig. 15.8. Profile (b) is better since larger flows rate are conveyed before its crest (vee's point) is submerged. For 2-year flows, depths are small, and a small triangular area conveys outflow. Its equation for a 90 degree weir is:

$$Q = 2.5\,H^{2.5} \tan \theta/2 \tag{15.5}$$

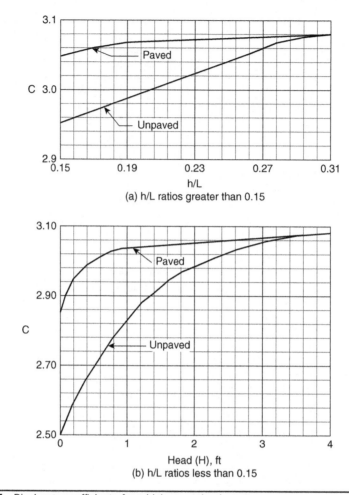

FIGURE 15.5 Discharge coefficients for a highway embankment.

where Q = flow rate, cfs
H = head on weir, feet, water surface, or EGL minus weir crest
θ = total central angle, degrees

Values of C for other central angles are shown in Fig. 5-4 of Brater and King (1976). Figure 15.9 is a V-notch weir with three central angles. This allows greater outflows when storage volumes are small. This weir with multiple angles is useful as shown in Fig. 15.8 or Fig. 15.9.

Question: if water depth in Fig. 15.9 is greater than the first change in central angle, is the lower V-notch a weir or an orifice?

Answer: orifice.

Question: When water is near the top of Fig. 15.9, how many weirs and orifices are there?

Answer: Two orifices and a top Cippoletti weir. If your answers are different and not sure why my answers are correct, you are not yet ready to design outlets.

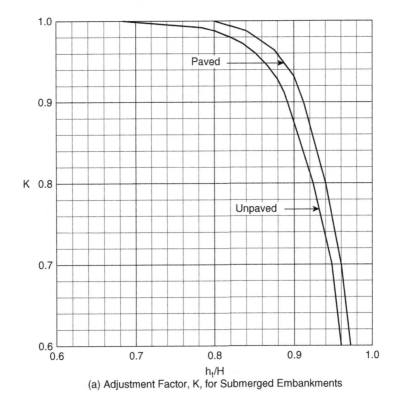

(a) Adjustment Factor, K, for Submerged Embankments

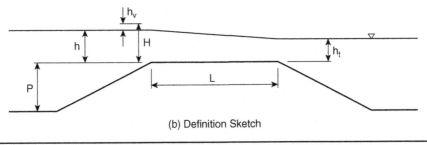

(b) Definition Sketch

Figure 15.6 Flow over a highway embankment.

How are the two orifices' heads measured? A submerged orifice's head is the difference between the up and downstreams' water surfaces—if it is above the orifice's centroid. If not, then head is the difference of upstream water surface and the orifice's centroid. Question? How deep must water be in a next higher weir before a lower one changes from a weir to an orifice? Unfortunately, I did not think to ask this question while doing experiments.

15.4.2 Length

A weir's length is its vena contracta's length as shown in Fig. 15.10, also showing water drops' trajectories in three locations approaching a weir. A middle drop flows directly through it. Two drops near its sides do not know a wall is there until they are diverted

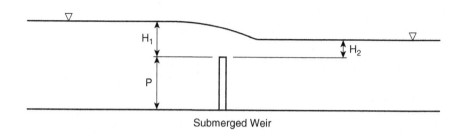

Submerged Weir

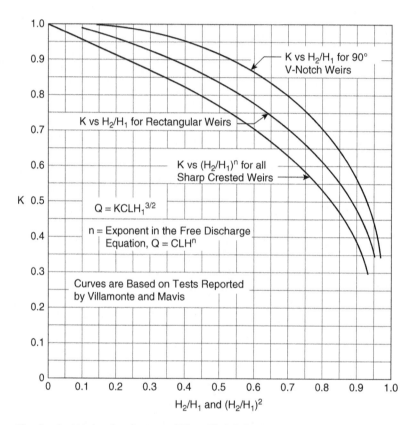

K vs H_2/H_1 for 90°
V-Notch Weirs

K vs H_2/H_1 for Rectangular Weirs

K vs $(H_2/H_1)^n$ for all
Sharp Crested Weirs

$Q = KCLH_1^{3/2}$

n = Exponent in the Free Discharge
Equation, $Q = CLH^n$

Curves are Based on Tests Reported
by Villamonte and Mavis

K

H_2/H_1 and $(H_2/H_1)^2$

Handbook of Hydraulics, Brater and King, Sixth Edition

FIGURE 15.7 Increase in head (H1) due to submergence of a weir.

by it towards its opening. These trajectories form the vena contracta. Its length is estimated from Eq. (15.6).

$$L' = L - 0.1\,nH \tag{15.6}$$

where L' = length to be used in weir equation, ft
L = actual length of weir, ft
n = number of end contractions, dimensionless
H = head on weir, ft

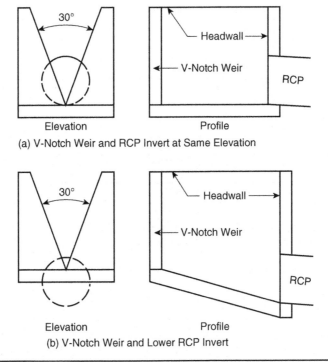

(a) V-Notch Weir and RCP Invert at Same Elevation

(b) V-Notch Weir and Lower RCP Invert

Figure 15.8   V–Notch weir in a headwall.

Assume an 8-ft long weir in a channel (Fig. 15.10) has two end contractions and is operating with a head of 3-ft. What length should be used in the weir equation?

$$L' = L - 0.1\,n\,H$$
$$L' = 8.0 - 0.1 \times 2 \times 3$$
$$L' = 7.4 \text{ ft}$$

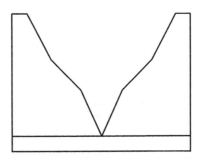

Figure 15.9   V–Notch weir with multiple angles.

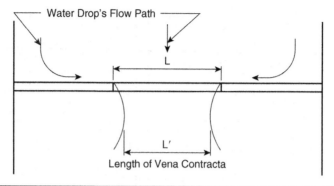

Figure 15.10   Length of a vena contractra.

This equation usually works but consider the weir shown in Fig. 15.11. What length should be used in the weir equation when head is 5.0 ft?

$$L' = L - 0.1\, n H$$
$$L' = 1.0 - 0.1 \times 2 \times 5$$
$$L' = 0.0 \text{ ft}$$

The answer is zero so there is no flow. End contractions in a 1.0-ft wide weir meet so a vena contracta's length is zero. However, anyone standing in front of the weir will get wet—some water <u>is</u> flowing through it. Equation (15.6) breaks down in this situation.

15.4.3   Head

Head has two definitions in a weir equation (see Fig. 15.12): difference between its crest and upstream hydraulic grade line (HGL) (water surface) and difference between its crest and upstream energy grade line (EGL). Measure head 10 times the head upstream

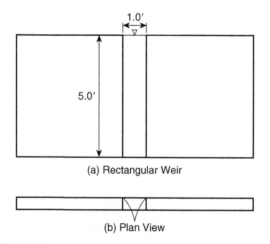

(a) Rectangular Weir

(b) Plan View

Figure 15.11   Narrow rectangular weir.

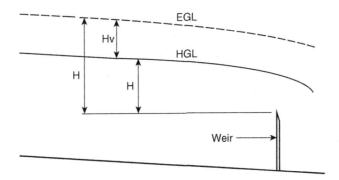

FIGURE 15.12   Two definitions of head on a weir.

of its crest because a water surface's drawdown begins upstream of it. Either definition is the head used in Eq. (15.2), but my preference is the EGL for reasons shown in the next problems. A vena contracta's length is 5.0 ft, and depth is 3.0 ft.

Problem 1 uses head's second definition with an approach velocity of 1.0 fps.

$$V^2/2g = (1.0)^2/(2 \times 32.16) = 0.02 \text{ ft}$$

$$Q = CLH^{3/2} = 3.32 \times 5.0 \times (3.0 + 0.02)^{3/2} = 87 \text{ cfs}$$

Problem 2 is similar to problem 1 except velocity of approach is 8.0 fps.

$$V^2/2g = (8.0)^2/(2 \times 32.16) = 1.00 \text{ ft}$$

$$Q = CLH^{3/2} = 3.32 \times 5.0 \times (3.0 + 1.0)^{3/2} = 133 \text{ cfs}$$

Problem 3 is similar to problem 1 except the first definition of head is used.

$$Q = CLH^{3/2} = 3.32 \times 5.0 \times (3.0)^{3/2} = 86 \text{ cfs}$$

Results show that either head's definition yields similar flows when velocities are small. When velocities are high, the EGL definition yields a flow 54 percent greater than when using the HGL definition. Thus, always use the EGL definition because it always yields good results.

15.5   Is a Weir Always a Weir Just as an Orifice Is Always an Orifice?

The answer is no. A weir becomes an orifice at deeper depths in one of two ways. See Fig. 15.13. A drop inlet's open top is shown in Figs. 15.13a and b. It acts as a weir under low heads as water flows over its crest along a drop inlet's perimeter. Flow over its sides do not interfere with each other. At higher heads, a vortex forms as it attempts to aerate the inlet. At even higher heads, the vortex disappears and it becomes an orifice.

The second way is explained as follows. Assume a 3-ft square orifice is 3.0-ft high in a wall as shown in a side view in Figs. 15.13c and d. It acts as a weir, broad- then

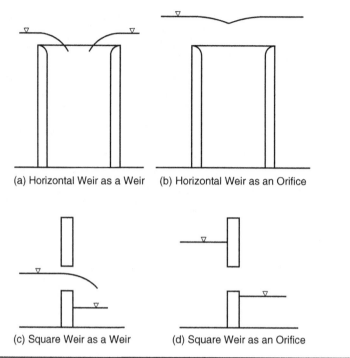

(a) Horizontal Weir as a Weir   (b) Horizontal Weir as an Orifice

(c) Square Weir as a Weir   (d) Square Weir as an Orifice

FIGURE 15.13   Weirs becoming orifices.

sharp-crested, under lower heads (Fig. 15.13c) because water is unaware the opening has a top. When it is submerged (Fig. 15.13d), an upstream water surface is deeper than the opening's top, so it acts as an orifice. Flow increases at higher upstream depths when a downstream water surface is lower than the opening's centroid. Spreadsheets for orifices and weirs are included on the McGraw-Hill's website (www.mhprofessional .com/sdsd).

15.6   Water Quality Outlets

Figure 15.14 shows three outlets. Figures 15.14a and c are used in detention and sedimentation basins to trap sediment and larger debris to prevent downstream movement. To prevent clogging their holes, wrap the pipes with geofabric surrounded by a gravel cone. Remove geofabric and gravel, wash and replace it every 5 to 10 years to prevent clogging. Settled sediment should be removed every 10 to 20 years depending on accumulation rate.

A pipe in Fig. 15.14b is used in facilities with a pond. It traps floatable debris and prevents them from moving downstream. Debris such as paper, cans, leaves, and other elements are removed as they are an aesthetic nuisance. A 3-ft dimension is minimum to prevent a vortex from forming and sucking in debris. It is sized to trap all quality runoff volume from exiting the basin except through the small pipe. The riser's open top handles runoff from larger storm events.

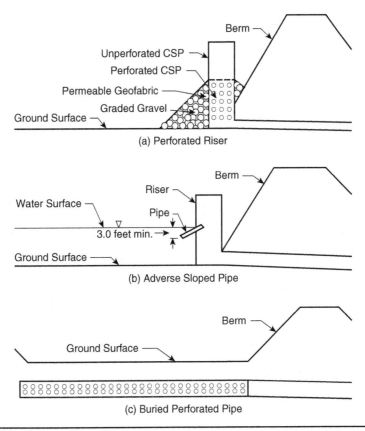

Figure 15.14   Water quality outlets.

Figure 15.15 is a Pacific Northwest outlet. The riser has a water quantity inlet near the top and a water quality inlet near the bottom. The pipe near the riser's bottom has a circular orifice in the vertical portion's bottom. This retains floatables inside the riser from exiting it; settleables in the riser fall to its bottom. Both are removed periodically. Removal times are based on written instructions to inspectors as to amount of floatables and settleable debris depths.

15.7 Examples

15.7.1 Example 15.1: Short, Steep 24-Inch RCP with a 50-Degree V-Notch Weir

A riser upstream of a short 24-in reinforced concrete pipe (RCP) (groove end at the riser) has a 50 degree V-notch opening 5.0 ft above the RCP's invert. The 24-in RCP is on a steep slope in inlet control. Its depth-discharge calculations are shown in Table 15.2 and plotted in Fig. 15.16. Values for HW/D were taken from Chart 2 of HEC-5 (USDOT,

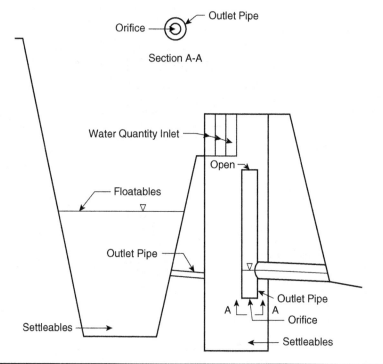

FIGURE 15.15 Pacific Northwest water quality outlet.

Depth	H	HW/D	Q_{18}
1.0	1.0	0.5	4.0
2.0	2.0	1.0	14.0
3.0	3.0	1.5	22.0
4.0	4.0	2.0	28.0
5.0	5.0	2.5	34.0
6.0	6.0	3.0	39.0
7.0	7.0	3.5	43.0
8.0	8.0	4.0	46.0
9.0	9.0	4.5	49.0
10.0	10.0	5.0	52.0
11.0	11.0	5.5	55.0
12.0	12.0	6.0	58.0

TABLE 15.2 Depth-Discharge Curve for a 24" RCP

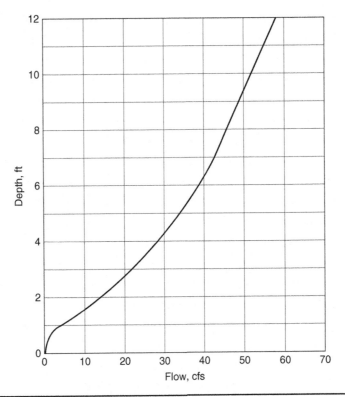

FIGURE 15.16 Depth-discharge curve for a 24" RCP.

reprinted 1980). The opening has a free discharge as long as the RCP's HW is less than 5.0 ft. Head is pond depth above its invert. Calculate the opening's depth-flow curve using appropriate head definitions and Eq. (15.10)

$$Qw = 2.5\, H^{2.5}\ \tan \theta/2 = 2.5 \times 0.466 \times H^{2.5} = 1.166\, H^{2.5} \qquad (15.10)$$

Its head is the temporary pond's depth – using the appropriate definition of head. Results are shown in Table 15.3. The weir becomes submerged above a pond depth of 3.85 ft. As water increases in the pond by 1 ft, the head inside the riser increases but only by 0.1 ft. This small difference is related to the relative sizes of the weir and pipe.

15.7.2 Example 15.2: Rectangular Weir in a Riser

A 4-ft wide weir is located in a 10-ft square riser, 12-ft high with an open top. See Fig. 15.17. Its crest is 4.0 ft below the top. Develop a depth-discharge curve for the weir using 1.0-ft increments and Eq. (15.11). Riser walls are 9-in thick with C obtained from Table 15.1.

$$Q = C\,Le\,H^{3/2} \qquad (15.11)$$

Depth	Head	H^2.5	Q	24" RCP HW	Comment
0.0	0.00	0.00	0.0	0.0	ok
1.0	1.00	1.00	1.2	0.5	ok
2.0	2.00	5.66	6.6	1.4	ok
3.0	3.00	15.59	18.2	2.5	ok
4.0	3.85	29.08	33.9	5.0	ok
5.0	4.05	33.01	38.5	5.95	ok
6.0	4.20	36.15	42.2	6.8	ok
7.0	4.31	38.56	45.0	7.69	ok
8.0	4.40	40.61	47.4	8.6	ok
9.0	4.50	42.96	50.1	9.5	ok
10.0	4.60	45.38	52.9	10.4	ok
11.0	4.69	47.64	55.5	11.31	ok
12.0	4.79	50.22	58.6	12.2	ok

TABLE 15.3   Results for Example 15.1–50° V-Notch Weir

where Q = flow rate, cfs
C = weir coefficient, dimensionless
Le = equivalent weir length due to end contractions, ft
H = head above weir crest, including velocity head of approach, if any, ft

Equivalent length and weir coefficient are taken from Eq. (15.6) and Table 15.1, respectively. Length is 4.0 ft with two end contractions. Velocity in the pond is assumed to be zero. Velocity through a weir is Q/A. Crest depth is Dc. Determine head by using flow depth plus Hv.

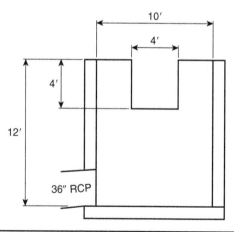

FIGURE 15.17   Rectangular weir in a riser.

Depth	C	L	Q	Vch	V²/2g	H′	H′³/²	Q
H, feet		feet	cfs	fps	feet	feet	feet	cfs
0.0	0.00	4.00	0.0	0.00	0.000	0.000	0.000	0.0
1.0	3.14	3.80	11.9	0.13	0.000	1.000	1.000	11.9
2.0	3.32	3.60	33.8	0.34	0.002	2.002	2.833	33.9
3.0	3.32	3.40	58.6	0.53	0.004	3.004	5.206	58.8
4.0	3.32	3.20	85.0	0.71	0.008	4.008	8.024	85.2

TABLE 15.4   Depth-Discharge Calculations for a 4-ft Wide Rectangular Weir

Calculations for Example 15.2 are listed in Table 15.4 and its curve plotted in Fig. 15.18. A first Q is obtained from Eq. (15.2) using H. Velocity in Table 14.10 equals Q divided by area (10 × (depth + 8). A second Q is obtained from Eq. (15.2) using H'. In this case, inclusion of approach velocity head increases flow an average of 0.2 percent. Table 15.5 indicates that for a flow of 85 cfs, HW for a 36-in RCP is 7.8 ft. Thus, this weir always has a free discharge.

15.7.3   Example 15.3: Vertical Perforated Riser

Assume an 18-in RCP is perforated with 2-in holes at 6-in centers around its periphery with six rows on its 2.75-ft height in a sediment basin's outlet (Fig. 15.14a). First holes are at the bottom. Other rows are 6 in center to center. Its open top functions as a horizontal weir, then as an orifice. Its circumference and area are 4.71 ft and 1.77 ft², respectively. Area of a 2-in hole is 0.02182 ft². With holes at 6-in centers, each row has nine holes. Head on each row is difference between water surface and each row's center. If water is 0.50-ft deep, row head is 0.42 ft.

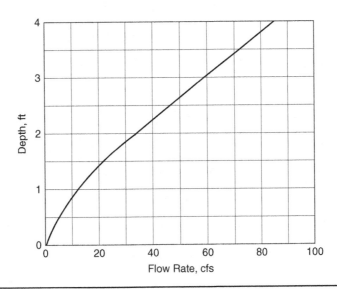

FIGURE 15.18   Rectangular weir flows.

Project: Rectangular Weir

Designer:
Date:

Hydrologic and Channel Information

Q1 = 85 cfs   TW1 = low ft

Q2 = cfs   TW2 = ft

(Q1 = Design Discharge, say Q25)

(Q2 = Check Discharge, say Q50 or Q100)

Sketch   **Station:**

Elevation =

AHW =

Elevation =   So = ft/ft   TW = low

L = ft   Elevation =

HEADWATER COMPUTATION

Culvert Description (Entrance Type)	Q cfs	Inlet Control			Outlet Control HW = H + Ho − LSo								Controlling HW ft	Outlet Velocity fps	Cost Dollars	Comment
		Size ft	HW/D	HW ft	Ke	H ft	Dc ft	(Dc + D)/2 ft	TW ft	Ho ft	LSo ft	HW ft				
RCP w/ square edged headwall	85	36″	2.60	7.8	0.5	4.7	3.0	3.0	low	3.0	1.0	6.7	7.8			ok

SUMMARY & RECOMMENDATIONS

TABLE 15.5   HW for a 36-inch RCP in Example 15.2

	L = length of tube		p = perimeter of cross section of tube		
	Condition of edges at entrance				
L/p	All corners square	Contractions suppressed on bottom only	Contractions suppressed on bottom and one side	Contractions suppressed on bottom and two sides	Contractions suppressed on on bottom, two sides, and top
0.02	0.61	0.63	0.68	0.77	0.95
0.04	0.62	0.64	0.68	0.77	0.94
0.06	0.63	0.65	0.69	0.76	0.94
0.08	0.65	0.66	0.69	0.74	0.93
0.10	0.66	0.67	0.69	0.73	0.93
0.12	0.67	0.68	0.70	0.72	0.93
0.14	0.69	0.69	0.71	0.72	0.92
0.16	0.71	0.70	0.72	0.72	0.92
0.18	0.72	0.71	0.73	0.72	0.92
0.20	0.74	0.73	0.74	0.73	0.92
0.22	0.75	0.74	0.75	0.75	0.91
0.24	0.77	0.75	0.76	0.78	0.91
0.26	0.78	0.76	0.77	0.81	0.91
0.28	0.78	0.76	0.78	0.82	0.91
0.30	0.79	0.77	0.79	0.83	0.91
0.35	0.79	0.78	0.80	0.84	0.90
0.40	0.80	0.79	0.80	0.84	0.90
0.60	0.80	0.80	0.81	0.84	0.90
0.80	0.80	0.80	0.81	0.85	0.90
1.00	0.80	0.81	0.82	0.85	0.90

Table 15.6   Coefficients of Discharge C for Submerged Tubes

Equation (15.1) is used for flow through rows of holes. Each hole is a short tube through a 5-in thick pipe. C in Table 15.6 is taken from (Brater and King, 1976) is 0.80 for a square-edged hole with $L/p = 5/(3.14159x2) = 0.80$. Eq. (15.1) and Eq. (15.2) are flows over the pipe's top, whichever yields a lesser value at each head. Table 15.1 contains weir coefficients. Example equations follow:

$$Q\text{holes} = 9\,C\,a\,(2g)^{1/2}\,(H)^{1/2} = 9 \times 0.80 \times 0.02182$$
$$\times (64.32)^{1/2}\,(H)^{1/2} = 1.26\,(H)^{1/2} \qquad (15.11)$$

1	2	3	4	5	6	7	8	9	10
Depth ft	Row 1 cfs	Row 2 cfs	Row 3 cfs	Row 4 cfs	Row 5 cfs	Row 6 cfs	Weir cfs	Orifice cfs	Total Q cfs
0.00	0.00	0.00	0.00	0.00	0.00	0.00	0.00	0.00	0.0
0.25	0.63	0.00	0.00	0.00	0.00	0.00	0.00	0.00	0.6
0.50	0.89	0.00	0.00	0.00	0.00	0.00	0.00	0.00	0.9
0.75	1.09	0.63	0.00	0.00	0.00	0.00	0.00	0.00	1.7
1.00	1.26	0.89	0.00	0.00	0.00	0.00	0.00	0.00	2.2
1.25	1.41	1.09	0.63	0.00	0.00	0.00	0.00	0.00	3.1
1.50	1.54	1.26	0.89	0.00	0.00	0.00	0.00	0.00	3.7
1.75	1.67	1.41	1.09	0.63	0.00	0.00	0.00	0.00	4.8
2.00	1.78	1.54	1.26	0.89	0.00	0.00	0.00	0.00	5.5
2.25	1.89	1.67	1.41	1.09	0.63	0.00	0.00	0.00	6.7
2.50	1.99	1.78	1.54	1.26	0.89	0.00	0.00	0.00	7.5
2.75	2.09	1.89	1.67	1.41	1.09	0.63	0.00	0.00	8.8
3.00	2.18	1.90	1.78	1.54	1.26	0.89	1.62	4.25	11.2
3.25	2.27	2.09	1.89	1.67	1.41	1.09	4.74	6.01	15.2
3.50	2.36	2.18	1.90	1.78	1.54	1.26	9.21	7.36	18.4
3.75	2.44	2.27	2.09	1.89	1.67	1.41	14.80	8.50	20.3
4.00	2.52	2.36	2.18	1.99	1.78	1.54	21.21	9.51	21.9
4.25	2.60	2.44	2.27	2.09	1.89	1.67	28.35	10.41	23.4
4.50	2.67	2.52	2.36	2.18	1.99	1.78	36.11	11.25	24.8
4.75	2.75	2.60	2.44	2.27	2.09	1.89	44.25	12.02	26.1
5.00	2.82	2.67	2.52	2.36	2.18	1.99	52.80	12.76	27.3

TABLE **15.7** Depth-Outflow Curve for an 18-inch Perforated Pipe Riser

$$Q_{weir} = C L H^{3/2} = 4.71 \, C \, H^{3/2} \tag{15.12}$$

$$Q_{orifice} = C \, a \, (2g)^{1/2} \, (H)^{1/2} = 0.60 \times 1.767 \times (64.32)^{1/2} \, (H)^{1/2} = 8.50 \, (H)^{1/2} \tag{15.13}$$

Outflow from each row is the same for a given head (Table 15.7) with values offset by one-half ft. Weir and orifice equations for an open riser top are shown in Cols. 8 and 9. The lesser of the two values is added to the sum of Cols. 2 through 7 to obtain total

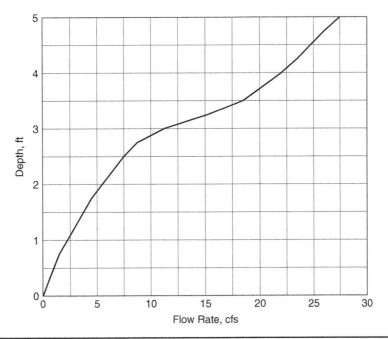

FIGURE **15.19**   Depth–outflow curve for a perforated riser pipe.

outflow. Columns 1 and 10 of Table 15.7 are the depth-outflow curve for the perforated pipe and are plotted in Fig. 15.19.

15.7.4   Example 15.4: Riser with Multiple Outlets

Assume an 8-ft square riser with an 8-ft height forms an inlet structure to a 54-in RCP (Fig. 15.20a). The RCP is 100-ft long, has a slope of 1.0 percent, and Manning's n of 0.012. Riser openings are shown in Fig. 15.20b. A 6-in circular orifice is located one ft above the RCP's invert. On another face, a 1.0-ft by 2.0-ft high rectangular orifice has its crest 2-ft above the RCP's invert. On a third face, a 2.0-ft square opening has its crest 4-ft above the RCP's invert. The riser's top has a cover on it with a 2-ft diameter opening.

Water depth inside this riser is the headwater (HW) for a 54-in RCP as shown in Fig. 15.21. Determine a depth-discharge curve for this riser to a depth of 12 ft using 1-ft increments. Calculations for the 54-in RCP are shown in Table 15.8 and its curve plotted in Fig. 15.21.

Coloumn 1 of Table 15.9 is flow depth. Cols. 2 through 6 are calculations for a 6-in diameter orifice which always acts as an orifice. A next set of calculations, Cols. 7 through 14, is for a 1-ft wide by 2-ft high orifice. When depth is 3 ft, there is 1 ft of water in this 2-ft high orifice. At this depth, the opening is a weir. It acts as an orifice at heads greater than 2 feet.

The next calculation, Cols. 15 through 22, is for a 2-ft by 2-ft orifice. When depth is 5-ft, there is 1-ft of water in this 2-ft high orifice acting as a weir. It is an orifice at heads

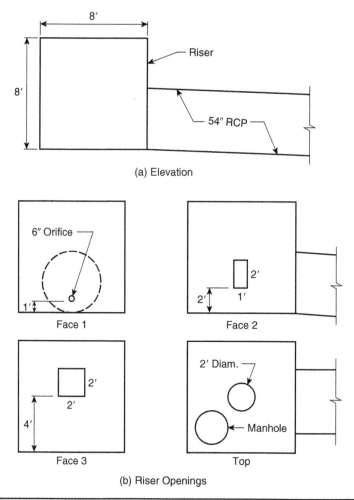

FIGURE 15.20 Square riser with weirs and orifices.

greater than 2-ft. The last calculation, Cols. 23 through 27, is for a horizontal 2.0-ft diameter horizontal orifice with the temporary pond depths greater than 8-ft. Flows through the four openings is total flow to the RCP and are listed in Col. 28. Column 29 is HW depth for the RCP from Fig. 15.21 for flows listed in Col. 28.

Total flow through all riser openings is flow into the RCP. If depth inside it is greater than a weir's crest, it is submerged and flow rate is reduced by the factor shown in Fig. 15.7. In orifice flow, depth inside it from Fig. 15.21 is compared to Col. 1 depth. If submerged, head is difference between the water surfaces. If not, Col. 1 depth is repeated and another head assumed. On each line, assume if an opening is submerged. If the RCP's HW is sum of head on each opening plus its crest or centroid or is equal to difference between two water surfaces if submerged, then the line is labeled "ok." If not, the line is labeled "ng." for no good. Each opening's heads are changed on the next line and more calculations made. Columns 1 and 28 are plotted in Fig. 15.22.

Designer:
Date:

Project: Riser with Multiple Openings

Sketch Station:

Hydrologic and Channel Information

Q1 = _____ cfs   TW1 = low ft
Q2 = _____ cfs   TW2 = _____ ft

Elevation = 100.0

AHW = _____

TW = low

(Q1 = Design Discharge, say Q25)
(Q2 = Check Discharge, say Q50 or Q100)

Elevation = 87.0   Elevation = 86.0

So = 0.01 ft/ft
L = 100 ft

HEADWATER COMPUTATION

Culvert Description (Entrance Type)	Q cfs	Size ft	Inlet Control HW/D	Inlet Control HW ft	Outlet Control HW = H + Ho − LSo: Ke	H ft	Dc ft	(Dc + D)/2 ft	TW ft	Ho ft	LSo ft	HW ft	Controlling HW ft	Outlet Velocity fps	Cost Dollars	Elevation
RCP w/ grooved	25	4.5	0.44	2.0	0.2	0.1	1.5	3.0	low	3.0	1.0	2.1	2.0			89
end in a headwall	90		0.89	4.0	0.2	0.7	2.9	3.7	low	3.7	1.0	3.4	4.0			91
	140		1.33	6.0	0.2	1.8	3.5	4.0	low	4.0	1.0	4.8	6.0			93
	195		1.78	8.0	0.2	3.4	4.1	4.3	low	4.3	1.0	6.7	8.0			95
	240		2.22	10.0	0.2	5.4	4.3	4.4	low	4.4	1.0	8.8	10.0			97
	270		2.67	12.0	0.2	6.9	4.5	4.5	low	4.5	1.0	10.3	12.0			99

SUMMARY & RECOMMENDATIONS

TABLE 15.8 Culvert Elevation-Discharge Calculations for Example 15.4

Depth	6-inch Diameter					1.0-foot by 2.0-foot								
ft 1	C 2	A sq. ft. 3	H ft. 4	H^1/2 5	Q cfs 6	C 7	L or A "ft., sq. ft." 8	H ft. 9	H^3/2 10	H^1/2 11	Q cfs 12	Factor 13	Q cfs 14	C 15
0.0	0.00	0.000	0.000	0.000	0.00	0.00	0.0	0.00	0.000	0.000	0.00	1.00	0.00	0.00
1.0	0.00	0.000	0.000	0.000	0.00	0.00	0.0	0.00	0.000	0.000	0.00	1.00	0.00	0.00
2.0	0.60	0.196	0.750	0.866	0.82	0.00	0.0	0.00	0.000	0.000	0.00	1.00	0.00	0.00
3.0	0.60	0.196	1.750	1.323	1.25	3.14	0.8	1.00	1.000	1.000	2.51	1.00	2.51	0.00
4.0	0.60	0.196	2.750	1.658	1.56	3.32	0.6	2.00	2.828	1.414	5.63	1.00	5.63	0.00
5.0	0.60	0.196	3.500	1.871	1.76	0.60	2.0	2.00	0.000	1.414	13.61	1.00	13.61	3.14
5.0	0.60	0.196	3.200	1.789	1.69	0.60	2.0	2.00	0.000	1.414	13.61	1.00	13.61	3.14
6.0	0.60	0.196	4.000	2.000	1.89	0.60	2.0	3.00	0.000	1.732	16.67	1.00	16.67	3.32
6.0	0.60	0.196	3.800	1.949	1.84	0.60	2.0	3.00	0.000	1.732	16.67	1.00	16.67	3.32
7.0	0.60	0.196	3.900	1.975	1.86	0.60	2.0	4.00	0.000	2.000	19.25	1.00	19.25	0.60
7.0	0.60	0.196	4.400	2.098	1.98	0.60	2.0	4.00	0.000	2.000	19.25	1.00	19.25	0.60
8.0	0.60	0.196	4.500	2.121	2.00	0.60	2.0	4.50	0.000	2.121	20.42	1.00	20.42	0.60
8.0	0.60	0.196	5.200	2.280	2.15	0.60	2.0	5.20	0.000	2.280	21.95	1.00	21.95	0.60
9.0	0.60	0.196	5.200	2.280	2.15	0.60	2.0	5.20	0.000	2.280	21.95	1.00	21.95	0.60
9.0	0.60	0.196	5.300	2.302	2.17	0.60	2.0	5.30	0.000	2.302	22.16	1.00	22.16	0.60
10.0	0.60	0.196	5.500	2.345	2.21	0.60	2.0	5.50	0.000	2.345	22.57	1.00	22.57	0.60
10.0	0.60	0.196	6.000	2.449	2.31	0.60	2.0	6.00	0.000	2.449	23.57	1.00	23.57	0.60

TABLE 15.9   Depth-Discharge Calculations for Various Openings in a Riser

In some cases, larger circular orifices are only partially filled at certain outflow depths. In these cases, Tables 15.10 and 15.11 are used to determine the areas and centroids of partially filled circles, respectively. Both tables were taken from Brater and King (1976).

15.8 Summary

This chapter used outlet conduit design plus weir and orifice equations to estimate outflow rates for various riser outlets. These structures can be any size and prismatic shape and limited only by designers' imaginations. One point to remember is in a riser,

2.0-foot by 2.0-foot						2-foot Diameter						Total Q	HW Depth	Comment
L or A "ft., sq. ft." 16	H ft. 17	H^3/2 18	H^1/2 19	Q cfs 20	Factor 21	Q cfs 22	C 23	A sq. ft. 24	H ft. 25	H^1/2 26	Q cfs 27	28	54-in. 29	30
0.0	0.00	0.000	0.000	0.00	0.00	0.00	0.00	0.00	0.00	0.000	0.00	0.0	0.0	ok
0.0	0.00	0.000	0.000	0.00	0.00	0.00	0.00	0.00	0.00	0.000	0.00	0.0	0.0	ok
0.0	0.00	0.000	0.000	0.00	0.00	0.00	0.00	0.00	0.00	0.000	0.00	0.8	0.1	ok
0.0	0.00	0.000	0.000	0.00	0.00	0.00	0.00	0.00	0.00	0.000	0.00	3.8	0.4	ok
0.0	0.00	0.000	0.000	0.00	0.00	0.00	0.00	0.00	0.00	0.000	0.00	7.2	0.8	ok
1.8	1.00	1.000	0.000	5.65	1.00	5.65	0.00	0.00	0.00	0.000	0.00	21.0	1.8	ng
1.8	1.00	1.000	0.000	5.65	1.00	5.65	0.00	0.00	0.00	0.000	0.00	20.9	1.8	ok
1.6	2.00	2.828	0.000	15.02	1.00	15.02	0.00	0.00	0.00	0.000	0.00	33.6	2.2	ng
1.6	2.00	2.828	0.000	15.02	1.00	15.02	0.00	0.00	0.00	0.000	0.00	33.5	2.2	ok
4.0	2.00	0.000	1.414	27.22	1.00	27.22	0.00	0.00	0.00	0.000	0.00	48.3	2.6	ng
4.0	2.00	0.000	1.414	27.22	1.00	27.22	0.00	0.00	0.00	0.000	0.00	48.4	2.6	ok
4.0	3.00	0.000	1.732	33.34	1.00	33.34	0.00	0.00	0.00	0.000	0.00	55.8	2.8	ng
4.0	3.00	0.000	1.732	33.34	1.00	33.34	0.00	0.00	0.00	0.000	0.00	57.4	2.8	ok
4.0	4.00	0.000	2.000	38.50	1.00	38.50	0.60	3.14	1.00	1.000	15.11	77.7	3.7	ng
4.0	4.00	0.000	2.000	38.50	1.00	38.50	0.60	3.14	1.00	1.000	15.11	77.9	3.7	ok
4.0	5.00	0.000	2.236	43.04	1.00	43.04	0.60	3.14	2.00	1.414	21.37	89.2	4.0	ng
4.0	5.00	0.000	2.236	43.04	1.00	43.04	0.60	3.14	2.00	1.414	21.37	90.3	4.0	ok

TABLE 15.9   Depth-Discharge Calculations for Various Openings in a Riser (Continued)

care must be taken to ensure whatever types, sizes, and elevations of weirs are used, water inside a riser must be below a weir's crest. If not, it is submerged and a reduction factor must be used. If an opening acts as an orifice, water inside the riser depth must be below its centroid. If not, definition of head changes to difference in the two water surfaces. At times a weir becomes an orifice and then is a weir again on a hydrograph's recession limb.

The weir and orifice equations nuances must be taken into account every foot of depth in a pond during a runoff event as depths increase and decrease and compared to depths inside a riser at that point in time. Remember flow in an outlet conduit and HW depth inside the riser is the total flow through those outlets whose crests are

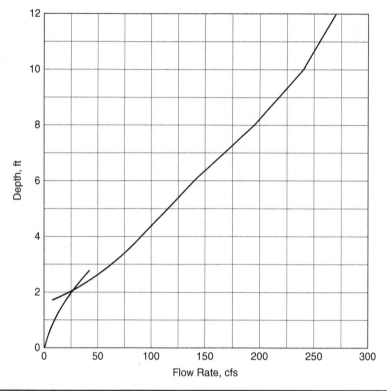

FIGURE 15.21   Depth–outflow curve for a 54–inch RCP.

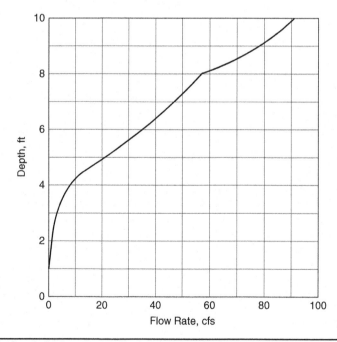

FIGURE 15.22   Depth-discharge curve for riser in Example 15.4.

D/d	.00	.01	.02	.03	.04	.05	.06	.07	.08	.09
0.0	.0000	.0013	.0037	.0069	.0105	.0147	.0192	.0242	.0294	.0350
0.1	.0409	.0470	.0534	.0600	.0668	.0739	.0811	.0885	.0961	.1039
0.2	.1118	.1199	.1281	.1365	.1449	.1535	.1623	.1711	.1800	.1890
0.3	.1982	.2074	.2167	.2260	.2355	.2450	.2546	.2642	.2739	.2836
0.4	.2934	.3032	.3130	.3229	.3328	.3428	.3527	.3627	.3727	.3827
0.5	.393	.403	.413	.423	.433	.443	.453	.462	.472	.482
0.6	.492	.502	.512	.521	.531	.540	.550	.559	.569	.578
0.7	.587	.596	.605	.614	.623	.632	.640	.649	.657	.666
0.8	.674	.681	.689	.697	.704	.712	.719	.725	.732	.738
0.9	.745	.750	.756	.761	.766	.771	.775	.779	.782	.784

TABLE 15.10   Values for Area of a Circle Flowing Partially Full

D/d	.00	.01	.02	.03	.04	.05	.06	.07	.08	.09
0.0	.000	.004	.008	.012	.016	.0320	.024	.028	.032	.036
0.1	.040	.044	.049	.053	.057	.061	.065	.069	.073	.077
0.2	.082	.086	.090	.094	.098	.103	.107	.111	.115	.119
0.3	.124	.128	.132	.137	.141	.145	.150	.154	.158	.163
0.4	.167	.172	.176	.181	.185	.189	.194	.199	.203	.208
0.5	.212	.217	.221	.226	.231	.235	.240	.245	.250	.254
0.6	.259	.254	.269	.274	.279	.284	.289	.294	.299	.304
0.7	.309	.314	.320	.325	.330	.336	.341	.347	.352	.358
0.8	.363	.369	.375	.381	.387	.393	.399	.405	.411	.418
0.9	.424	.431	.438	.445	.452	.459	.466	.474	.482	.491

TABLE 15.11   Vertical Distance y to Centroid of a Circle Flowing Partly Full

below the pond's water surface at each point in time as the flood flows into and out of a basin over the time water is above the lowest outlet's crest elevation. If an elevation-outflow relationship does not correctly depict the factors mentioned above, reservoir routings will be incorrect. The result will be maximum temporary pond elevations that are too low or too high and maximum outflow rates will be too low or too high for the flood magnitudes investigated. A good elevation-outflow curve is the most difficult part of the hydraulic design of any type of BMP to develop successfully– so be extremely careful.

CHAPTER 16

Hydrograph Routing

16.1 Introduction

Hydrographs were described in Chap. 12, storage in Chap. 11, and outflow in Chaps. 13 through 15. This chapter puts them all together, develops a routing equation, routing curve, routing procedure, and uses them in the Catfish Creek example.

This chapter is included for two reasons. First, many computer programs only develop and route hydrographs. Depth-storage and depth-outflow calculations are not included; so they must be calculated separately in other computer programs, then results somehow changed into a routing curve and its results entered into the hydrograph program. If all is done with programs, then the designer is left to the mercies of the programmers' assumptions and inclusions.

The second reason is contained in the above. Designers must remain in control of their designs. Using the spreadsheets contained on the McGraw-Hill website (www.mhprofessional.com/sdsd) and the explanations contained in this book, designers remain in control. They select each number to input, and then review the results to ensure that they achieve what the designer had in mind. The spreadsheets merely make the calculations without the assumptions embedded in programs.

16.2 Routing Equation

A routing equation's definition sketch is shown in Fig. 16.1. Routing begins with the continuity equation, Eq. (16.1):

$$\hat{I} - \hat{O} = \Delta S / \Delta T \tag{16.1}$$

where $\hat{I}$ is average inflow in cfs during ΔT, $\hat{O}$ is average outflow in cfs during ΔT, ΔS is storage change in ac ft or cu ft during ΔT in hours. ΔT is short enough so straight lines drawn from Is or Os at a period's beginning and end fall directly on curvilinear I and O hydrograph plots. Equation (16.1) variables are as follows: A hydrograph is divided into ΔT time-steps. For small areas, ΔT is 0.1 or 0.2 hour. For larger areas, ΔT is 0.5 or 1.0 hour. On the Mississippi, ΔT is a week. Use at least 5 time-steps before the peak flow occurs. I_1 and I_2 are inflows in cfs, O_1 and O_2 are outflows in cfs, S_1 and S_2 are storage volumes at beginning and end of time periods, ac ft or cu ft.

When inflow is greater than outflow in Eq. (16.1), left side is positive, so right side is positive. Storage is positive, and a best management practice's (BMP's) water depth increases. When inflow is less than outflow, left side is negative, so right side is negative. This means storage is negative, and a BMP is emptying.

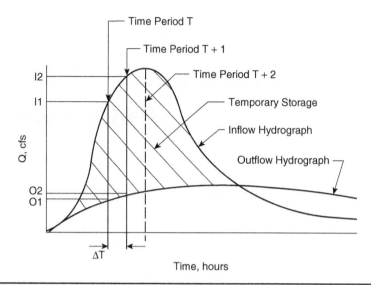

FIGURE 16.1   Definition sketch for the routing equation.

A hydrograph is routed each time you brush your teeth. BMP is the sink for storage. Inflow hydrograph is turning the faucet on. The hole at a sink's bottom is the outflow. Water fills the sink until O equals I. If turned on too hard, water overtops the sink because of insufficient storage. Water goes onto the floor. Plug the sink, and there is a mess because the outlet is clogged.

For BMPs, S and O rate are balanced with an inflow hydrograph so no overflow results in a mess. Maintenance is a must so outlets do not plug. Turning a faucet on too hard is akin to a greater-than-design storm occurring and must be checked during design.

Equation 16.1 is not usable as is because its flow and storage units are different and must be changed. With this being the case, $\hat{I}$ and $\hat{O}$ can be rewritten as average inflow and outflow at a time period's beginning and end. Likewise, ΔS can be written as difference between storage amounts at a time period's end and beginning. Substituting these into Eq. (16.1) yields Eq. (16.2).

$$(I_1 + I_2)/2 - (O_1 + O_2)/2 = (S_2 - S_1)/\Delta T \qquad (16.2)$$

Rearranging Eq. (16.2) to get all known values on the equation's left side and all unknowns on its right side and multiplying through by two yields Eq. (16.3).

$$I_1 + I_2 + 2S_1/\Delta T - O_1 = 2S_2/\Delta T + O_2 \qquad (16.3)$$

Equation 16.3 is correct. We know I_1 and I_2 because we know all ordinates. We select ΔT. Referring to Fig. 16.1, O and S values at end of period ΔT are O and S values at beginning of period $\Delta T + 1$. They are the same points. Thus we know all numbers on Eq. (16.3)'s left side for all periods. Equation (16.3) is the routing equation. However, it has two unknowns on the right-hand side, S_2 and O_2, so there are an infinite number of solutions. By developing a second equation that relates S_2 and O_2, these two equations are solved simultaneously to arrive at a single solution. Combining depth-storage and depth-outflow relationships into a routing curve yields this second equation.

16.3   Routing Curve

Combining them at similar elevations develops a routing curve by filling out Table 16.1. A routing period of 0.1 hour was used. Columns. 3 and 4 of Table 16.1 are plotted as a routing curve in Fig. 16.2 on arithmetic paper. It is easier to plot and read outflows if

Routing Curve Calculations			
Volume in Acre Feet			
Routing Period, $\Delta T = 0.1$ hr			
$2S/\Delta T + O = (2S$ (ac ft)$/\Delta T$ hr$) + O$ (cfs) $\times$ 1 cfs-day$/2$ ac ft $\times$ 24 hr$/1$ day $= 240 * S + O$			
Elev. feet 1	S acre feet 2	O cfs 3	$2S/\Delta T + O$ cfs 4
			0
			0
			0
			0
			0
			0
			0
			0
			0
			0

TABLE 16.1   Typical Routing Curve Calculation Spreadsheet

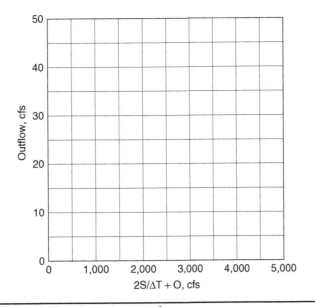

FIGURE 16.2   Typical arithmetic routing curve form.

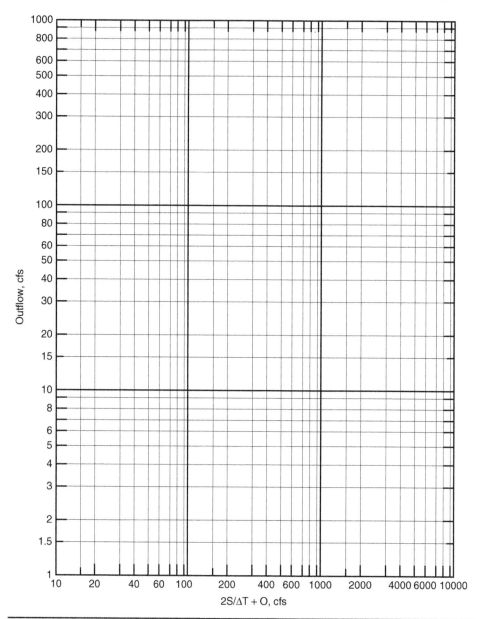

FIGURE 16.3 Typical logarithmic routing curve form.

Cols. 3 and 4 of Table 16.1 are on log-log paper. It is also better to use log-log paper if your range of outflow and $2S/\Delta T + O$ is large, two or three log scales. Figure 16.3 is the second equation. Flow units are cfs with volume in ac ft or cu ft. Adding cfs to acre-feet/hour or cubic feet/hour to get $2S/\Delta T + O$ in cfs does not work. Units of $2S/\Delta T$ must be changed to cfs and is done as shown near the top of Table 16.1.

16.4   Routing Procedure

A routing procedure uses an inflow hydrograph, routing equation, and routing curve. Computer programs use a routing curve as a lookup table. Equation (16.3)'s left-hand side is solved for a number, is inserted into a lookup table, and an outflow value is interpreted. Then a table similar to Table 16.2 is output. Table 16.2 is used if routing is done by hand or with a spreadsheet. Note that Col. 4 plus Col. 5 equals Col. 6, the routing equation. Equations are buried in cols. 4, 5, and 6.

Hydrograph Routing					Type II Rainfall		$\Delta T = 0.1$ hour	
Line No. 1	Time hr 2	Inflow, I_2 cfs 3	$I_1 + I_2$ cfs 4	+	$2S_1/\Delta T - O_1$ cfs 5	= $2S_2/\Delta T + O_2$ cfs 6	Outflow cfs 7	Elevation ft 8
1	11.0							
2	11.1		0.0		0.0	0.0		
3	11.2		0.0		0.0	0.0		
4	11.3		0.0		0.0	0.0		
5	11.4		0.0		0.0	0.0		
6	11.5		0.0		0.0	0.0		
7	11.6		0.0		0.0	0.0		
8	11.7		0.0		0.0	0.0		
9	11.8		0.0		0.0	0.0		
10	11.9		0.0		0.0	0.0		
11	12.0		0.0		0.0	0.0		
12	12.1		0.0		0.0	0.0		
13	12.2		0.0		0.0	0.0		
14	12.3		0.0		0.0	0.0		
15	12.4		0.0		0.0	0.0		
16	12.5		0.0		0.0	0.0		
17	12.6		0.0		0.0	0.0		
18	12.7		0.0		0.0	0.0		
19	12.8		0.0		0.0	0.0		
20	12.9		0.0		0.0	0.0		
21	13.0		0.0		0.0	0.0		
22	13.1		0.0		0.0	0.0		
23	13.2		0.0		0.0	0.0		
24	13.3		0.0		0.0	0.0		
25	13.4		0.0		0.0	0.0		
26	13.5		0.0		0.0	0.0		

TABLE 16.2   Typical Hydrograph Routing Form

Hydrograph Routing				Type II Rainfall			$\Delta T = 0.1$ hour	
Line No. 1	Time hr 2	Inflow, I_2 cfs 3	$I_1 + I_2$ cfs 4	+	$2S_1/\Delta T - O_1$ cfs 5	=	$2S_2/\Delta T + O_2$ cfs 6	Outflow cfs 7	Elevation ft 8
26	13.5		0.0		0.0		0.0		
27	13.6		0.0		0.0		0.0		
28	13.7		0.0		0.0		0.0		
29	13.8		0.0		0.0		0.0		
30	13.9		0.0		0.0		0.0		
31	14.0		0.0		0.0		0.0		
32	14.1		0.0		0.0		0.0		
33	14.2		0.0		0.0		0.0		
34	14.3		0.0		0.0		0.0		
35	14.4		0.0		0.0		0.0		
36	14.5		0.0		0.0		0.0		
37	14.6		0.0		0.0		0.0		
38	14.7		0.0		0.0		0.0		
39	14.8		0.0		0.0		0.0		
40	14.9		0.0		0.0		0.0		
41	15.0		0.0		0.0		0.0		
42	15.1		0.0		0.0		0.0		
43	15.2		0.0		0.0		0.0		
44	15.3		0.0		0.0		0.0		
45	15.4		0.0		0.0		0.0		
46	15.5		0.0		0.0		0.0		
47	15.6		0.0		0.0		0.0		
48	15.7		0.0		0.0		0.0		
49	15.8		0.0		0.0		0.0		
50	15.9		0.0		0.0		0.0		
51	16.0		0.0		0.0		0.0		

TABLE 16.2 Typical Hydrograph Routing Form (*Continued*)

Columns in Table 16.2 are defined as follows. Column 1 is a line and routing period number. Column 2 is time from beginning of routing. Values in Col. 2 exceed previous values by ΔT. Column 3 is a listing of inflow hydrograph ordinates at a period's end. Col. 4 is sum of hydrograph values in Col. 3 at a time period's beginning and end listed in Col. 1. One assumption is outflow equals inflow in the first routing period. From Fig. 16.1, at the beginning of runoff, outflow is always less than or equal to inflow. If a program is used, it starts at time zero and first inflows and outflows are zero.

For Col. 4 of Line 1 in Table 16.2: assume a value slightly less than the number in Col. 3 of Line 1, then add this number to the number in Col. 3 of Line 1 and place this sum in col. 4 of Line 1. An assumption I make with spreadsheets is to use an outflow value from Table 16.1 that is less than or equal to inflow on line 1 of Table 16.2. Place it in Col. 7 of Line 1, Table 16.2. A value in Col. 6, Line 1 is taken from the routing curve, Fig. 16.3, based on outflow in Col. 7, Line 1. Value in Col. 5, Line 1 is obtained knowing Col. 4 plus Col. 5 equals Col. 6, completing Line 1. Col. 5 can be negative for several time periods, but becomes positive as runoff is temporarily stored in a BMP.

The last step to become familiar with is how to proceed from one line to the next. From Fig. 16.1, end of period outflow, O_2, and storage, S_2, in time period, N, become beginning of period outflow, O_1, and storage, S_1, for the next routing period, $N+1$, or

$$O_{2,N} = O_{1,N+1} \tag{16.4}$$

$$S_{2,N} = S_{1,N+1} \tag{16.5}$$

thus,

$$(2S_1/\Delta T - O_1)_{N+1} = (2S_2/\Delta T - O_2)_N \tag{16.6}$$

or

$$(2S_1/\Delta T - O_1)_{N+1} = (2S_2/\Delta T + O_2)_N - (2O_2)_N \tag{16.7}$$

N is the line number. On Line 3, $(2S_1/\Delta T - O_1)$ equals $(2S_2/\Delta T + O_2) - 2O_2$ on Line 2, i.e., Col. 6 on Line 2 minus two times Col. 7 on Line 2 is equal to Col. 5 on Line 3.

Use this procedure to complete a routing after Line 1 is completed. Use Eq. (16.7) to obtain Col. 5 in Line 2. Add Cols. 4 and 5 to obtain Col. 6 in Line 2. Use a routing curve with Col. 6 to obtain Col. 7 in Line 2. Use Eq. (16.7) to obtain Col. 5 in Line 3. Do this until the routing is completed. You need to input values in Col. 3, then finish Line 1 as described previously. Input outflows in Col. 7 based on Col. 6 values and the routing curve. If more than one page is needed to complete the routing, change the values in Col. 1 through Col. 3. The first line on page 2 is page 1's last line. Several pages of routing may be needed to estimate time needed to allow all water from a water-quality storm to exit the basin. It should remain in storage for an average of 2 to 3 days. Spreadsheets on McGraw-Hill's website (www.mhprofessional.com/sdsd) perform all calculations.

16.5 Example 16.1

16.5.1 Description

In previous chapters, I used an existing culvert on a 126-acre Catfish Creek tributary to illustrate the three sets of data needed to route an inflow hydrograph through a detention basin.

16.5.2 Inflow Hydrographs

Basic watershed variables were listed in Tables 12.18 and 12.19. These values were determined by the Soil Conservation Service (SCS) over 40 years ago as a demonstration project. At that point in time only homes on a single cul-de-sac existed in the Catfish Creek tributary watershed. These tables showed calculations for CN for both undeveloped and developed conditions.

Basic hydrograph variables for the 2- and 100-year hydrographs were listed in Table 12.20. Inflow hydrographs were developed using the methodology described in TR-55 (USDA, 1986) in Chap. 12. The 2- and 100-year, 24-hour rainfall amounts in Northeastern Iowa are 3.1 and 6.3 inches using maps in App. B of TR-55 (USDA, 1986). A composite-developed CN was 81.

Developed condition hydrographs had peak flow rates of 190 and 539 cfs, respectively, for the 2- and 100-year, 24-hour storm events (see Tables 12.21 and 12.22). The 100-year runoff was 4.1 inches. Values of Ia/P of 0.10 were used for all subareas for the 2- and 100-year events even though Ia/P for subareas 1 and 2 for the 2-year event were over 0.20. This was done so the inflow hydrograph ordinates were larger and would be a more severe test of the site.

16.5.3 Elevation Storage

Available storage was determined from data listed in Table 11.3 and shown in Fig. 11.2. No changes have been made to the ravine over the years. They indicate that 61.5 AF of storage is available at elevation 848. Total runoff volume from a 100-year, 24-hour storm event is 126 ac × 4.1 in. × 1 ft/12 in. = 43.0 AF. From Fig. 11.2, this corresponds to elevation 845. Thus, all 100-year runoff volume could be stored with depth to spare below the homes at elevation 848. There is no need for a culvert, and a pond forms. However, a pipe is used to empty the pond to prevent children from playing in this attractive nuisance, to prevent other storm runoff from further raising the water surface, and to drain the pond.

16.5.4 Elevation Outflow

A 72-in reinforced concrete pipe (RCP) was installed when the four-lane divided Northwest Arterial was constructed. Elevation-outflow calculations for this culvert are contained in Table 16.3 using nomographs presented in HEC-5 (FHWA, reprinted 1980) and plotted in Fig. 16.4. Always develop storage volumes and outflow rates above those needed to route a 100-year, 24-hour inflow hydrograph through BMPs. This gives you additional values to route an even greater flow through the site.

16.5.5 Routing Curve

A routing curve is calculated in Table 16.4 using a ΔT of 0.2 hours, using values of the elevation-storage and elevation-outflow relationships. It is plotted in Fig. 16.5.

16.5.6 Routing the 2-Year Hydrograph through Catfish Creek Tributary

The 2-year, 24-hour flood event was routed through the existing 72-in RCP as shown in Table 16.5. The peak inflow rate of 190 cfs was reduced to just 160 cfs, a reduction of 16 percent because of the large outlet pipe. Peak flow was not delayed from hour 12.2. Peak elevation reached by the temporary pond was 829.7, a total of 5.7 ft of depth. Inflow equaled outflow for over 20 hours. At hour 19.0, flow rate was reduced to 4.0 cfs and only 0.3 ft of water remained in the temporary pond. Thus, runoff water quality was improved very little, if at all.

16.5.7 Routing the 100-Year Hydrograph through Catfish Creek Tributary

The 100-year, 24-hour flood event was routed through the existing 72-in RCP as shown in Table 16.6. The peak inflow of 539 cfs was reduced to a peak outflow rate of 404 cfs, a reduction of 25 percent. Peak flow was delayed from hour 12.4 to hour 12.6, a total of 0.2

Pipe Culvert Elevation–Discharge Calculations

Project: Catfish Creek Tributary
Road/Berm Elev: 856   AHW: 28
Size = 72 inches

Location: Dubuque, Iowa
Inlet Elev: 824   Outl. Elev: 821
Ke = 0.5

Designer:
Length: 300   Slope: 0.01000
n = 0.015   If n ≠ 0.012, in outlet control L' = 469

Date:

1	2	3	4	5	6	7	8	9	10	11	12	13	14	15	16
		Inlet Control					Outlet Control (HW = H + Ho – L*So)								
Culvert Description	Q cfs	HW/D	HW ft	Ke	H ft	Dc ft	(Dc + D/2) ft	TW ft	Ho ft	L*So ft	HW ft	Controlling Headwater ft	Outlet Velocity fps	Cost $	Elevation ft
RCP w/ square edge and flared headwall	38	0.33	2.00	0.5	0.1	1.6	3.80	0.4	3.80	3.00	0.9	2.0	6.2	30,000	826.0
	105	0.67	4.00	0.5	0.5	2.8	4.40	1.0	4.40	3.00	1.9	4.0	8.0		828.0
	240	1.33	8.00	0.5	2.9	4.2	5.10	1.5	5.10	3.00	5.0	8.0	11.4		832.0
	400	2.00	12.00	0.5	8.7	5.4	5.70	2.0	5.70	3.00	11.4	12.0	14.9		836.0
	490	2.67	16.00	0.5	13.0	5.8	5.90	2.5	5.90	3.00	15.9	16.0	17.5		840.0
	560	3.33	20.00	0.5	17.0	6.0	6.00	3.0	6.00	3.00	20.0	20.0	19.8		844.0
	610	4.00	24.00	0.5	20.0	6.0	6.00	3.5	6.00	3.00	23.0	24.0	21.6		848.0
		0.00	0.00	0.0			0.00		0.00	0.00	0.0	0.0			0.0
		0.00	0.00	0.0			0.00		0.00	0.00	0.0	0.0			0.0
		0.00	0.00	0.0			0.00		0.00	0.00	0.0	0.0			0.0

Summary and Recommendations

TABLE 16.3   Seventy-Two-Inch RCP Culvert Design for Example 16.1

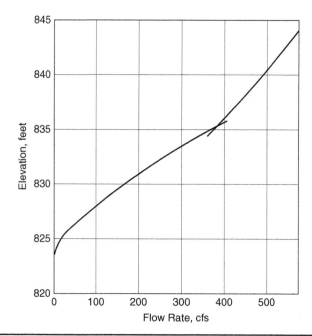

FIGURE 16.4   Elevation-outflow curve for Example 16.1.

hour. Peak elevation reached by the temporary pond was 836.2, a total of 12.2 ft. Inflow equaled outflow for about 19 hours. At hour 19.0, the flow rate was reduced to 13 cfs and only 0.9 ft of water remained in the temporary pond. Thus, runoff water quality was improved very little, if any.

Routing Period = 0.2 hr			
2S/ΔT + O = (2S (ac ft)/0.1 hr) + O (cfs) × 1 cfs-day/2 ac ft × 24 hr/1 day = 120 *S + O			
Elev. ft 1	S ac ft 2	O cfs 3	2S/ΔT + O cfs 4
824	0.0	0.0	0
825	0.04	15.0	20
826	0.1	38.0	50
828	0.3	105.0	141
832	2.4	240.0	528
836	7.7	400.0	1324
840	18.4	490.0	2698
844	35.9	560.0	4868
848	61.5	610.0	7990

TABLE 16.4   Routing Curve Calculations for Example 16.1

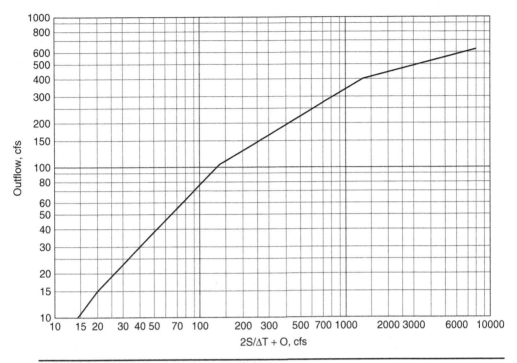

FIGURE 16.5   Routing Curve for Example 16.1.

Line No. 1	Time hr 2	Inflow, I_2 cfs 3	$I_1 + I_2$ cfs 4	$2S_1/\Delta T - O_1$ cfs 5	$2S_2/\Delta T - O_2$ cfs 6	Outflow cfs 7	Elevation ft 8
1	11.0	5.6	10.2	−4.9	5.3	4.0	824.3
2	11.2	6.6	12.2	−2.7	9.5	6.0	
3	11.4	8.0	14.6	−2.5	12.1	7.0	
4	11.6	11.5	19.5	−1.9	17.6	12.0	
5	11.8	26.9	38.4	−6.4	32.0	25.0	825.7
6	12.0	66.1	93.0	−18.0	75.0	58.0	
7	12.2	173.3	239.4	−41.0	198.4	130.0	
8	12.4	189.5	362.8	−61.6	301.2	160.0	829.7 peak
9	12.6	128.0	317.5	−18.8	298.7	159.0	
10	12.8	56.2	184.2	−19.3	164.9	110.0	
11	13.0	31.6	87.8	−55.1	32.7	26.0	
12	13.2	22.2	53.8	−19.3	34.5	26.0	825.5
13	13.4	18.6	40.8	−17.5	23.3	18.6	
14	13.6	15.1	33.7	−13.9	19.8	14.9	

TABLE 16.5   2-Year Hydrograph Routing for Example 16.1

Line No. 1	Time hr 2	Inflow, I_2 cfs 3	$I_1 + I_2$ cfs 4	$2S_1/\Delta T - O_1$ cfs 5	$2S_2/\Delta T - O_2$ cfs 6	Outflow cfs 7	Elevation ft 8
15	13.8	13.4	28.5	−10.0	18.5	14.0	
16	14.0	12.2	25.6	−9.5	16.1	12.0	824.8
17	14.2	11.2	23.4	−7.9	15.5	11.0	
18	14.4	10.4	21.6	−6.5	15.1	10.4	
19	14.6	9.7	20.1	−5.7	14.4	9.7	
20	14.8	9.2	18.9	−5.0	13.9	9.2	
21	15.0	8.8	18.0	−4.5	13.5	8.8	
22	15.2	8.5	17.3	−4.1	13.2	8.5	
23	15.4	8.2	16.7	−3.8	12.9	8.2	
24	15.6	7.9	16.1	−3.5	12.6	7.9	
25	15.8	7.5	15.4	−3.2	12.2	7.5	
26	16.0	7.1	14.6	−2.8	11.8	7.1	
27	16.2	6.8	13.9	−2.4	11.5	6.8	
28	16.4	6.6	13.4	−2.1	11.3	6.6	
29	16.6	6.4	13.0	−1.9	11.1	6.4	
30	16.8	6.2	12.6	−1.7	10.9	6.2	
31	17.0	6.0	12.2	−1.5	10.7	6.0	
32	17.2	5.8	11.8	−1.3	10.5	5.8	
33	17.4	5.6	11.4	−1.1	10.3	5.6	
34	17.6	5.4	11.0	−0.9	10.1	5.4	
35	17.8	5.2	10.6	−0.7	9.9	5.2	
36	18.0	5.0	10.2	−0.5	9.7	5.0	
37	18.2	4.8	9.8	−0.3	9.5	4.8	
38	18.4	4.6	9.4	−0.1	9.3	4.6	
39	18.6	4.4	9.0	0.1	9.1	4.4	
40	18.8	4.2	8.6	0.3	8.9	4.2	
41	19.0	4.0	8.2	0.5	8.7	4.0	824.3

TABLE 16.5 Two-Year Hydrograph Routing for Example 16.1 (*Continued*)

16.6 Cost and Summary

The cost of a 300-ft long, 72-in RCP some 40 years ago was about $30,000. Only two things were accomplished when a 72-in RCP was constructed at this site: $30,000 was spent (plus earthwork), and its peak flow was reduced only 16 percent and 25 percent, free to continue to flood downstream people and property with no reduction in pollutants conveyed in the runoff. But that occurred over 40 years ago before the city became involved with the concept of sustainability and the triple bottom line: people, planet, and profit. See App. A.

Line No. 1	Time hr 2	Inflow, I_2 cfs 3	$I_1 + I_2$ cfs 4	$2S_1/\Delta T - O_1$ cfs 5	$2S_2/\Delta T + O_2$ cfs 6	Outflow cfs 7	Elevation ft 8
1	11.0	16	31	−11	20	15	825.0
2	11.2	20	36	−10	26	20	
3	11.4	25	45	−14	31	24	
4	11.6	34	59	−17	42	32	826.0
5	11.8	70	104	−22	82	64	
6	12.0	201	271	−46	225	120	
7	12.2	501	702	−15	687	240	832.0
8	12.4	539	1040	207	1247	380	
9	12.6	355	894	487	1381	404	836.2 peak
10	12.8	158	513	573	1086	340	
11	13.0	90	248	406	654	255	832.0
12	13.2	64	154	144	298	160	
13	13.4	53	117	−22	95	72	
14	13.6	44	97	−49	48	37	
15	13.8	39	83	−26	57	42	826.0
16	14.0	35	74	−27	47	37	
17	14.2	33	68	−27	41	31	
18	14.4	30	63	−21	42	32	
19	14.6	28	58	−22	36	28	
20	14.8	27	55	−20	35	26	
21	15.0	26	53	−17	36	26	
22	15.2	25	51	−16	35	26	
23	15.4	24	49	−17	32	25	
24	15.6	23	47	−18	29	23	
25	15.8	22	45	−17	28	22	
26	16.0	21	43	−16	27	21	
27	16.2	20	41	−15	26	20	
28	16.4	19	39	−14	25	19	
29	16.6	18	37	−13	24	18	
30	16.8	18	36	−12	24	18	
31	17.0	17	35	−12	23	17	
32	17.2	17	34	−11	23	17	
33	17.4	16	33	−11	22	16	
34	17.6	16	32	−10	22	16	

TABLE 16.6   100-Year Hydrograph Routing for Example 16.1

Line No. 1	Time hr 2	Inflow, I_2 cfs 3	$I_1 + I_2$ cfs 4	$2S_1/\Delta T - O_1$ cfs 5	$2S_2/\Delta T + O_2$ cfs 6	Outflow cfs 7	Elevation ft 8
35	17.8	15	31	−10	20	15	825.0
36	18.0	15	30	−11	20	15	
37	18.2	15	30	−10	19	15	
38	18.4	14	29	−10	19	14	
39	18.6	14	28	−9	19	14	
40	18.8	13	27	−9	18	13	
41	19.0	13	26	−8	18	13	824.9

TABLE 16.6   100-Year Hydrograph Routing for Example 16.1 (*Continued*)

Appendices

The appendices contain several worked examples plus other information. The first five contain text, figures, and tables containing all calculations for five common BMPs. These are:

   A. Retrofitted Culvert

   B. Greenroof

   C. Rain Garden

   D. Bioswale

   E. Parking Lot

The next three appendices contain only figures and text containing the final results for sites that utilize several on-site BMPs to control both runoff quantity and quality. No calculations are included since they would just be repetitious of those in the first five appendices. These are:

   F. Single Family

   G. Office Park

   H. Industrial Site

The final appendices include:

   I. List of Source Control BMPs

   J. List of Manning's Roughness Coefficients

   K. List of References

Appendix A shows how an existing culvert can be retrofitted at low cost to serve both runoff quantity and runoff quality requirements. Neither was included when it was constructed decades ago. This book's purposes were listed as part of Chap. 1. One purpose is to include the triple bottom line (TBL) (people, planet, profits) in projects. Retroftting culverts and older detention basins includes all three.

Retrofit for Catfish Creek

A.1 Problem Statement

Retrofitting existing culverts or detention basins is the cheapest way to meet the Federal Emergency Management Agency (FEMA) and the Environmental Protection Agency (EPA) rules for reducing flooding and enhancing runoff water quality. For relatively few dollars invested, subwatersheds can be retrofitted for most storm events. If certain topographic conditions are present, storms from a 6-month through a 100-year, 24-hour event can be controlled.

An existing Catfish Creek culvert in Northeast Iowa is retrofited by removing its inlet end section and replacing it with a 8-ft by 6-ft reinforced concrete box (RCB) riser as shown in Fig. A.1. A short reverse slope pipe is placed with its invert at a pond's surface and its lower end set three feet below it to keep floatables in the pond. The riser's top is covered with grates, one portion being a hinged access grate. If a portion of it is plugged, the remainder allows water to drop through. A walkway is built between the fill's foreslope and riser with chain link fences and a locked gate.

In Fig. A.1, vertical 6-in by 6-in protrusions exist to the riser's full height. This is done to counteract the coriolis force caused by the earth's rotation. As water enters the top, the force causes it to rotate, resulting in water forced to the riser's perimeter with no water in its center. The orifice equation assumes an orifice's area is full of water and these protusions make the assumption true.

Assume only a 3-ft by 3-ft top opening remains unplugged. This forces runoff to pond deeper during all storm events other than the 2- to 5- or 10-year events. Set the riser's top elevation such that all 2-year event runoff is stored in a temporary pond. Its only outlet is an orifice sized to store all 2-year runoff for an average of at least two days. The 10- and 100-year storms outlet through the riser's top. Sufficient freeboard below homes at elevation 848 remains.

A.2 Permanent Pond Elevation

To offset evaporation, a pond's area should be 1 to 3 percent of its drainage area in the Midwest. The tributary watershed is 126 ac. Thus a pond's area ranges from 1.26 to 3.78 ac. Figure 11.2 in Chap. 11 showed the area is 1.26 ac at elevation 834.0 and 3.78 ac at 839.5. Pond area is 2.3 ac at 837.0. and 2.7 ac at 838. Set a small pipe's invert at elevation 837.0 and its lower pond end at 834.

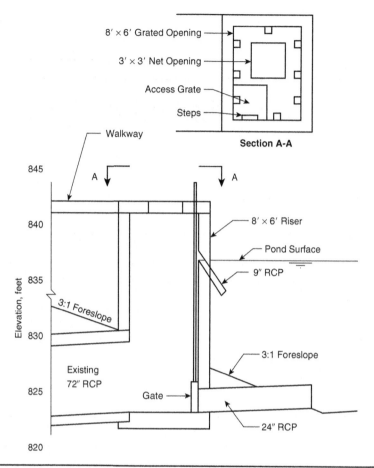

FIGURE A.1 Retrofitted outlet structure for Catfish Creek.

A.3 Top of Riser Elevation

I set the riser's top so all runoff from a 2-year storm would be stored below it. Runoff from this event is 1.4 in or 14.7 AF ($126 \times 1.4/12$) based on TR-55 (USDA, 1986). From Fig. 11.2, pond volume at elevation 837.0 is 9.9 AF. This plus 14.7 is 24.6 AF at elevation 841.5, so use 842.

A.4 Inflow Hydrographs

Inflow hydrographs for the 2- and 100-year, 24-hour storms are the same as those listed in Chap. 12. We have not changed anything about the rainfall or watershed. I have just added a reinforced concrete riser and its openings at the upstream end of the existing 72-in reinforced concrete pipe (RCP).

A.5 Revised Storage Volumes

With a pond, volumes to store runoff during a storm are reduced. These net storage volumes are developed in Table A.1 and plotted in Fig. A.2. Total volume at 848.0 is now 51.7 AF.

Elevation/Depth-Storage Calculations							
Frustum of a Cone Method							
Permanent Pond Elevation = 837.0 Total Pond Vol. = 9.9							
Elev/Depth ft 1	Area ac 2	(A1*A2)^.5 ac 3	Δ Depth ft 4	Δ Volume ac ft 5	Tot. Vol. ac ft 6	Pond Vol. ac ft 7	Net Vol. ac ft 8
824.0	0.00				0.0	0.0	0.0
		0.0	2.0	0.05			
826.0	0.08				0.1	0.0	0.0
		0.1	2.0	0.31			
828.0	0.24				0.4	0.0	0.0
		0.3	2.0	0.72			
830.0	0.50				1.1	0.0	0.0
		0.6	2.0	1.33			
832.0	0.84				2.4	0.0	0.0
		1.1	2.0	2.18			
834.0	1.36				4.6	0.0	0.0
		1.6	2.0	3.24			
836.0	1.90				7.8	0.0	0.0
		2.1	1.0	2.10			
837.0	2.30				9.9	9.9	0.0
		2.5	1.0	2.50			
838.0	2.70				12.4	9.9	2.5
		2.9	1.0	2.90			
839.0	3.10				15.3	9.9	5.4
		3.3	1.0	3.32			
840.0	3.54				18.6	9.9	8.7
		3.7	1.0	3.74			
841.0	3.94				22.4	9.9	12.5
		3.9	1.0	3.96			
842.0	4.40				26.3	9.9	16.4
		4.6	1.0	4.60			
843.0	4.80				30.9	9.9	21.0
		5.0	1.0	5.04			
844.0	5.28				36.0	9.9	26.1
		5.5	1.0	5.55			
845.0	5.82				41.5	9.9	31.6
		6.1	1.0	6.13			
846.0	6.44				47.7	9.9	37.8
		6.7	1.0	6.70			
847.0	6.96				54.4	9.9	44.5
		7.3	1.0	7.26			
848.0	7.56				61.6	9.9	51.7

TABLE A.1 Retrofitted Storage Volumes for Catfish Creek Tributary

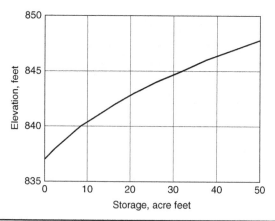

FIGURE A.2   Net storage volumes for Catfish Creek.

A.6   Attenuation of Inflow Hydrographs

Table A.2 is filled out as follows. Peak inflow rates are taken from Tables 12.21 and 12.22. Select peak outflow values but no less than 10 percent of peak inflows. Qin/Qout values are calculated for you. Select values of Vs/Vr from Fig. 12.5. Enter runoff values of 1.4 and 4.1 in for the 2- and 100-year storms, respectively. Runoff and storage volumes are calculated for you. Use Fig. A.2 to enter values of the temporary pond's maximum water surface elevations.

These calculations indicate that the site has sufficient storage volume to reduce peak inflow rates by at least 90 percent. Maximum water surface in its temporary pond during a 100-year, 24-hour rainfall event is estimated to be elevation 843.6, 4.4 ft below the homes. This needs to be checked by routing the 100-year storm.

A.7   Elevation-Outflow Calculations for the 2-Year Orifice

The elevation-outflow relationship for the 72-in. RCP is the same as that developed in Chap. 16. It is repeated here as Fig. A.3.

Flow Rate Reduction Calculations		
Drainage Area = 0.197 sq. mi. Rainfall Distri. = II		
Frequency	2-Year	100-Year
Peak Inflow	190	539
Peak Outflow	20	55
Qout/Qin	0.11	0.10
Vs/Vr	0.54	0.55
Runoff, Q	1.40	4.10
Runoff, Vol.	14.7	43.1
Storage Vol.	7.9	23.7
Max. Elev.	839.8	843.6

TABLE A.2   Flow Reductions for Catfish Creek Tributary

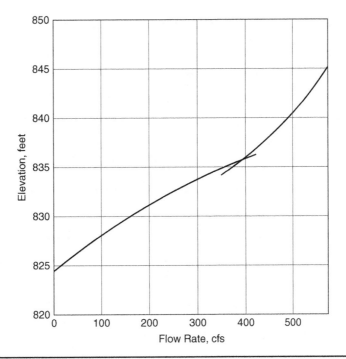

FIGURE A.3   Elevation-outflow curve for a 72-in RCP

We now need to develop the riser's elevation-outflow relationship. A cylindrical tube having a length of two to three diameters with the inner end flat with a flush wall so as to form a sharp-cornered entrance is called a standard short tube. The contraction coefficient is assumed to be 1.0 because the jet first contracts and then expands to fill the tube. Discharge coefficients range from 0.78 to 0.83 with a mean value of 0.82. Inward-projecting short tubes have coefficients which range from 0.72 to 0.80 with a mean value of 0.75. The discharge coefficient for submerged short tubes ranges from 0.61 to 0.95.

Values for sharp-edged entrances were listed in Table 14.12. In Col. 1, "L" is tube length and "p" is tube's perimeter. In our case, the pipe will not be submerged because of its height above the 72-in RCP's invert. A quick routing to size the orifice assumes the following. Use a 9-in diameter orifice and an average head of 2.12 ft ((842 − 837/2) − 0.75/2).

$$Q = Ca(2gH)^{1/2} = 0.6 \times 0.7854\ (0.75)^2 \times 8.02 \times (2.12)^{0.5} = 3.10 \text{ cfs} \tag{A.1}$$

There are $14.7 \times 43,560 = 640,330$ ft³ in 14.7 AF. Time = $640,330/3.10 = 206,560$ seconds or 2.4 days to drain runoff from a 2-year, 24-hour storm event. This is undoubtedly short because outflow is not a straight line reduction due to its falling head as the temporary pond empties. We need to verify time required to empty this temporary storage volume from the 2-year storm. This is done by routing it through the detention basin.

An elevation-outflow curve is developed for the 2-year water quality event in Table A.3. These are calculations for a 9-in diameter vertical orifice with its crest elevation at 837.0. Its equation is listed in Eq. (A.2).

$$Q = Ca(2gH)^{1/2} = 0.6 \times 0.7854\ (0.75)^2 \times 8.02 \times H^{0.5} = 2.13\ H^{0.5} \text{ cfs} \tag{A.2}$$

Outflow Curve for a Vertical Circular Orifice Flowing Partially Full and Full						
Head, H, listed in Col. 4, assumes that an orifice is acting with a free discharge. If an orifice is submerged, then head in Col. 4 must reflect the second definition of head. Insert a number in Col. 2 and revise it until head plus downstream water surface elevation equals elevation in Col. 1.						
Q = C a (2g)^0.5 (H)^0.5　Obtain Cŷ from Table 15.11　Obtain Ca from Table 15.10						
Using a 9-inch vertical circular orifice at elevation 837.0						
D = 0.75 ft in diameter						
Q = 0.6 * Ca * D^2 * (64.32)^0.5 * (H)^0.5 = 4.812 Ca * D^2 * H^0.5						
If C is other than 0.6, change equation in spreadsheet grid space G20.						
Elevation ft 1	Depth ft 2	Cŷ 3	H ft 4	H^0.5 5	Ca 6	Q cfs 7
837.0	0.00	0.500	0.000	0.000	0.7854	0.00
838.0	1.00	0.500	0.625	0.791	0.7854	1.68
839.0	2.00	0.500	1.625	1.275	0.7854	2.71
840.0	3.00	0.500	2.625	1.620	0.7854	3.44
841.0	4.00	0.500	3.625	1.904	0.7854	4.05
842.0	5.00	0.500	4.625	2.151	0.7854	4.57
843.0	6.00	0.500	5.625	2.372	0.7854	5.04
844.0	7.00	0.500	6.625	2.574	0.7854	5.47
845.0	8.00	0.500	7.625	2.761	0.7854	5.87
846.0	9.00	0.500	8.625	2.937	0.7854	6.24
847.0	10.00	0.500	9.625	3.102	0.7854	6.60
848.0	11.00	0.500	10.625	3.260	0.7854	6.93

TABLE A.3　Elevation-Outflow Curve for a 9-in Orifice

A.8　Elevation-Outflow Calculations for the Top of Riser

Calculations for its elevation-outflow curve are developed in Table A.4. They take into account whether a 3-ft by 3-ft horizontal orifice in the riser's top is a weir or an orifice as well as the correct definition of head in the orifice equation. An assumption was made that it always functioned with a free discharge.

A.9　Total Riser Inflows

Total riser inflows are the sum of inflows to the 9-in orifice and 3-ft square orifice. Flows through the two orifices from elevations 837 through 848 were obtained from Table A.3 and Table A.4. Total flows are shown in Table A.5. Elevations inside the riser for total flow entering it were obtained from Fig. A.3 as headwaters for the 72-in outlet conduit. These elevations indicated that water surfaces inside the riser are always less than elevation 837; therefore, the first definition of head for an orifice was a correct assumption.

Depth-Outflow Curve for a Horizontal Square Orifice

The head, H, listed in Col. 3 assumes that the orifice is acting with a free discharge. If the orifice is submerged, then the head in Col. 3 must reflect the second definition of head. Insert a number in Col. 2 and revise it until the head plus the downstream water elevation equals the elevation in Col. 1.

$Q = C a (2g)^{0.5} (H)^{0.5}$ or $Q = C L H^{3/2}$

Using a 3-ft by 3-ft square horizontal orifice

Width = 3.0 ft   Length = 3.0 ft

Weir Breadth = 1.5 ft

$Q = 0.6 * a * (2g)^{0.5} * (H)^{0.5} = 43.31 *(H)^{0.5}$ If C is other than 0.6, change equation in spreadsheet grid space E21.

Perimeter = 12.00 ft

Elevation ft 1	Depth ft 2	H ft 3	H^0.5 4	Qo cfs 5	C Table 14.1 6	H^3/2 7	Qw cfs 8	Q cfs 9
842.0	0.0	0.00	0.000	0.0	0.00	0.000	0.0	0.0
843.0	1.0	1.00	1.000	43.3	2.75	1.000	33.0	33.0
844.0	2.0	2.00	1.414	61.2	3.14	2.828	106.6	61.2
845.0	3.0	3.00	1.732	75.0	3.32	5.196	207.0	75.0
846.0	4.0	4.00	2.000	86.6	3.32	8.000	318.7	86.6
847.0	5.0	5.00	2.236	96.8	3.32	11.180	445.4	96.8
848.0	6.0	6.00	2.449	106.1	3.32	14.697	585.5	106.1

TABLE A.4   Outflow Calculations for the Top of the Riser

Total Elevation-Outflow Curve for a Vertical Riser in a Detention Basin

1 9-in vertical circular orifice at elevation 837
2 3-ft square horizontal orifice at elevation 842

1 Elevation ft	2 Outlet 1 cfs	3 Outlet 2 cfs	4 Outlet 3 cfs	5 Outlet 4 cfs	6 Outlet 5 cfs	7 Total cfs	8 Riser Elev.
837	0.0	0.0				0.0	824.0
838	1.7	0.0				1.7	824.1
839	2.7	0.0				2.7	824.2
840	3.4	0.0				3.4	824.3
841	4.0	0.0				4.0	824.4
842	4.6	0.0				4.6	824.5
843	5.0	33.0				38.0	826.0
844	5.5	61.2				66.7	837.0
845	5.9	75.0				80.9	827.5
846	6.2	86.6				92.8	827.8
847	6.6	96.8				103.4	828.2
848	6.9	106.1				113.0	828.5

TABLE A.5   Total Inflows to Catfish Creek Retrofit Riser

A.10   Routing Curve

Its calculations using ΔT of 0.2 hour are presented in Table A.6 and plotted in Fig. A.4. They take into account total flow through a 9-in circular orifice and a partially open riser top. Storage volumes were taken from Table A.1 and total outflow rates were used from Table A.5.

A.11   Hydrograph Routing for the Water Quality Storm Event

The inflow hydrograph for a 2-year storm event was developed in Chap. 12 in Table 12.21. Ia/P is about 0.2. However, Ia/P equals 0.1 was used since this gives the detention basin a more severe test (higher flow rates) than Ia/P equals 0.3. 0.1 and 0.3 are values from TR-55 (USDA, 1986).

A 2-year storm is routed in Table A.7a through A.7j, but only these two tables are shown. Runoff is stored for a total of five days for an average of 2.5 days. Water is a foot deep in about 2.9 days. Maximum elevation reached is 841.4, 0.6-ft below the riser's top. Again, this is a conservative estimate because Ia/P used was 0.1. Maximum inflow rate of 190 cfs was reduced to a peak outflow rate of 4.5 cfs with a delay of 6 hours in the peak outflow rate.

Routing Curve Calculations Volume in Acre Feet			
Routing Period, $\Delta T = 0.2$ hr			
$2S/\Delta T + O = (2S$ (ac ft)$/\Delta T$ hr$) + O$ (cfs$) \times 1$ cfs-day$/2$ ac ft $\times 24$ hr$/1$ day $= 120 * S + O$			
Elev. feet 1	S acre feet 2	O cfs 3	$2S/\Delta T + O$ cfs 4
837.0	0.0	0.0	0
837.5	1.0	0.5	121
838.0	2.5	1.7	302
839.0	5.4	2.7	651
840.0	8.7	3.4	1047
841.0	12.5	4.0	1504
842.0	16.4	4.6	1973
843.0	21.0	38.0	2558
844.0	26.1	66.7	3199
845.0	31.6	80.9	3873
846.0	37.8	92.8	4629
847.0	44.5	103.4	5443
848.0	51.7	113.0	6317

TABLE A.6   Routing Curve for the Catfish Creek Retrofit

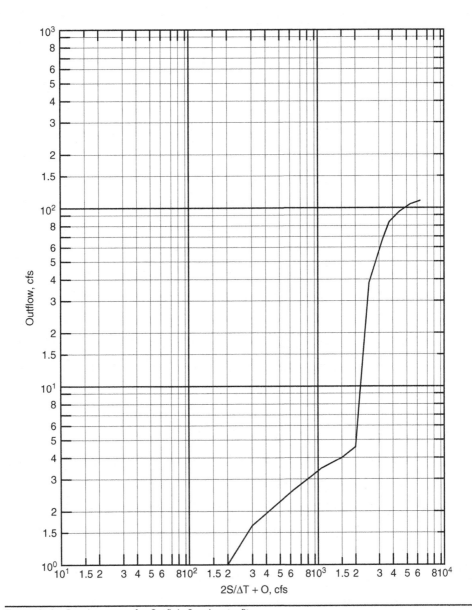

Figure A.4 Routing curve for Catfish Creek retrofit.

A.12 Hydrograph Routing for the 100-Year Storm Event

The 100-year hydrograph is the same as that developed in Chap. 12 in Table 12.22. A 100-year hydrograph is routed in Tables A.8a and A.8b. Runoff is temporarily ponded to 844.7, still 3.3 ft below any homes. When rain stops at hour 24, water is still at about 842.1, slightly above the riser's top. Runoff then continues to drain from the detention basin similar to that for the 2-year storm event. Peak flow rate is reduced from 540 cfs

Line No. 1	Time hr 2	Inflow, I_2 cfs 3	$I_1 + I_2$ cfs 4	$2S_1/\Delta T - 0_1$ cfs 5	$2S_2/\Delta T + 0_2$ cfs 6	Outflow cfs 7	Elevation ft 8
1	11.0	5.6	10.2	110.8	121.0	0.5	837.5
2	11.2	6.6	12.2	120.0	132.2	0.6	
3	11.4	8.0	14.6	131.0	145.6	0.7	
4	11.6	11.5	19.5	144.2	163.7	0.8	
5	11.8	26.9	38.4	162.1	200.5	1.0	
6	12.0	66.1	93.0	198.5	291.5	1.6	837.9
7	12.2	173.3	239.4	288.3	527.7	2.3	
8	12.4	189.5	362.8	523.1	885.9	3.1	
9	12.6	128.0	317.5	879.7	1197.2	3.5	840.2
10	12.8	56.2	184.2	1190.2	1374.4	3.8	
11	13.0	31.6	87.8	1366.8	1454.6	3.9	
12	13.2	22.2	53.8	1446.8	1500.6	4.0	841.0
13	13.4	18.6	40.8	1492.6	1533.4	4.0	
14	13.6	15.1	33.7	1525.4	1559.1	4.1	
15	13.8	13.4	28.5	1550.9	1579.4	4.1	
16	14.0	12.2	25.6	1571.2	1596.8	4.2	
17	14.2	11.2	23.4	1588.4	1611.8	4.2	
18	14.4	10.4	21.6	1603.4	1625.0	4.3	
19	14.6	9.7	20.1	1616.4	1636.5	4.3	
20	14.8	9.2	18.9	1627.9	1646.8	4.3	
21	15.0	8.8	18.0	1638.2	1656.2	4.3	
22	15.2	8.5	17.3	1647.6	1664.9	4.3	
23	15.4	8.2	16.7	1656.3	1673.0	4.4	
24	15.6	7.9	16.1	1664.2	1680.3	4.4	
25	15.8	7.5	15.4	1671.5	1686.9	4.4	
26	16.0	7.1	14.6	1678.1	1692.7	4.4	
27	16.2	6.8	13.9	1683.9	1697.8	4.4	
28	16.4	6.6	13.4	1689.0	1702.4	4.4	
29	16.6	6.4	13.0	1693.6	1706.6	4.4	
30	16.8	6.2	12.6	1697.8	1710.4	4.4	
31	17.0	6.0	12.2	1701.6	1713.8	4.4	
32	17.2	5.8	11.8	1705.0	1716.8	4.4	
33	17.4	5.6	11.4	1708.0	1719.4	4.4	
34	17.6	5.4	11.0	1710.6	1721.6	4.5	
35	17.8	5.2	10.6	1712.6	1723.2	4.5	
36	18.0	5.0	10.2	1714.2	1724.4	4.5	
37	18.2	4.8	9.8	1715.4	1725.2	4.5	
38	18.4	4.6	9.4	1716.2	1725.6	4.5	841.4 peak
39	18.6	4.4	9.0	1716.6	1725.6	4.5	
40	18.8	4.2	8.6	1716.6	1725.2	4.5	
41	19.0	4.0	8.2	1716.2	1724.4	4.5	841.4

TABLE A.7a   2-Year Hydrograph Routing for Catfish Creek

Line No. 1	Time hr 2	Inflow, I_2 cfs 3	$I_1 + I_2$ cfs 4	$2S_1/\Delta T - O_1$ cfs 5	$2S_2/\Delta T + O_2$ cfs 6	Outflow cfs 7
361	83.0	0.0	0.0	122.4	122.4	0.6
362	83.2	0.0	0.0	121.2	121.2	0.5
363	83.4	0.0	0.0	120.2	120.2	0.5
364	83.6	0.0	0.0	119.2	119.2	0.5
365	83.8	0.0	0.0	118.2	118.2	0.5
366	84.0	0.0	0.0	117.2	117.2	0.5
367	84.2	0.0	0.0	116.2	116.2	0.5
368	84.4	0.0	0.0	115.2	115.2	0.5
369	84.6	0.0	0.0	114.2	114.2	0.5
370	84.8	0.0	0.0	113.2	113.2	0.5
371	85.0	0.0	0.0	112.2	112.2	0.5
372	85.2	0.0	0.0	111.2	111.2	0.5
373	85.4	0.0	0.0	110.2	110.2	0.5
374	85.6	0.0	0.0	109.2	109.2	0.4
375	85.8	0.0	0.0	108.4	108.4	0.4
376	86.0	0.0	0.0	107.6	107.6	0.4
377	86.2	0.0	0.0	106.8	106.8	0.4
378	86.4	0.0	0.0	106.0	106.0	0.4
379	86.6	0.0	0.0	105.2	105.2	0.4
380	86.8	0.0	0.0	104.4	104.4	0.4
381	87.0	0.0	0.0	103.6	103.6	0.4
382	87.2	0.0	0.0	102.8	102.8	0.4
383	87.4	0.0	0.0	102.0	102.0	0.4
384	87.6	0.0	0.0	101.2	101.2	0.4
385	87.8	0.0	0.0	100.4	100.4	0.4
386	88.0	0.0	0.0	99.6	99.6	0.4
387	88.2	0.0	0.0	98.8	98.8	0.3
388	88.4	0.0	0.0	98.2	98.2	0.3
389	88.6	0.0	0.0	97.6	97.6	0.3
390	88.8	0.0	0.0	97.0	97.0	0.3
391	89.0	0.0	0.0	96.4	96.4	0.3
392	89.2	0.0	0.0	95.8	95.8	0.3
393	89.4	0.0	0.0	95.2	95.2	0.3
394	89.6	0.0	0.0	94.6	94.6	0.3
395	89.8	0.0	0.0	94.0	94.0	0.3
396	90.0	0.0	0.0	93.4	93.4	0.3
397	90.2	0.0	0.0	92.8	92.8	0.3
398	90.4	0.0	0.0	92.2	92.2	0.3
399	90.6	0.0	0.0	91.6	91.6	0.3
400	90.8	0.0	0.0	91.0	91.0	0.3
401	91.0	0.0	0.0	90.4	90.4	0.3

TABLE A.7j 2-Year Hydrograph Routing for Catfish Creek

Line No. 1	Time hr 2	Inflow, I_2 cfs 3	$I_1 + I_2$ cfs 4	$2S_1/\Delta T - O_1$ cfs 5	$2S_2/\Delta T + O_2$ cfs 6	Outflow cfs 7	Elevation ft 8
1	11.0	16	31	90	121	0.5	837.5
2	11.2	20	36	120	156	0.8	
3	11.4	25	45	154	199	1.0	
4	11.6	34	59	197	256	1.4	
5	11.8	70	104	254	358	1.8	
6	12.0	201	271	354	625	2.7	839.0
7	12.2	501	702	620	1322	3.7	
8	12.4	539	1040	1314	2354	27	
9	12.6	355	894	2300	3194	66	844.0
10	12.8	158	513	3062	3575	75	
11	13.0	90	248	3425	3673	76	
12	13.2	64	154	3521	3675	76	844.7 peak
13	13.4	53	117	3523	3640	76	
14	13.6	44	97	3488	3585	75	
15	13.8	39	83	3435	3518	74	
16	14.0	35	74	3370	3444	73	
17	14.2	33	68	3298	3366	71	
18	14.4	30	63	3224	3287	69	
19	14.6	28	58	3149	3207	67	844.0
20	14.8	27	55	3073	3128	64	
21	15.0	26	53	3000	3053	61	
22	15.2	25	51	2931	2982	58	
23	15.4	24	49	2866	2915	55	
24	15.6	23	47	2805	2852	52	
25	15.8	22	45	2748	2793	49	
26	16.0	21	43	2695	2738	46	
27	16.2	20	41	2646	2687	43	
28	16.4	19	39	2601	2640	41	
29	16.6	18	37	2558	2595	39	
30	16.8	18	36	2517	2553	38	843.0
31	17.0	17	35	2477	2511	36	
32	17.2	17	34	2439	2472	35	
33	17.4	16	33	2403	2436	33	
34	17.6	16	32	2370	2402	31	
35	17.8	15	31	2339	2370	30	
36	18.0	15	30	2311	2342	28	
37	18.2	15	30	2286	2315	26	
38	18.4	14	29	2263	2292	25	
39	18.6	14	28	2243	2271	23	
40	18.8	13	27	2225	2252	21	
41	19.0	13	26	2210	2236	20	842.2

TABLE A.8a   100-Year Hydrograph Routing for Catfish Creek

Line No. 1	Time hr 2	Inflow, I_2 cfs 3	$I_1 + I_2$ cfs 4	$2S1/\Delta T - 0_1$ cfs 5	$2S2/\Delta T + 0_2$ cfs 6	Outflow cfs 7	Elevation ft 8
41	19.0	13.0	26.0	2210.0	2236.0	20.0	842.2
42	19.2	12.6	25.6	2196.0	2221.6	19.4	
43	19.4	12.2	24.8	2182.8	2207.6	18.8	
44	19.6	11.8	24.0	2169.9	2193.9	18.2	
45	19.8	11.4	23.2	2157.5	2180.7	17.6	
46	20.0	11.0	22.4	2145.4	2167.8	17.1	
47	20.2	10.6	21.6	2133.7	2155.3	16.5	
48	20.4	10.2	20.8	2122.4	2143.2	15.9	
49	20.6	9.8	20.0	2111.4	2131.4	15.3	
50	20.8	9.4	19.2	2100.9	2120.1	14.7	
51	21.0	9.0	18.4	2090.7	2109.1	14.1	
52	21.2	8.6	17.6	2080.9	2098.5	13.5	
53	21.4	8.2	16.8	2071.5	2088.3	12.9	
54	21.6	7.8	16.0	2062.4	2078.4	12.3	
55	21.8	7.4	15.2	2053.8	2069.0	11.7	
56	22.0	7.0	14.4	2045.5	2059.9	11.2	
57	22.2	6.6	13.6	2037.6	2051.2	10.6	
58	22.4	6.2	12.8	2030.1	2042.9	10.0	
59	22.6	5.8	12.0	2022.9	2034.9	9.4	
60	22.8	5.4	11.2	2016.2	2027.4	8.8	
61	23.0	5.0	10.4	2009.8	2020.2	8.2	
62	23.2	4.6	9.6	2003.8	2013.4	7.6	
63	23.4	4.2	8.8	1998.2	2007.0	7.0	
64	23.6	3.8	8.0	1992.9	2000.9	6.4	
65	23.8	3.4	7.2	1988.1	1995.3	5.8	
66	24.0	3.0	6.4	1983.6	1990.0	5.3	
67	24.2	2.6	5.6	1979.5	1985.1	4.8	
68	24.4	2.2	4.8	1975.5	1980.3	4.7	
69	24.6	1.8	4.0	1970.9	1974.9	4.6	842.0
70	24.8	1.4	3.2	1965.7	1968.9	4.6	
71	25.0	1.0	2.4	1959.7	1962.1	4.6	
72	25.2	0.6	1.6	1952.9	1954.5	4.6	
73	25.4	0.4	1.0	1945.4	1946.4	4.6	
74	25.6	0.2	0.6	1937.2	1937.8	4.6	
75	25.8	0.0	0.2	1928.7	1928.9	4.6	
76	26.0	0.0	0.0	1919.8	1919.8	4.5	
77	26.2	0.0	0.0	1910.7	1910.7	4.5	
78	26.4	0.0	0.0	1901.7	1901.7	4.5	
79	26.6	0.0	0.0	1892.6	1892.6	4.5	
80	26.8	0.0	0.0	1883.6	1883.6	4.5	
81	27.0	0.0	0.0	1874.6	1874.6	4.5	841.8

TABLE A.8b 100-Year Hydrograph Routing for Catfish Creek

to only 76 cfs because of the available storage volume. This should eliminate flooding and overtopping of the downstream intersection. Runoff quality is also enhanced because of the extended detention orifice at elevation 837.

A.13   Cost

Riser cost, walkway, grates, fencing, and removal of one end section is about $15,000. The 72-in RCP in the 1960s and 1970s cost about $30,000 plus embankment fill. At that point in time, the city had not yet begun to utilize existing detention volumes or consider runoff water quality in their designs.

A.14   Summary

Quantity and quality aspects of drainage are so intermeshed today that both must be given equal consideration in design. Sites like this must be preserved and utilized for both water quantity and water quality control in compliance with local ordinances and state and federal laws. It should not be allowed to be developed for any other type of land use. If it were, flooding would either continue or another less desirable site would have to be found elsewhere.

This example and Chaps. 14 and 15 on culvert and riser design, respectively, are included in this book because they are the least-cost retrofit to achieve both runoff water quantity and quality control. While working for the Iowa State Highway Commission (ISHC) in the 1970s, we calculated that Iowa had about 400,000 culverts in existence. With 50 states, this is upwards of 20,000,000 culverts. If only 5 percent were suitable as retrofits, a million culverts would be usable as retrofit sites—at a minimal cost for each.

In addition, since FEMA and EPA came into existence, we have constructed hundreds of thousands of land-consuming detention basins. Some percentage of these have excess unused storage in them that could also be used as retrofit sites—again at a minimal cost for each.

In both cases, these facilities already exist. There should be no problems from local officials, special interest groups, or not in my back yard (NIMBY) from surrounding neighbors. In fact the reverse could be true because of the benefits that would accrue. Flow rates could be reduced, runoff quality enhanced, and sites refurbished to present a more pleasant appearance, enhancing all portions of the triple bottom line (TBL) of people, planet, and profit.

APPENDIX B

Greenroofs

B.1 Definition

A greenroof is a building roof that is partially or completely covered with vegetation and soil or a growing medium, planted over a waterproofing membrane. It includes additional layers such as root barriers and drainage and irrigation systems. Another distinction is between pitched and flat greenroofs. Pitched sod roofs, a traditional feature of many Scandinavian buildings, tend to be a much simpler design than flat greenroofs. This is because roof pitch reduces risk of water penetrating through a roof structure, allowing use of fewer waterproofing and drainage layers.

The three types of greenroofs (extensive, intensive, and modular) are used in all land uses. Commercial, industrial, and municipal buildings are the most common but are also used in single- and multi-family developments. Their advantages are reducing energy costs to heat and cool buildings by 20 to 30 percent, reducing runoff peaks and volumes, removing some pollutants from runoff, and adding color to usually bland rooftops. They add usable space to buildings by providing areas for coffee breaks, lunches, meetings, and flower and vegetable gardens. They support a variety of grasses, shrubs, plants, and trees. Soil depths range from a few inches to several feet. Leakage into upper floors is prevented by concrete slabs or other water barriers.

B.2 Apartment Building

The building is located near the west coast of Oregon (Fig. B.1). Local ordinances require that rainfall be retained on-site to replenish groundwater supplies and best management practices (BMPs) be used to enhance water quality. A 100-year, 24-hour storm event contains 4.8 in of rain. Figure B.1 shows the arrangement of the twenty-eight 1,680-ft² apartments. The central atrium is two stories high with skylights. A circular staircase leads to the second floor and roof.

Half of the greenroof has grass in 6-in of amended soil. Tables, chairs, and lounges allow residents to chat, play cards, and sun bathe. The other half is used for flower and vegetable gardens. It has a concrete slab and one layer is 6-in of gravel with a porosity of 0.40. Storage volume is 0.4×6 or 2.4 in. Thus, all rainfall is stored in the soil, gravel, or used for plant growth. Roof runoff is conveyed in a 2-in square downspout to a rain garden. Rout the hydrograph through the gravel and downspout to determine the inflow hydrograph to the rain garden.

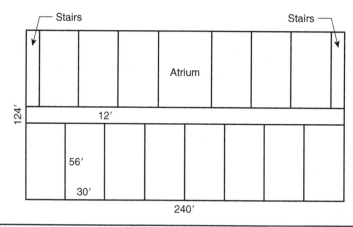

FIGURE B.1   Plan view of apartment building.

B.2.1   Drainage Area

It is 124 × 240 ft and has an area of 29,760 ft², 0.683 ac, or 0.00107 mi².

B.2.2   Time of Concentration

A downspout is located in a corner of the building's end. Flow length is $((240)^2 + (124)^2)^{0.5}$ or 270 ft. Tc is estimated in Table B.1 as 56.2 min. Use Tc as 1.0 hour. The asterisks in the first column indicate that a number must be entered in the other columns if the number in that column is different from the number already in the column for that flow path or sub-area.

B.2.3   100-Year Inflow Hydrograph

Use Ia/P as 0.1. The inflow hydrograph to the greenroof is estimated in Table B.2 and B.3. The peak flow rate of 0.32 cfs occurs at hour 8.8.

B.2.4   Depth-Storage Calculations

With two of the 4.8 in of rain retained in the soil, runoff can be 2.8 in. Runoff is 124 × 240 × 2.8/12 or 6,940 ft³. Storage in 6-in of gravel with a 40 percent porosity is 124 × 240 × 0.5 × 0.40 is 5,950 ft³. Storage volumes in each 0.1-ft is determined in Table B.4.

B.2.5   Depth-Outflow Calculations

A 2.0-in square downspout is a horizontal orifice with heads from zero to a half foot. These calculations are depicted in Table B.5. Maximum outflow rate is 0.09 cfs.

B.2.6   Routing Curve Calculations

A routing curve combines both depth-storage and depth-outflow curves into a single curve. These calculations are shown is Table B.6 and are plotted in Fig. B.2.

B.2.7   100-Year Hydrograph Routing

The 100-year hydrograph is routed in Table B.7a to c using Type 1A rainfall. An inflow peak flow of 0.32 cfs at hour 8.8 is reduced to a peak outflow of 0.07 cfs at hour 21.4, a delay of 12.6 hours. Maximum depth in the gravel is 0.27 ft. Outflow is 0.02 cfs at hour 40.0, with depth in the gravel of 0.05 ft. The gravel is empty at about hour 72.0, 2.0 days after the rain stops.

Worksheet 2				
Overland (Sheet)	1	2	3	Total
Pathway Length, ft*	270.0	0.0	0.0	
Upstream Elevation, ft*	100.4	0.0	0.0	
Downstream Elevation, ft*	100.0	0.0	0.0	
Pathway Slope, ft/ft	0.00148	4.00000	4.00000	
Manning's n*	0.10	0.10	0.10	
2-yr, 24-hr Rainfall, in.*	2.0	2.0	2.0	
Flow Velocity, fps	0.08	0.25	0.25	
Travel Time, min	56.2	0.0	0.0	56.2
Shallow Concentrated	1	2	3	
Pathway Length, ft*	0.1	0.1	0.1	
Upstream Elevation, ft*	0.1	0.1	0.1	
Downstream Elevation, ft*	0.0	0.0	0.0	
Pathway Slope, ft/ft	1.00000	1.00000	1.00000	
Equation Coefficient*	16.1	20.3	20.3	
Flow Velocity, fps	16.1	20.3	20.3	
Travel Time, min	0.0	0.0	0.0	0.0
Channel	1	2	3	
Pathway Length, ft*	0.0	0.1	0.1	
Upstream Elevation, ft*	0.0	0.1	0.1	
Downstream Elevation, ft*	0.0	0.0	0.0	
Pathway Slope, ft/ft	4.00000	1.00000	1.00000	
Bottom Width, ft*	1.0	1.0	1.0	
Flow Depth, ft*	1.0	1.0	1.0	
Side Slope, H:V*	1.0	1.0	1.0	
Area, sq ft	1.00	1.00	1.00	
Wetted Perimeter, ft	1.00	1.00	1.00	
Manning's n*	0.040	0.040	0.040	
Flow Velocity, fps	74.50	37.25	37.25	
Travel Time, min	0.0	0.0	0.0	0.0
Gutter	1	2	3	
Pathway Length, ft*	0.1	0.1	0.1	
Upstream Elevation, ft*	0.1	0.1	0.1	
Downstream Elevation, ft*	0.0	0.0	0.0	
Pathway Slope, ft/ft	1.0	1.0	1.0	
Street Cross Slope, ft/ft*	0.02	0.02	0.02	
Flow Depth, ft*	0.25	0.25	0.25	
Flow Top Width, ft	12.5	12.5	12.5	
Manning's n*	0.016	0.016	0.016	
Flow Velocity, fps	27.78	27.78	27.78	
Travel Time, min	0.0	0.0	0.0	0.0
Total Travel Time, Min	56.2	0.0	0.0	56.2

TABLE B.1 Time of Concentration for Apartment Greenroof

Hydrograph Development—Worksheet 5A

Project: Greenroof

Outline one: Present | Developed

Location: Oregon

Frequency (yr): 100

By: Date:

Checked by: Date:

Subarea Name 1	Drainage Area sq. mi. 2	Time of Concen. hr. 3	Travel Time thru Subarea hr. 4	Downstream Subarea Names 5	Travel Time Summation to Outlet 6	24-hr Rainfall in. 7	Runoff Curve Number 8	Runoff in. 9	AmQ sq. mi.-in. 10	Initial Abstraction in. 11	Ia/P 12
Greenroof	0.00107	1.00	0.00	—	0.00	2.8	100	2.8	0.00300	0.000	0.000
								#DIV/0!	#DIV/0!		#DIV/0!
								#DIV/0!	#DIV/0!		#DIV/0!
								#DIV/0!	#DIV/0!		#DIV/0!

TABLE B.2 Hydrograph Variables for Apartment Greenroof

Hydrograph Development—Worksheet 5B

Project: Greenroof Location: Oregon Date:

Outline one: Present [Developed] Frequency (yr): 100 By: Checked by: Date:

Select and enter hydrograph times in hours from exhibit 5-III

Discharge at selected hydrograph times — cfs

Subarea Name	Subarea Tc hr	ΣTt to Outlet hr	Ia/P	AmQ sq.mi.in.	7.0	7.3	7.6	7.9	8.0	8.2	8.4	8.6	8.8	9.0	9.2	9.4
1	2	3	4	5	6	7	8	9	10	11	12	13	14	15	16	17
Greenroof	1.0	0.0	0.1	0.00300	19	23	26	33	37	52	76	97	108	102	91	81
					0.06	0.07	0.08	0.10	0.11	0.16	0.23	0.29	0.32	0.31	0.27	0.24
					0.0	0.0	0.0	0.0	0.0	0.0	0.0	0.0	0.0	0.0	0.0	0.0
					0.0	0.0	0.0	0.0	0.0	0.0	0.0	0.0	0.0	0.0	0.0	0.0
					0.0	0.0	0.0	0.0	0.0	0.0	0.0	0.0	0.0	0.0	0.0	0.0
Total					0.06	0.07	0.08	0.10	0.11	0.16	0.23	0.29	0.32	0.31	0.27	0.24

Select and enter hydrograph times in hours from exhibit 5-III

Subarea Name	Subarea Tc hr	ΣTt to Outlet hr	Ia/P	AmQ sq.mi.in.	9.6	9.8	10.0	10.3	10.6	11.0	12.0	13.0	14.0	15.0	16.0	18.0
Greenroof	1.0	0.0	0.1	0.00300	72	65	59	52	48	43	38	33	32	31	30	28
					0.22	0.20	0.18	0.16	0.14	0.13	0.11	0.10	0.10	0.09	0.09	0.08
					0.0	0.0	0.0	0.0	0.0	0.0	0.0	0.0	0.0	0.0	0.0	0.0
					0.0	0.0	0.0	0.0	0.0	0.0	0.0	0.0	0.0	0.0	0.0	0.0
					0.0	0.0	0.0	0.0	0.0	0.0	0.0	0.0	0.0	0.0	0.0	0.0
Total					0.22	0.20	0.18	0.16	0.14	0.13	0.11	0.10	0.10	0.09	0.09	0.08

Table B.3 100-year Hydrograph Ordinates for Apartment Greenroof

239

Depth ft	Area sq ft	Aver. Area sq ft	Δ Depth ft	Δ Volume cu ft	Tot. Vol. cu ft
0.0	11900				0
		11900	0.10	1190	
0.1	11900				1190
		11900	0.10	1190	
0.2	11900				2380
		11900	0.10	1190	
0.3	11900				3570
		11900	0.10	1190	
0.4	11900				4760
		11900	0.10	1190	
0.5	11900				5950

TABLE **B.4**   Depth-Storage for Apartment Greenroof

Depth-Outflow Curve for a Horizontal Rectangular Orifice
The head, H, listed in Col. 3 assumes that the orifice is acting with a free discharge. If the orifice is submerged, then the head in Col. 3 must reflect the second definition of head. Insert a number in Col. 2 and revise it until the head plus the downstream water elevation equals the elevation in Col. 1.

$Q = C a (2g)*0.5 (H)^0.5$ or $Q = C L H^{3/2}$

Using a 2.0 × 2.0 inch horizontal orifice

Width = 0.17 ft   Length = 0.17 ft

Weir Breadth = 0.2 ft

$Q = 0.6 * a * (2g)^{0.5} * (H)^{0.5} = 0.1342 * (H)^{0.5}$ If C is other than 0.6, change equation in E21.

Perimeter = 0.67

Depth ft 1	Depth ft 2	H ft 3	H^0.5 4	Qo cfs 5	C Table 8 6	H^3/2 7	Qw cfs 8	Q cfs 9
0.0	0.000	0.000	0.000	0.00	0.00	0.000	0.00	0.00
0.1	0.100	0.100	0.316	0.04	3.32	0.032	0.07	0.04
0.2	0.200	0.200	0.447	0.06	3.32	0.089	0.20	0.06
0.3	0.300	0.300	0.548	0.07	3.32	0.164	0.36	0.07
0.4	0.400	0.400	0.632	0.08	3.32	0.253	0.56	0.08
0.5	0.500	0.500	0.707	0.09	3.32	0.354	0.78	0.09

TABLE **B.5**   Outflows for a 2.0 × 2.0 in Downspout

Routing Period = 0.2 hr			
$2S/\Delta T + O = (2\ S\ (cu\ ft)/0.20\ hr) + O\ (cfs) \times (1\ hr/60\ min \times)(1\ min/60\ sec)$ $= 0.002778\ *S + O$			
Depth **ft** **1**	**S** **cu ft** **2**	**O** **cfs** **3**	**$2S/\Delta T + O$** **cfs** **4**
0.0	0	0.00	0.00
0.1	1190	0.04	3.35
0.2	2380	0.06	6.67
0.3	3570	0.07	9.99
0.4	4760	0.08	13.30
0.5	5950	0.09	16.62

TABLE B.6   Routing Curve for Apartment Greenroof

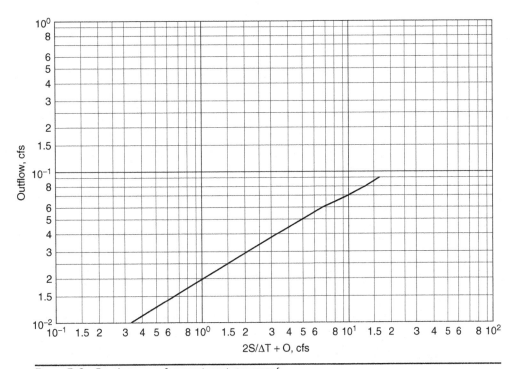

FIGURE B.2   Routing curve for apartment greenroof.

Line No. 1	Time hr 2	Inflow, I_2 cfs 3	$(I_1 + I_2)$ + cfs 4	$(2S_1/\Delta T - 0_1)$ cfs 5	=	$(2S_2/\Delta T + 0_2)$ cfs 6	Outflow cfs 7	Depth ft 8
1	7.0	0.06	0.11	0.73		0.84	0.01	0.03
2	7.2	0.06	0.12	0.82		0.94	0.01	
3	7.4	0.07	0.13	0.92		1.05	0.01	
4	7.6	0.08	0.15	1.03		1.18	0.01	
5	7.8	0.09	0.17	1.16		1.33	0.01	
6	8.0	0.11	0.20	1.31		1.51	0.02	
7	8.2	0.16	0.27	1.47		1.74	0.02	
8	8.4	0.23	0.39	1.70		2.09	0.03	
9	8.6	0.29	0.52	2.03		2.55	0.03	
10	8.8	0.32	0.61	2.49		3.10	0.04	
11	9.0	0.31	0.63	3.02		3.65	0.04	
12	9.2	0.27	0.58	3.57		4.15	0.04	
13	9.4	0.24	0.51	4.07		4.58	0.05	
14	9.6	0.22	0.46	4.48		4.94	0.05	
15	9.8	0.20	0.42	4.84		5.26	0.05	
16	10.0	0.18	0.38	5.16		5.54	0.05	
17	10.2	0.16	0.34	5.44		5.78	0.05	
18	10.4	0.15	0.31	5.68		5.99	0.06	
19	10.6	0.14	0.29	5.87		6.16	0.06	
20	10.8	0.13	0.27	6.04		6.31	0.06	
21	11.0	0.13	0.26	6.19		6.45	0.06	
22	11.2	0.13	0.26	6.33		6.59	0.06	
23	11.4	0.12	0.25	6.47		6.72	0.06	0.20
24	11.6	0.12	0.24	6.60		6.84	0.06	
25	11.8	0.12	0.24	6.72		6.96	0.06	
26	12.0	0.11	0.23	6.84		7.07	0.06	
27	12.2	0.11	0.22	6.95		7.17	0.06	
28	12.4	0.11	0.22	7.05		7.27	0.06	
29	12.6	0.11	0.22	7.15		7.37	0.06	
30	12.8	0.11	0.22	7.25		7.47	0.06	
31	13.0	0.10	0.21	7.35		7.56	0.06	
32	13.2	0.10	0.20	7.44		7.64	0.06	
33	13.4	0.10	0.20	7.52		7.72	0.06	
34	13.6	0.10	0.20	7.60		7.80	0.06	
35	13.8	0.10	0.20	7.68		7.88	0.06	
36	14.0	0.10	0.20	7.76		7.96	0.06	
37	14.2	0.10	0.20	7.84		8.04	0.06	
38	14.4	0.10	0.20	7.92		8.12	0.06	
39	14.6	0.10	0.20	8.00		8.20	0.06	
40	14.8	0.10	0.20	8.08		8.28	0.06	

TABLE B.7a   100-Year Hydrograph Routing for Apartment Greenroof

Line No. 1	Time hr 2	Inflow, I_2 cfs 3	$(I_1 + I_2)$ cfs 4	$+ (2S_1/\Delta T - O_1)$ cfs 5	$= (2S_2/\Delta T + O_2)$ cfs 6	Outflow cfs 7	Depth ft 8
41	15.0	0.09	0.19	8.16	8.35	0.06	
42	15.2	0.09	0.18	8.23	8.41	0.07	
43	15.4	0.09	0.18	8.27	8.45	0.07	
44	15.6	0.09	0.18	8.31	8.49	0.07	
45	15.8	0.09	0.18	8.35	8.53	0.07	
46	16.0	0.09	0.18	8.39	8.57	0.07	
47	16.2	0.09	0.18	8.43	8.61	0.07	
48	16.4	0.09	0.18	8.47	8.65	0.07	
49	16.6	0.09	0.18	8.51	8.69	0.07	
50	16.8	0.09	0.18	8.55	8.73	0.07	
51	17.0	0.08	0.17	8.59	8.76	0.07	
52	17.2	0.08	0.16	8.62	8.78	0.07	
53	17.4	0.08	0.16	8.64	8.80	0.07	
54	17.6	0.08	0.16	8.66	8.82	0.07	
55	17.8	0.08	0.16	8.68	8.84	0.07	
56	18.0	0.08	0.16	8.70	8.86	0.07	0.27

TABLE **B.7a**　100-Year Hydrograph Routing for Apartment Greenroof (*Continued*)

Line No. 1	Time hr 2	Inflow, I_2 cfs 3	$(I_1 + I_2)$ cfs 4	$+ (2S_1/\Delta T - O_1)$ cfs 5	$= (2S_2/\Delta T + O_2)$ cfs 6	Outflow cfs 7	Depth ft 8
56	18.0	0.08	0.16	8.70	8.86	0.07	0.27
57	18.2	0.08	0.16	8.72	8.88	0.07	
58	18.4	0.08	0.16	8.74	8.90	0.07	
59	18.6	0.08	0.16	8.76	8.92	0.07	
60	18.8	0.08	0.16	8.78	8.94	0.07	
61	19.0	0.08	0.16	8.80	8.96	0.07	
62	19.2	0.08	0.16	8.82	8.98	0.07	
63	19.4	0.08	0.16	8.84	9.00	0.07	
64	19.6	0.08	0.16	8.86	9.02	0.07	
65	19.8	0.08	0.16	8.88	9.04	0.07	
66	20.0	0.08	0.16	8.90	9.06	0.07	
67	20.2	0.07	0.15	8.92	9.07	0.07	
68	20.4	0.07	0.14	8.93	9.07	0.07	
69	20.6	0.07	0.14	8.93	9.07	0.07	
70	20.8	0.07	0.14	8.93	9.07	0.07	
71	21.0	0.07	0.14	8.93	9.07	0.07	
72	21.2	0.07	0.14	8.93	9.07	0.07	
73	21.4	0.07	0.14	8.93	9.07	0.07	0.27 peak
74	21.6	0.07	0.14	8.93	9.07	0.07	

TABLE **B.7b**　100-Year Hydrograph Routing for Apartment Greenroof

Line No. 1	Time hr 2	Inflow, I_2 cfs 3	$(I_1 + I_2)$ cfs 4	+ $(2S_1/\Delta T - 0_1)$ cfs 5	= $(2S_2/\Delta T + 0_2)$ cfs 6	Outflow cfs 7	Depth ft 8
75	21.8	0.07	0.14	8.93	9.07	0.07	
76	22.0	0.07	0.14	8.93	9.07	0.07	
77	22.2	0.07	0.14	8.93	9.07	0.07	
78	22.4	0.06	0.13	8.93	9.06	0.07	
79	22.6	0.06	0.12	8.92	9.04	0.07	
80	22.8	0.06	0.12	8.90	9.01	0.07	
81	23.0	0.05	0.11	8.87	8.98	0.07	
82	23.2	0.05	0.10	8.84	8.94	0.07	
83	23.4	0.05	0.09	8.80	8.90	0.07	
84	23.6	0.04	0.09	8.76	8.85	0.07	
85	23.8	0.04	0.08	8.71	8.79	0.07	
86	24.0	0.04	0.07	8.65	8.72	0.07	
87	24.2	0.03	0.07	8.58	8.65	0.07	
88	24.4	0.03	0.06	8.51	8.57	0.07	
89	24.6	0.02	0.05	8.43	8.48	0.07	
90	24.8	0.02	0.05	8.34	8.38	0.07	
91	25.0	0.02	0.04	8.24	8.28	0.07	
92	25.2	0.01	0.03	8.14	8.17	0.07	
93	25.4	0.01	0.02	8.03	8.06	0.07	
94	25.6	0.01	0.02	7.92	7.94	0.06	
95	25.8	0.00	0.01	7.82	7.83	0.06	
96	26.0	0.00	0.00	7.71	7.71	0.06	
97	26.2	0.00	0.00	7.59	7.59	0.06	
98	26.4	0.00	0.00	7.47	7.47	0.06	
99	26.6	0.00	0.00	7.35	7.35	0.06	
100	26.8	0.00	0.00	7.23	7.23	0.06	
101	27.0	0.00	0.00	7.11	7.11	0.06	
102	27.2	0.00	0.00	6.99	6.99	0.06	
103	27.4	0.00	0.00	6.87	6.87	0.06	
104	27.6	0.00	0.00	6.75	6.75	0.06	
105	27.8	0.00	0.00	6.63	6.63	0.06	0.20
106	28.0	0.00	0.00	6.51	6.51	0.06	
107	28.2	0.00	0.00	6.39	6.39	0.06	
108	28.4	0.00	0.00	6.27	6.27	0.06	
109	28.6	0.00	0.00	6.15	6.15	0.06	
110	28.8	0.00	0.00	6.03	6.03	0.06	
111	29.0	0.00	0.00	5.91	5.91	0.06	0.18

TABLE B.7b   100-Year Hydrograph Routing for Apartment Greenroof (*Continued*)

Line No. 1	Time hr 2	Inflow, I_2 cfs 3	$(I_1 + I_2)$ cfs 4	+	$(2S_1/\Delta T - 0_1)$ cfs 5	=	$(2S_2/\Delta T + 0_2)$ cfs 6	Outflow cfs 7	Depth ft 8
111	29.0	0.00	0.00		5.91		5.91	0.06	0.18
112	29.2	0.00	0.00		5.79		5.79	0.06	
113	29.4	0.00	0.00		5.67		5.67	0.06	
114	29.6	0.00	0.00		5.55		5.55	0.06	
115	29.8	0.00	0.00		5.43		5.43	0.06	
116	30.0	0.00	0.00		5.31		5.31	0.06	
117	30.2	0.00	0.00		5.19		5.19	0.06	
118	30.4	0.00	0.00		5.07		5.07	0.05	
119	30.6	0.00	0.00		4.97		4.97	0.05	
120	30.8	0.00	0.00		4.87		4.87	0.05	
121	31.0	0.00	0.00		4.77		4.77	0.05	
122	31.2	0.00	0.00		4.67		4.67	0.05	
123	31.4	0.00	0.00		4.57		4.57	0.05	
124	31.6	0.00	0.00		4.47		4.47	0.05	
125	31.8	0.00	0.00		4.37		4.37	0.05	
126	32.0	0.00	0.00		4.27		4.27	0.05	
127	32.2	0.00	0.00		4.17		4.17	0.04	
128	32.4	0.00	0.00		4.09		4.09	0.04	
129	32.6	0.00	0.00		4.01		4.01	0.04	
130	32.8	0.00	0.00		3.93		3.93	0.04	
131	33.0	0.00	0.00		3.85		3.85	0.04	
132	33.2	0.00	0.00		3.77		3.77	0.04	
133	33.4	0.00	0.00		3.69		3.69	0.04	
134	33.6	0.00	0.00		3.61		3.61	0.04	
135	33.8	0.00	0.00		3.53		3.53	0.04	
136	34.0	0.00	0.00		3.45		3.45	0.04	
137	34.2	0.00	0.00		3.37		3.37	0.04	0.10
138	34.4	0.00	0.00		3.29		3.29	0.04	
139	34.6	0.00	0.00		3.21		3.21	0.04	
140	34.8	0.00	0.00		3.13		3.13	0.03	
141	35.0	0.00	0.00		3.07		3.07	0.03	
142	35.2	0.00	0.00		3.01		3.01	0.03	
143	35.4	0.00	0.00		2.95		2.95	0.03	
144	35.6	0.00	0.00		2.89		2.89	0.03	
145	35.8	0.00	0.00		2.83		2.83	0.03	
146	36.0	0.00	0.00		2.77		2.77	0.03	
147	36.2	0.00	0.00		2.71		2.71	0.03	

TABLE **B.7c**   100-Year Hydrograph Routing for Apartment Greenroof

Line No.	Time hr	Inflow, I_2 cfs	$(I_1 + I_2)$ cfs	$+$	$(2S_1/\Delta T - O_1)$ cfs	$=$	$(2S_2/\Delta T + O_2)$ cfs	Outflow cfs	Depth ft
1	2	3	4		5		6	7	8
148	36.4	0.00	0.00		2.65		2.65	0.03	
149	36.6	0.00	0.00		2.59		2.59	0.03	
150	36.8	0.00	0.00		2.53		2.53	0.03	
151	37.0	-0.01	-0.01		2.47		2.46	0.03	
152	37.2	0.00	-0.01		2.40		2.39	0.03	
153	37.4	0.00	0.00		2.33		2.33	0.03	
154	37.6	0.00	0.00		2.27		2.27	0.03	
155	37.8	0.00	0.00		2.21		2.21	0.03	
156	38.0	0.00	0.00		2.15		2.15	0.03	
157	38.2	0.00	0.00		2.09		2.09	0.03	
158	38.4	0.00	0.00		2.03		2.03	0.03	
159	38.6	0.00	0.00		1.97		1.97	0.03	
160	38.8	0.00	0.00		1.91		1.91	0.03	
161	39.0	0.00	0.00		1.85		1.85	0.03	
162	39.2	0.00	0.00		1.79		1.79	0.03	
163	39.4	0.00	0.00		1.73		1.73	0.03	
164	39.6	0.00	0.00		1.67		1.67	0.02	
165	39.8	0.00	0.00		1.63		1.63	0.02	
166	40.0	0.00	0.00		1.59		1.59	0.02	0.05

TABLE B.7c   100-Year Hydrograph Routing for Apartment Greenroof (*Continued*)

B.3   Summary

A peak inflow rate of 0.32 cfs at hour 8.8 is reduced to a peak outflow rate of 0.07 cfs at hour 21.4. Maximum depth in the gravel is 0.27 ft. At hour 40.0, this depth is decreased to 0.05 ft and is zero 2.0 days later. This outflow hydrograph becomes part of an inflow hydrograph to a rain garden. To the roof hydrograph is added a hydrograph from rain falling directly onto the rain garden. However, runoff from this greenroof is delayed until after the intense portion of the 100-year rain event is past. Depending on native soil type and infiltration rate under the rain garden, a layer of gravel will probably be needed under the rain garden as well to meet the zero surface runoff condition required by a local ordinance.

Residential Rain Gardens

C.1 Definition

Rain gardens are attractive landscaped areas that capture and filter stormwater from various surfaces. They duplicate a natural forest's hydology by collecting water in free-form vegetated areas and allow it to infiltrate. They reduce erosion potential and pollutants flowing into storm drains and creeks by taking advantage of runoff in their design and plant selection. They are designed to withstand moisture extremes and use nutrients, e.g., nitrogen and phosphorus, found in runoff.

C.2 Introduction

Besides existing detention basin and culvert retrofits, rain gardens give the most bang for the buck and add color to their surroundings. Announcements by the Environmental Protection Agency (EPA) show a preference for green rather than gray infrastucture and explicitly mention rain gardens as a best management practice (BMP) to be included in new developments, retrofits, and redevelopment projects. They are cheap, need little maintenance, are of any size, and replace lawn areas. Soils are amended to a mixture of sand, organic matter, and wood chips. This encourages infiltration and use of water and pollutants for plant growth.

Some city programs encourage owners to add rain gardens to their homes and businesses. They have prepared lists of plants, grasses, shrubs, and trees well suited to their climates.

C.3 Description of Rain Garden Example

The rain garden receives runoff from a garage, driveway, sidewalk and half a street's width plus front yard runoff as shown in Fig. C.1. A porous curb and gutter allow street runoff to flow into the garden under the sidewalk. A roof is covered with asphalt shingles, and its downspout conveys runoff to the garden in a swale along a driveway. Soils are Natural Resources Conservation Service (NRCS) hydrologic soil group (HSG) Type A with an infiltration rate of 0.40 in/hour. Rainfall for the 2-, 10-, and 100-year, 24-hour storms are 3.3, 5.1, and 7.2 in, respectively, near the East Coast with the time distribution shown in Table C.1 for the 100-year event.

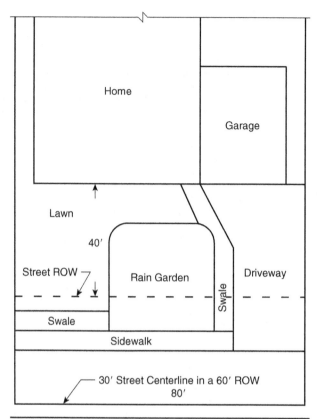

FIGURE C.1   Single–family residential lot.

Hour	% of Total	Total P	Increm. P	Hour	% of Total	Total P	Increm. P
1	0.014	0.101	0.10	13	0.727	5.234	0.34
2	0.032	0.230	0.13	14	0.766	5.515	0.28
3	0.052	0.374	0.14	15	0.802	5.774	0.26
4	0.073	0.526	0.15	16	0.836	6.019	0.24
5	0.095	0.684	0.16	17	0.863	6.214	0.19
6	0.119	0.857	0.17	18	0.887	6.386	0.17
7	0.157	1.130	0.27	19	0.909	6.545	0.16
8	0.199	1.433	0.30	20	0.930	6.696	0.15
9	0.248	1.786	0.35	21	0.951	6.847	0.15
10	0.491	3.535	1.75	22	0.970	6.984	0.14
11	0.629	4.529	0.99	23	0.987	7.106	0.12
12	0.68	4.896	0.37	24	1.000	7.200	0.09

TABLE C.1   Hourly Amounts During a 24-hour Rainfall Event, Inches

C.3.1   Front Yard and Street Runoff Volume

Runoff from a 24×32 ft garage is $768 \times 0.98 \times 7.2 / 12$ or 450 ft^3, allowing for initial abstraction (Ia). Runoff from a 20×55 ft driveway is $1,100 \times 0.98 \times 7.2/12$ or 650 cu ft. Street and sidewalk runoff is $(80 \times 15 + 60 \times 5) \times 0.98 \times 7.2/12$ or 880 ft^3. Excess rain occurs in hours 10 and 11. This amounts to $1.75 + 0.99 - 2 \times 0.40 = 1.94$ in. Front yard dimensions are $60 \times (40 + 10)$ ft. Runoff is $3,000 \times 1.94/12 = 480$ ft^3. Total runoff is $450 + 650 + 880 + 480 = 2,460$ ft^3. Total impervious area is $768 + 1,100 + 1,500$ or $3,370$ ft^2.

Use a 30×50-ft rain garden, 1-ft deep, with 5:1 side slopes. Total volume is $0.5 \times 1.0 \times (20 \times 40 + 30 \times 50)$ or $1,150$ ft^3 plus 1.5-ft of gravel with 40 percent porosity is $30 \times 50 \times 1.5 \times 0.4$ is 900 ft^3, a total of $2,050$ ft^3. Excavate 3 ft of soil within a 30×50 ft area. The first 0.5 ft is amended soil; second 1.5 ft is gravel with 40 percent porosity; the upper one foot is the amended soil.

C.3.2   Time of Concentration

Runoff from a roof takes place in just a few minutes. Street runoff flows through the porous concrete curb and gutter in less than a minute. Therefore, use Tc of 6 minutes or 0.1 hour.

C.3.3   Runoff Hydrograph

Runoff curve number (CN) is 40 for grass in good condition and 98 for impervious areas, Table 2-2a of TR-55 (USDA, 1986). CN is $(40 \times 3,000 + 98 \times 3,370)/6,370$ or 71. Drainage area is $6,370/43,560/640 = 0.000228$ sq mi. Variables are listed in Table C.2.

C.3.3.1   Two-Year Storm

Hydrograph ordinates are listed in Table C.3 with Tc of 0.1 hour and Ia/P of 0.1. Peak inflow rate is 0.14 cfs at hour 12.2 from runoff into the rain garden gravel.

C.3.3.2   10-Year Storm

Hydrograph ordinates are listed in Table C.4 with Tc of 0.1 hour and Ia/P of 0.1. Peak inflow rate is 0.33 cfs at hour 12.2 from runoff into the rain garden gravel.

C.3.3.3   100-Year Storm

Hydrograph ordinates are listed in Table C.5 with Tc of 0.1 hour and Ia/P of 0.1. Peak inflow rate is 0.59 cfs at hour 12.2 from runoff into the rain garden gravel.

C.3.4   Storage Volume

Total storage is enough to temporarily retain all runoff including water used by the rain garden plants for growth. Storage areas are calculated in Table C.6. Storage volumes each 0.2-ft are developed in Table C.7. Below ground storage area is $30 \times 60 \times 1.5 \times 0.4 = 1,080$ sq ft. Total storage volume below and above ground is $2,206$ ft^3.

C.3.5   Outflow Rate

Outflow is a function of the soil's infiltration rate of 0.40 in./h over the rain garden's area. Outflow rate is a constant value of $50 \times 30 \times 0.40/12 = 50.0$ ft^3 per hour or $50.0/60/60$ or 0.014 cfs.

C.3.6   Routing Curve

A routing curve for the rain garden is calculated in Table C.8. There is no need to plot the routing curve because outflow is a constant 0.014 cfs.

Hydrograph Development—Worksheet 5a

Project: Front Yard Rain Garden
Location: East Coast
By:
Date:

Outline one: Present | **Developed**
Frequency (yr): 2, 10, 100
Checked by:
Date:

Subarea Name 1	Drainage Area sq. mi. 2	Time of Concen. hr. 3	Travel Time thru Subarea hr. 4	Downstream Subarea Names 5	Travel Time Summation to Outlet 6	24-hr Rainfall in. 7	Runoff Curve Number 8	Runoff in. 9	AmQ sq. mi.-in. 10	Initial Abstraction in. 11	Ia/P 12
Rain Garden	0.000228	0.10	0.00	—	0.00	3.3	71	0.9	0.000214	0.817	0.248
Rain Garden	0.000228	0.10	0.00	—	0.00	5.1	71	2.2	0.000500	0.817	0.160
Rain Garden	0.000228	0.10	0.00	—	0.00	7.2	71	3.9	0.000887	0.817	0.113

Table C.2 Front Yard Rain Garden Hydrograph Variables

250

Hydrograph Development—Worksheet 5b

Project: Rain Garden   Location: East Coast   By: Date:

Outline one: Present   Frequency (yr): 2   Checked by: Date:

[Developed]

	Basic Watershed Data Used				Select and enter hydrograph times in hours from exhibit 5-											
Subarea Name	Subarea Tc	ΣTt to Outlet	Ia/P	AmQ sq. mi.–in.	11.0	11.3	11.6	11.9	12.0	12.2	12.4	12.6	12.8	13.0	13.2	13.4
					Discharge at selected hydrograph times, cfs											
1	2	3	4	5	6	7	8	9	10	11	12	13	14	15	16	17
Rain Garden	0.10	0.00	0.10	0.000214	29	38	57	172	241	662	345	191	101	83	68	62
					0.01	0.01	0.01	0.04	0.05	0.14	0.07	0.04	0.02	0.02	0.01	0.01
					0.0	0.0	0.0	0.0	0.0	0.0	0.0	0.0	0.0	0.0	0.0	0.0
					0.0	0.0	0.0	0.0	0.0	0.0	0.0	0.0	0.0	0.0	0.0	0.0
Total					0.01	0.01	0.01	0.04	0.05	0.14	0.07	0.04	0.02	0.02	0.01	0.01

	Basic Watershed Data Used				Select and enter hydrograph times in hours from exhibit 5-											
Subarea Name	Subarea Tc	ΣTt to Outlet	Ia/P	AmQ sq. mi.–in.	13.6	13.8	14.0	14.3	14.6	15.0	15.5	16.0	17.0	18.0	20.0	22.0
					Discharge at selected hydrograph times, cfs											
1	2	3	4	5	6	7	8	9	10	11	12	13	14	15	16	17
Rain Garden	0.10	0.00	0.10	0.000214	58	54	50	44	41	37	37	32	21	16	13	11
					0.01	0.01	0.01	0.01	0.01	0.01	0.01	0.01	0.00	0.00	0.00	0.00
					0.0	0.0	0.0	0.0	0.0	0.0	0.0	0.0	0.0	0.0	0.0	0.0
					0.0	0.0	0.0	0.0	0.0	0.0	0.0	0.0	0.0	0.0	0.0	0.0
Total					0.01	0.01	0.01	0.01	0.01	0.01	0.01	0.01	0.00	0.00	0.00	0.00

TABLE C.3   2-Year Hydrograph Ordinates for Front Yard Rain Garden

Hydrograph Development—Worksheet 5b

Project: Rain Garden   Location: East Coast   By: Date:

Outline one: Present   [Developed]   Frequency (yr): 10   Checked by: Date:

Select and enter hydrograph times in hours from exhibit 5-

Subarea Name	Subarea Tc	ΣTt to Outlet	Ia/P	AmQ sq. mi.-in.	11.0	11.3	11.6	11.9	12.0	12.2	12.4	12.6	12.8	13.0	13.2	13.4
1	2	3	4	5	6	7	8	9	10	11	12	13	14	15	16	17
					Discharge at selected hydrograph times, cfs											
Rain Garden	0.10	0.00	0.10	0.000500	29	38	57	172	241	662	345	191	101	83	68	62
					0.01	0.02	0.03	0.09	0.12	0.33	0.17	0.10	0.05	0.04	0.03	0.03
					0.0	0.0	0.0	0.0	0.0	0.0	0.0	0.0	0.0	0.0	0.0	0.0
					0.0	0.0	0.0	0.0	0.0	0.0	0.0	0.0	0.0	0.0	0.0	0.0
Total					0.01	0.02	0.03	0.09	0.12	0.33	0.17	0.10	0.05	0.04	0.03	0.03

Select and enter hydrograph times in hours from exhibit 5-

Subarea Name	Subarea Tc	ΣTt to Outlet	Ia/P	AmQ sq. mi.-in.	13.6	13.8	14.0	14.3	14.6	15.0	15.5	16.0	17.0	18.0	20.0	22.0
Rain Garden	0.10	0.00	0.10	0.000500	58	54	50	44	41	37	37	32	21	16	13	11
					0.03	0.03	0.03	0.02	0.02	0.02	0.02	0.02	0.01	0.01	0.01	0.01
					0.0	0.0	0.0	0.0	0.0	0.0	0.0	0.0	0.0	0.0	0.0	0.0
					0.0	0.0	0.0	0.0	0.0	0.0	0.0	0.0	0.0	0.0	0.0	0.0
Total					0.03	0.03	0.03	0.02	0.02	0.02	0.02	0.02	0.01	0.01	0.01	0.01

TABLE C.4   10-Year Hydrograph Ordinates for Front Yard Rain Garden

Hydrograph Development—Worksheet 5b

Project: Rain Garden   Location: East Coast   By: Date:

Outline one: Present   [Developed]   Frequency (yr): 100   Checked by: Date:

	Basic Watershed Data Used				Select and enter hydrograph times in hours from exhibit 5-											
	Subarea Tc	ΣTt to Outlet	Ia/P	AmQ sq. mi.-in.	11.0	11.3	11.6	11.9	12.0	12.2	12.4	12.6	12.8	13.0	13.2	13.4
Subarea Name	2	3	4	5	6	7	8	9	10	11	12	13	14	15	16	17
1					Discharge at selected hydrograph times, cfs											
Rain Garden	0.10	0.00	0.10	0.000887	29	38	57	172	241	662	345	191	101	83	68	62
					0.03	0.03	0.05	0.15	0.21	0.59	0.31	0.17	0.09	0.07	0.06	0.05
					0.0	0.0	0.0	0.0	0.0	0.0	0.0	0.0	0.0	0.0	0.0	0.0
					0.0	0.0	0.0	0.0	0.0	0.0	0.0	0.0	0.0	0.0	0.0	0.0
Total					0.03	0.03	0.05	0.15	0.21	0.59	0.31	0.17	0.09	0.07	0.06	0.05

					Select and enter hydrograph times in hours from exhibit 5-											
	Subarea Tc	ΣTt to Outlet	Ia/P	AmQ sq. mi.-in.	13.6	13.8	14.0	14.3	14.6	15.0	15.5	16.0	17.0	18.0	20.0	22.0
Subarea Name	2	3	4	5	6	7	8	9	10	11	12	13	14	15	16	17
Rain Garden	0.10	0.00	0.10	0.000887	58	54	50	44	41	37	37	32	21	16	13	11
					0.05	0.05	0.04	0.04	0.04	0.03	0.03	0.03	0.02	0.01	0.01	0.01
					0.0	0.0	0.0	0.0	0.0	0.0	0.0	0.0	0.0	0.0	0.0	0.0
					0.0	0.0	0.0	0.0	0.0	0.0	0.0	0.0	0.0	0.0	0.0	0.0
Total					0.05	0.05	0.04	0.04	0.04	0.03	0.03	0.03	0.02	0.01	0.01	0.01

TABLE C.5   100-Year Hydrograph Ordinates for Front Yard Rain Garden

Depth ft	Length ft	Width ft	Area sq ft
−1.50	30	60	720
−1.40	30	60	720
−1.20	30	60	720
−1.00	30	60	720
−0.80	30	60	720
−0.60	30	60	720
−0.40	30	60	720
−0.20	30	60	720
0.00	30	60	720
0.00	20	40	800
0.20	22	42	924
0.40	24	44	1056
0.60	26	46	1196
0.80	28	48	1344
1.00	30	50	1500

TABLE **C.6**   Depth-Area Calculations for Rain Garden

Depth ft	Area sq ft	Aver. Area sq ft	Δ Depth ft	Δ Volume cu ft	Tot. Vol. cu ft
−1.5	720				0
		720	0.1	72	
−1.4	720				72
		720	0.2	144	
−1.2	720				216
		720	0.2	144	
−1.0	720				360
		720	0.2	144	
−0.8	720				504
		720	0.2	144	
−0.6	720				648
		720	0.2	144	
−0.4	720				792
		720	0.2	144	

TABLE **C.7**   Depth-Storage Calculations for Rain Garden

Depth ft	Area sq ft	Aver. Area sq ft	Δ Depth ft	Δ Volume cu ft	Tot. Vol. cu ft
−0.2	720				936
		720	0.2	144	
0.0	720				1080
		760	0.0	0	
0.0	800				0
		822	0.2	164	
0.2	924				1244
		990	0.2	198	
0.4	1056				1442
		1126	0.2	225	
0.6	1196				1668
		1270	0.2	254	
0.8	1344				1922
		1422	0.2	284	
1.0	1500				2206

TABLE C.7   Depth-Storage Calculations for Rain Garden (*Continued*)

Routing Period = 0.2 hr

$2S/\Delta T + O = (2S \text{ (cu ft)}/0.1 \text{ hr}) + O \text{ (cfs)} \times (1 \text{ hr}/60 \text{ min}) \times (1 \text{ min}/60 \text{ sec}) = 0.00278 \quad *S + O$

Depth ft 1	S cu ft 2	O cfs 3	2S/ΔT + O cfs 4
−1.5	0	0.014	0.01
−1.4	72	0.014	0.21
−1.2	216	0.014	0.61
−1.0	360	0.014	1.01
−0.8	504	0.014	1.41
−0.6	648	0.014	1.81
−0.4	792	0.014	2.21
−0.2	936	0.014	2.61
0.0	1080	0.014	3.01
0.2	1244	0.014	3.47
0.4	1442	0.014	4.02
0.6	1668	0.014	4.65
0.8	1922	0.014	5.35
1.0	2206	0.014	6.14

TABLE C.8   Rain Garden Routing Curve Calculations

C.3.7   Hydrograph Routing

C.3.7.1   Two-Year Storm

This storm's hydrograph is routed in Table C.9. Maximum water depth in the gravel is 0.27 ft at hour 13.0. The gravel is empty of water at hour 18.4.

Line No. 1	Time hr 2	Inflow, I_2 cfs 3	$I_1 + I_2$ cfs 4	$2S_1/\Delta T - O_1$ cfs 5	$2S_2/\Delta T + O_2$ cfs 6	Outflow cfs 7	Depth ft 8
1	11.0	0.01	0.02	0.000	0.030	0.010	−1.48
2	11.2	0.01	0.02	0.010	0.030	0.010	
3	11.4	0.01	0.02	0.010	0.030	0.010	
4	11.6	0.01	0.02	0.010	0.030	0.010	
5	11.8	0.02	0.03	0.010	0.040	0.014	
6	12.0	0.05	0.07	0.012	0.082	0.014	
7	12.2	0.14	0.19	0.054	0.244	0.014	
8	12.4	0.07	0.21	0.216	0.426	0.014	
9	12.6	0.04	0.11	0.398	0.508	0.014	
10	12.8	0.02	0.06	0.480	0.540	0.014	
11	13.0	0.02	0.04	0.512	0.552	0.014	−1.23 peak
12	13.2	0.01	0.03	0.524	0.554	0.014	
13	13.4	0.01	0.02	0.526	0.546	0.014	
14	13.6	0.01	0.02	0.518	0.538	0.014	
15	13.8	0.01	0.02	0.510	0.530	0.014	
16	14.0	0.01	0.02	0.502	0.522	0.014	
17	14.2	0.01	0.02	0.494	0.514	0.014	
18	14.4	0.01	0.02	0.486	0.506	0.014	
19	14.6	0.01	0.02	0.478	0.498	0.014	
20	14.8	0.01	0.02	0.470	0.490	0.014	
21	15.0	0.01	0.02	0.462	0.482	0.014	
22	15.2	0.01	0.02	0.454	0.474	0.014	
23	15.4	0.01	0.02	0.446	0.466	0.014	
24	15.6	0.01	0.02	0.438	0.458	0.014	
25	15.8	0.01	0.02	0.430	0.450	0.014	
26	16.0	0.01	0.02	0.422	0.442	0.014	
27	16.2	0.01	0.02	0.414	0.434	0.014	
28	16.4	0.01	0.02	0.406	0.426	0.014	
29	16.6	0.01	0.02	0.398	0.418	0.014	

TABLE C.9   2-Year Rain Garden Routing

Line No. 1	Time hr 2	Inflow, I_2 cfs 3	$I_1 + I_2$ cfs 4	$2S_1/\Delta T - 0_1$ cfs 5	$2S_2/\Delta T + 0_2$ cfs 6	Outflow cfs 7	Depth ft 8
30	16.8	0.01	0.02	0.390	0.410	0.014	
31	17.0	0.00	0.01	0.382	0.392	0.014	
32	17.2	0.00	0.00	0.364	0.364	0.014	
33	17.4	0.00	0.00	0.336	0.336	0.014	
34	17.6	0.00	0.00	0.308	0.308	0.014	
35	17.8	0.00	0.00	0.280	0.280	0.014	
36	18.0	0.00	0.00	0.252	0.252	0.014	
37	18.2	0.00	0.00	0.224	0.224	0.112	−1.40
38	18.4	0.00	0.00	0.000	0.000	0.000	−1.50
39	18.6	0.00	0.00	0.000	0.000	0.000	
40	18.8	0.00	0.00	0.000	0.000	0.000	
41	19.0	0.00	0.00	0.000	0.000	0.000	
42	19.2	0.00	0.00	0.000	0.000	0.000	
43	19.4	0.00	0.00	0.000	0.000	0.000	
44	19.6	0.00	0.00	0.000	0.000	0.000	
45	19.8	0.00	0.00	0.000	0.000	0.000	
46	20.0	0.00	0.00	0.000	0.000	0.000	
47	20.2	0.00	0.00	0.000	0.000	0.000	
48	20.4	0.00	0.00	0.000	0.000	0.000	
49	20.6	0.00	0.00	0.000	0.000	0.000	
50	20.8	0.00	0.00	0.000	0.000	0.000	
51	21.0	0.00	0.00	0.000	0.000	0.000	
52	21.2	0.00	0.00	0.000	0.000	0.000	
53	21.4	0.00	0.00	0.000	0.000	0.000	
54	21.6	0.00	0.00	0.000	0.000	0.000	
55	21.8	0.00	0.00	0.000	0.000	0.000	
56	22.0	0.00	0.00	0.000	0.000	0.000	
57	22.2	0.00	0.00	0.000	0.000	0.000	
58	22.4	0.00	0.00	0.000	0.000	0.000	
59	22.6	0.00	0.00	0.000	0.000	0.000	
60	22.8	0.00	0.00	0.000	0.000	0.000	
61	23.0	0.00	0.00	0.000	0.000	0.000	−1.50

TABLE C.9 2-Year Rain Garden Routing (*Continued*)

C.3.7.2 Ten-Year Storm

This storm's hydrograph is routed in Table C.10a and b. Maximum water depth in the gravel is 0.98 ft at hour 17.0. The gravel is empty of water at hour 40.0.

Line No. 1	Time hr 2	Inflow, I_2 cfs 3	$I_1 + I_2$ cfs 4	$2S_1/\Delta T - O_1$ cfs 5	$2S_2/\Delta T + O_2$ cfs 6	Outflow cfs 7	Depth ft 8
1	11.0	0.01	0.04	0.000	0.040	0.010	−1.46
2	11.2	0.01	0.02	0.020	0.040	0.010	
3	11.4	0.02	0.03	0.020	0.050	0.014	
4	11.6	0.03	0.05	0.022	0.072	0.014	
5	11.8	0.07	0.10	0.044	0.144	0.014	
6	12.0	0.12	0.19	0.116	0.306	0.014	
7	12.2	0.33	0.45	0.278	0.728	0.014	−1.14
8	12.4	0.17	0.50	0.700	1.200	0.014	
9	12.6	0.10	0.27	1.172	1.442	0.014	−0.80
10	12.8	0.05	0.15	1.414	1.564	0.014	
11	13.0	0.04	0.09	1.536	1.626	0.014	
12	13.2	0.03	0.07	1.598	1.668	0.014	
13	13.4	0.03	0.06	1.640	1.700	0.014	
14	13.6	0.03	0.06	1.672	1.732	0.014	
15	13.8	0.03	0.06	1.704	1.764	0.014	
16	14.0	0.03	0.06	1.736	1.796	0.014	
17	14.2	0.02	0.05	1.768	1.818	0.014	−0.60
18	14.4	0.02	0.04	1.790	1.830	0.014	
19	14.6	0.02	0.04	1.802	1.842	0.014	
20	14.8	0.02	0.04	1.814	1.854	0.014	
21	15.0	0.02	0.04	1.826	1.866	0.014	
22	15.2	0.02	0.04	1.838	1.878	0.014	
23	15.4	0.02	0.04	1.850	1.890	0.014	
24	15.6	0.02	0.04	1.862	1.902	0.014	
25	15.8	0.02	0.04	1.874	1.914	0.014	
26	16.0	0.02	0.04	1.886	1.926	0.014	
27	16.2	0.02	0.04	1.898	1.938	0.014	
28	16.4	0.02	0.04	1.910	1.950	0.014	
29	16.6	0.02	0.04	1.922	1.962	0.014	

TABLE C.10a   10-Year Rain Garden Routing

Line No. 1	Time hr 2	Inflow, I_2 cfs 3	$I_1 + I_2$ cfs 4	$2S_1/\Delta T - O_1$ cfs 5	$2S_2/\Delta T + O_2$ cfs 6	Outflow cfs 7	Depth ft 8
30	16.8	0.02	0.04	1.934	1.974	0.014	
31	17.0	0.01	0.03	1.946	1.976	0.014	−0.52 peak
32	17.2	0.01	0.02	1.948	1.968	0.014	
33	17.4	0.01	0.02	1.940	1.960	0.014	
34	17.6	0.01	0.02	1.932	1.952	0.014	
35	17.8	0.01	0.02	1.924	1.944	0.014	
36	18.0	0.01	0.02	1.916	1.936	0.014	
37	18.2	0.01	0.02	1.908	1.928	0.014	
38	18.4	0.01	0.02	1.900	1.920	0.014	
39	18.6	0.01	0.02	1.892	1.912	0.014	
40	18.8	0.01	0.02	1.884	1.904	0.014	
41	19.0	0.01	0.02	1.876	1.896	0.014	
42	19.2	0.01	0.02	1.868	1.888	0.014	
43	19.4	0.01	0.02	1.860	1.880	0.014	
44	19.6	0.01	0.02	1.852	1.872	0.014	
45	19.8	0.01	0.02	1.844	1.864	0.014	
46	20.0	0.01	0.02	1.836	1.856	0.014	
47	20.2	0.01	0.02	1.828	1.848	0.014	
48	20.4	0.01	0.02	1.820	1.840	0.014	
49	20.6	0.01	0.02	1.812	1.832	0.014	
50	20.8	0.01	0.02	1.804	1.824	0.014	
51	21.0	0.01	0.02	1.796	1.816	0.014	
52	21.2	0.01	0.02	1.788	1.808	0.014	−0.60
53	21.4	0.01	0.02	1.780	1.800	0.014	
54	21.6	0.01	0.02	1.772	1.792	0.014	
55	21.8	0.01	0.02	1.764	1.784	0.014	
56	22.0	0.01	0.02	1.756	1.776	0.014	
57	22.2	0.01	0.02	1.748	1.768	0.014	
58	22.4	0.01	0.02	1.740	1.760	0.014	
59	22.6	0.01	0.02	1.732	1.752	0.014	
60	22.8	0.01	0.02	1.724	1.744	0.014	
61	23.0	0.01	0.02	1.716	1.736	0.014	−0.64

TABLE C.10a   10-Year Rain Garden Routing (*Continued*)

Line No. 1	Time hr 2	Inflow, I_2 cfs 3	$I_1 + I_2$ cfs 4	$2S_1/\Delta T - O_1$ cfs 5	$2S_2/\Delta T + O_2$ cfs 6	Outflow cfs 7	Depth ft 8
61	23.0	0.01	0.02	1.716	1.736	0.014	−0.64
62	23.2	0.01	0.02	1.708	1.728	0.014	
63	23.4	0.01	0.02	1.700	1.720	0.014	
64	23.6	0.01	0.02	1.692	1.712	0.014	
65	23.8	0.01	0.02	1.684	1.704	0.014	
66	24.0	0.01	0.02	1.676	1.696	0.014	
67	24.2	0.01	0.02	1.668	1.688	0.014	
68	24.4	0.01	0.02	1.660	1.680	0.014	
69	24.6	0.01	0.02	1.652	1.672	0.014	
70	24.8	0.01	0.02	1.644	1.664	0.014	
71	25.0	0.01	0.02	1.636	1.656	0.014	
72	25.2	0.01	0.02	1.628	1.648	0.014	
73	25.4	0.01	0.02	1.620	1.640	0.014	
74	25.6	0.01	0.02	1.612	1.632	0.014	
75	25.8	0.01	0.02	1.604	1.624	0.014	
76	26.0	0.01	0.02	1.596	1.616	0.014	
77	26.2	0.00	0.01	1.588	1.598	0.014	
78	26.4	0.00	0.00	1.570	1.570	0.014	
79	26.6	0.00	0.00	1.542	1.542	0.014	
80	26.8	0.00	0.00	1.514	1.514	0.014	
81	27.0	0.00	0.00	1.486	1.486	0.014	
82	27.2	0.00	0.00	1.458	1.458	0.014	
83	27.4	0.00	0.00	1.430	1.430	0.014	
84	27.6	0.00	0.00	1.402	1.402	0.014	−0.80
85	27.8	0.00	0.00	1.374	1.374	0.014	
86	28.0	0.00	0.00	1.346	1.346	0.014	
87	28.2	0.00	0.00	1.318	1.318	0.014	
88	28.4	0.00	0.00	1.290	1.290	0.014	
89	28.6	0.00	0.00	1.262	1.262	0.014	
90	28.8	0.00	0.00	1.234	1.234	0.014	
91	29.0	0.00	0.00	1.206	1.206	0.014	
92	29.2	0.00	0.00	1.178	1.178	0.014	
93	29.4	0.00	0.00	1.150	1.150	0.014	
94	29.6	0.00	0.00	1.122	1.122	0.014	

TABLE C.10b   10-Year Rain Garden Routing

Line No. 1	Time hr 2	Inflow, I_2 cfs 3	$I_1 + I_2$ cfs 4	$2S_1/\Delta T - O_1$ cfs 5	$2S_2/\Delta T + O_2$ cfs 6	Outflow cfs 7	Depth ft 8
95	29.8	0.00	0.00	1.094	1.094	0.014	
96	30.0	0.00	0.00	1.066	1.066	0.014	
97	30.2	0.00	0.00	1.038	1.038	0.014	
98	30.4	0.00	0.00	1.010	1.010	0.014	−1.00
99	30.6	0.00	0.00	0.982	0.982	0.014	
100	30.8	0.00	0.00	0.954	0.954	0.014	
101	31.0	0.00	0.00	0.926	0.926	0.014	
102	31.2	0.00	0.00	0.898	0.898	0.014	
103	31.4	0.00	0.00	0.870	0.870	0.014	
104	31.6	0.00	0.00	0.842	0.842	0.014	
105	31.8	0.00	0.00	0.814	0.814	0.014	
106	32.0	0.00	0.00	0.786	0.786	0.014	
107	32.2	0.00	0.00	0.758	0.758	0.014	
108	32.4	0.00	0.00	0.730	0.730	0.014	
109	32.6	0.00	0.00	0.702	0.702	0.014	
110	32.8	0.00	0.00	0.674	0.674	0.014	
111	33.0	0.00	0.00	0.646	0.646	0.014	
112	33.2	0.00	0.00	0.618	0.618	0.014	−1.20
113	33.4	0.00	0.00	0.590	0.590	0.014	
114	33.6	0.00	0.00	0.562	0.562	0.014	
115	33.8	0.00	0.00	0.534	0.534	0.014	
116	34.0	0.00	0.00	0.506	0.506	0.014	
117	34.2	0.00	0.00	0.478	0.478	0.014	
118	34.4	0.00	0.00	0.450	0.450	0.014	
119	34.6	0.00	0.00	0.422	0.422	0.014	
120	34.8	0.00	0.00	0.394	0.394	0.014	
121	35.0	0.00	0.00	0.366	0.366	0.014	−1.32

TABLE **C.10b**   10-Year Rain Garden Routing (*Continued*)

C.3.7.3   100-Year Storm

This storm's hydrograph is routed in Tables C.11a to c. Maximum water depth on the rain garden's surface is 0.42 ft at hour 18.0. From hours 13.0 to 33.2 water is on the surface, a total of 20.2 hours. The gravel is empty of water at hour 62.5.

Line No. 1	Time hr 2	Inflow, I_2 cfs 3	$I_1 + I_2$ cfs 4	$2S_1/\Delta T - 0_1$ cfs 5	$2S_2/\Delta T + 0_2$ cfs 6	Outflow cfs 7	Depth ft 8
1	11.0	0.03	0.03	0.010	0.040	0.014	−1.46
2	11.2	0.03	0.06	0.012	0.072	0.014	
3	11.4	0.04	0.07	0.044	0.114	0.014	
4	11.6	0.05	0.09	0.086	0.176	0.014	
5	11.8	0.12	0.17	0.148	0.318	0.014	−1.35
6	12.0	0.21	0.33	0.290	0.620	0.014	
7	12.2	0.59	0.80	0.592	1.392	0.014	−0.80
8	12.4	0.31	0.90	1.364	2.264	0.014	
9	12.6	0.17	0.48	2.236	2.716	0.014	
10	12.8	0.09	0.26	2.688	2.948	0.014	
11	13.0	0.07	0.16	2.920	3.080	0.014	0.00
12	13.2	0.06	0.13	3.052	3.182	0.014	
13	13.4	0.05	0.11	3.154	3.264	0.014	
14	13.6	0.05	0.10	3.236	3.336	0.014	
15	13.8	0.05	0.10	3.308	3.408	0.014	
16	14.0	0.04	0.09	3.380	3.470	0.014	0.20
17	14.2	0.04	0.08	3.442	3.522	0.014	
18	14.4	0.04	0.08	3.494	3.574	0.014	
19	14.6	0.04	0.08	3.546	3.626	0.014	
20	14.8	0.04	0.08	3.598	3.678	0.014	
21	15.0	0.03	0.07	3.650	3.720	0.014	
22	15.2	0.03	0.06	3.692	3.752	0.014	
23	15.4	0.03	0.06	3.724	3.784	0.014	
24	15.6	0.03	0.06	3.756	3.816	0.014	
25	15.8	0.03	0.06	3.788	3.848	0.014	
26	16.0	0.03	0.06	3.820	3.880	0.014	
27	16.2	0.03	0.06	3.852	3.912	0.014	
28	16.4	0.03	0.06	3.884	3.944	0.014	
29	16.6	0.03	0.06	3.916	3.976	0.014	
30	16.8	0.03	0.06	3.948	4.008	0.014	
31	17.0	0.02	0.05	3.980	4.030	0.014	0.40
32	17.2	0.02	0.04	4.002	4.042	0.014	
33	17.4	0.02	0.04	4.014	4.054	0.014	
34	17.6	0.02	0.04	4.026	4.066	0.014	

TABLE C.11a   100-Year Rain Garden Routing

Line No. 1	Time hr 2	Inflow, I_2 cfs 3	$I_1 + I_2$ cfs 4	$2S_1/\Delta T - O_1$ cfs 5	$2S_2/\Delta T + O_2$ cfs 6	Outflow cfs 7	Depth ft 8
35	17.8	0.02	0.04	4.038	4.078	0.014	
36	18.0	0.01	0.03	4.050	4.080	0.014	0.42 peak
37	18.2	0.01	0.02	4.052	4.072	0.014	
38	18.4	0.01	0.02	4.044	4.064	0.014	
39	18.6	0.01	0.02	4.036	4.056	0.014	
40	18.8	0.01	0.02	4.028	4.048	0.014	
41	19.0	0.01	0.02	4.020	4.040	0.014	
42	19.2	0.01	0.02	4.012	4.032	0.014	
43	19.4	0.01	0.02	4.004	4.024	0.014	
44	19.6	0.01	0.02	3.996	4.016	0.014	0.40
45	19.8	0.01	0.02	3.988	4.008	0.014	
46	20.0	0.01	0.02	3.980	4.000	0.014	
47	20.2	0.01	0.02	3.972	3.992	0.014	
48	20.4	0.01	0.02	3.964	3.984	0.014	
49	20.6	0.01	0.02	3.956	3.976	0.014	
50	20.8	0.01	0.02	3.948	3.968	0.014	
51	21.0	0.01	0.02	3.940	3.960	0.014	
52	21.2	0.01	0.02	3.932	3.952	0.014	
53	21.4	0.01	0.02	3.924	3.944	0.014	
54	21.6	0.01	0.02	3.916	3.936	0.014	
55	21.8	0.01	0.02	3.908	3.928	0.014	
56	22.0	0.01	0.02	3.900	3.920	0.014	
57	22.2	0.01	0.02	3.892	3.912	0.014	
58	22.4	0.01	0.02	3.884	3.904	0.014	
59	22.6	0.01	0.02	3.876	3.896	0.014	
60	22.8	0.01	0.02	3.868	3.888	0.014	
61	23.0	0.01	0.02	3.860	3.880	0.014	0.35

TABLE C.11a 100-Year Rain Garden Routing (*Continued*)

Line No. 1	Time hr 2	Inflow, I_2 cfs 3	$I_1 + I_2$ cfs 4	$2S_1/\Delta T - O_1$ cfs 5	$2S_2/\Delta T + O_2$ cfs 6	Outflow cfs 7	Depth ft 8
61	23.0	0.01	0.02	3.860	3.880	0.014	0.35
62	23.2	0.02	0.03	3.852	3.882	0.014	
63	23.4	0.02	0.04	3.854	3.894	0.014	
64	23.6	0.02	0.04	3.866	3.906	0.014	
65	23.8	0.02	0.04	3.878	3.918	0.014	
66	24.0	0.02	0.04	3.890	3.930	0.014	
67	24.2	0.01	0.03	3.902	3.932	0.014	
68	24.4	0.01	0.02	3.904	3.924	0.014	
69	24.6	0.01	0.02	3.896	3.916	0.014	
70	24.8	0.01	0.02	3.888	3.908	0.014	
71	25.0	0.01	0.02	3.880	3.900	0.014	
72	25.2	0.01	0.02	3.872	3.892	0.014	
73	25.4	0.01	0.02	3.864	3.884	0.014	
74	25.6	0.01	0.02	3.856	3.876	0.014	
75	25.8	0.01	0.02	3.848	3.868	0.014	
76	26.0	0.01	0.02	3.840	3.860	0.014	
77	26.2	0.01	0.02	3.832	3.852	0.014	
78	26.4	0.01	0.02	3.824	3.844	0.014	
79	26.6	0.01	0.02	3.816	3.836	0.014	
80	26.8	0.01	0.02	3.808	3.828	0.014	
81	27.0	0.01	0.02	3.800	3.820	0.014	
82	27.2	0.01	0.02	3.792	3.812	0.014	
83	27.4	0.01	0.02	3.784	3.804	0.014	
84	27.6	0.01	0.02	3.776	3.796	0.014	
85	27.8	0.00	0.01	3.768	3.778	0.014	
86	28.0	0.00	0.00	3.750	3.750	0.014	
87	28.2	0.00	0.00	3.722	3.722	0.014	
88	28.4	0.00	0.00	3.694	3.694	0.014	
89	28.6	0.00	0.00	3.666	3.666	0.014	
90	28.8	0.00	0.00	3.638	3.638	0.014	
91	29.0	0.00	0.00	3.610	3.610	0.014	
92	29.2	0.00	0.00	3.582	3.582	0.014	
93	29.4	0.00	0.00	3.554	3.554	0.014	
94	29.6	0.00	0.00	3.526	3.526	0.014	

TABLE **C.11b**   100-Year Rain Garden Routing

Line No. 1	Time hr 2	Inflow, I_2 cfs 3	$I_1 + I_2$ cfs 4	$2S_1/\Delta T - O_1$ cfs 5	$2S_2/\Delta T + O_2$ cfs 6	Outflow cfs 7	Depth ft 8
95	29.8	0.00	0.00	3.498	3.498	0.014	
96	30.0	0.00	0.00	3.470	3.470	0.014	
97	30.2	0.00	0.00	3.442	3.442	0.014	
98	30.4	0.00	0.00	3.414	3.414	0.014	
99	30.6	0.00	0.00	3.386	3.386	0.014	
100	30.8	0.00	0.00	3.358	3.358	0.014	
101	31.0	0.00	0.00	3.330	3.330	0.014	
102	31.2	0.00	0.00	3.302	3.302	0.014	
103	31.4	0.00	0.00	3.274	3.274	0.014	
104	31.6	0.00	0.00	3.246	3.246	0.014	
105	31.8	0.00	0.00	3.218	3.218	0.014	
106	32.0	0.00	0.00	3.190	3.190	0.014	
107	32.2	0.00	0.00	3.162	3.162	0.014	
108	32.4	0.00	0.00	3.134	3.134	0.014	
109	32.6	0.00	0.00	3.106	3.106	0.014	
110	32.8	0.00	0.00	3.078	3.078	0.014	
111	33.0	0.00	0.00	3.050	3.050	0.014	
112	33.2	0.00	0.00	3.022	3.022	0.014	0.00
113	33.4	0.00	0.00	2.994	2.994	0.014	
114	33.6	0.00	0.00	2.966	2.966	0.014	
115	33.8	0.00	0.00	2.938	2.938	0.014	
116	34.0	0.00	0.00	2.910	2.910	0.014	
117	34.2	0.00	0.00	2.882	2.882	0.014	
118	34.4	0.00	0.00	2.854	2.854	0.014	
119	34.6	0.00	0.00	2.826	2.826	0.014	
120	34.8	0.00	0.00	2.798	2.798	0.014	
121	35.0	0.00	0.00	2.770	2.770	0.014	−0.12

TABLE C.11b   100-Year Rain Garden Routing (*Continued*)

Line No. 1	Time hr 2	Inflow, I_2 cfs 3	$I_1 + I_2$ cfs 4	$2S_1/\Delta T - O_1$ cfs 5	$2S_2/\Delta T + O_2$ cfs 6	Outflow cfs 7	Depth ft 8
121	35.0	0.00	0.00	2.770	2.770	0.014	−0.12
122	35.2	0.00	0.00	2.742	2.742	0.014	
123	35.4	0.00	0.00	2.714	2.714	0.014	
124	35.6	0.00	0.00	2.686	2.686	0.014	
125	35.8	0.00	0.00	2.658	2.658	0.014	
126	36.0	0.00	0.00	2.630	2.630	0.014	
127	36.2	0.00	0.00	2.602	2.602	0.014	−0.20
128	36.4	0.00	0.00	2.574	2.574	0.014	
129	36.6	0.00	0.00	2.546	2.546	0.014	
130	36.8	0.00	0.00	2.518	2.518	0.014	
131	37.0	0.00	0.00	2.490	2.490	0.014	
132	37.2	0.00	0.00	2.462	2.462	0.014	
133	37.4	0.00	0.00	2.434	2.434	0.014	
134	37.6	0.00	0.00	2.406	2.406	0.014	
135	37.8	0.00	0.00	2.378	2.378	0.014	
136	38.0	0.00	0.00	2.350	2.350	0.014	
137	38.2	0.00	0.00	2.322	2.322	0.014	
138	38.4	0.00	0.00	2.294	2.294	0.014	
139	38.6	0.00	0.00	2.266	2.266	0.014	
140	38.8	0.00	0.00	2.238	2.238	0.014	
141	39.0	0.00	0.00	2.210	2.210	0.014	−0.40
142	39.2	0.00	0.00	2.182	2.182	0.014	
143	39.4	0.00	0.00	2.154	2.154	0.014	
144	39.6	0.00	0.00	2.126	2.126	0.014	
145	39.8	0.00	0.00	2.098	2.098	0.014	
146	40.0	0.00	0.00	2.070	2.070	0.014	
147	40.2	0.00	0.00	2.042	2.042	0.014	
148	40.4	0.00	0.00	2.014	2.014	0.014	
149	40.6	0.00	0.00	1.986	1.986	0.014	
150	40.8	0.00	0.00	1.958	1.958	0.014	
151	41.0	0.00	0.00	1.930	1.930	0.014	
152	41.2	0.00	0.00	1.902	1.902	0.014	
153	41.4	0.00	0.00	1.874	1.874	0.014	
154	41.6	0.00	0.00	1.846	1.846	0.014	

TABLE C.11c　100-Year Rain Garden Routing

Line No. 1	Time hr 2	Inflow, I_2 cfs 3	$I_1 + I_2$ cfs 4	$2S_1/\Delta T - O_1$ cfs 5	$2S_2/\Delta T + O_2$ cfs 6	Outflow cfs 7	Depth ft 8
155	41.8	0.00	0.00	1.818	1.818	0.014	−0.60
156	42.0	0.00	0.00	1.790	1.790	0.014	
157	42.2	0.00	0.00	1.762	1.762	0.014	
158	42.4	0.00	0.00	1.734	1.734	0.014	
159	42.6	0.00	0.00	1.706	1.706	0.014	
160	42.8	0.00	0.00	1.678	1.678	0.014	
161	43.0	0.00	0.00	1.650	1.650	0.014	
162	43.2	0.00	0.00	1.622	1.622	0.014	
163	43.4	0.00	0.00	1.594	1.594	0.014	
164	43.6	0.00	0.00	1.566	1.566	0.014	
165	43.8	0.00	0.00	1.538	1.538	0.014	
166	44.0	0.00	0.00	1.510	1.510	0.014	
167	44.2	0.00	0.00	1.482	1.482	0.014	
168	44.4	0.00	0.00	1.454	1.454	0.014	
169	44.6	0.00	0.00	1.426	1.426	0.014	
170	44.8	0.00	0.00	1.398	1.398	0.014	−0.80
171	45.0	0.00	0.00	1.370	1.370	0.014	
172	45.2	0.00	0.00	1.342	1.342	0.014	
173	45.4	0.00	0.00	1.314	1.314	0.014	
174	45.6	0.00	0.00	1.286	1.286	0.014	
175	45.8	0.00	0.00	1.258	1.258	0.014	
176	46.0	0.00	0.00	1.230	1.230	0.014	
177	46.2	0.00	0.00	1.202	1.202	0.014	
178	46.4	0.00	0.00	1.174	1.174	0.014	
179	46.6	0.00	0.00	1.146	1.146	0.014	
180	46.8	0.00	0.00	1.118	1.118	0.014	
181	47.0	0.00	0.00	1.090	1.090	0.014	−0.96

TABLE **C.11c** 100-Year Rain Garden Routing (*Continued*)

C.3.8 Summary

A rain garden acts as a detention basin and easily stores and infiltrates runoff into it from a roof, street, driveway, sidewalk, front yard, and itself. This meets requirements for runoff quantity and quality. It adds color by using native grasses, flowering plants, and shrubs. It allows amended soils and gravel to drain over some amount of time.

All runoff is stored in the gravel during the 2- and 10-year storms. However, a surface drain time of 20 hours during a 100-year event could be too long for some owners. Also, a rain garden that occupys most of a front yard may not please some owners. One alternative is to construct a storm sewer and design the garden for a less rare storm event. Another alternative is to use porous pavement for the street, driveway, and sidewalk with no runoff from them. This reduces a rain garden's size as shown in the next example.

C.4 Revised Rain Garden Example

How does its size change if the street, driveway, and sidewalk are constructed of porous concrete underlain by 9-in of gravel? Concrete has about 20 percent void space so 6-in stores 0.2×6 or 1.2 in of rain. The gravel stores $0.4 \times 9 = 3.6$ in of rain, a total of 4.8 in. They store all excess rain when its intensity is greater than a soil's infiltration capacity. Runoff into the rain garden will be from a garage roof and rain falling on a front lawn and rain garden.

C.4.1 Runoff Volume

Volume from a 24×32 ft garage is $768 \times 0.98 \times 7.2/12$ or 450 ft^3. Rain exceeding infiltration occurs in hours 10 and 11 as shown in Table C.1 is $1.75 + 0.99 - 2 \times 0.40 = 1.94$ in. Front-yard runoff volume is $60 \times 50 \times 1.94/12$ or 485 ft^3. Thus, total runoff volume is 935 ft^3.

C.4.2 Time of Concentration

Tc is estimated in Table C.12 as 0.09 hour from the front yard grass. Use Tc of 0.10 hour.

C.4.3 Runoff Hydrograph

CN is 40 for grass and 98 for roofs (USDA, 1986). Hydrograph variables are listed in Table C.13. In this revision, roof and front lawn are treated as separate areas.

Worksheet 2	
Overland (Sheet) Flow	**1**
Pathway Length, feet	30
Upstream Elevation, feet	100.6
Downstream Elevation, feet	100.0
Pathway Slope, feet per foot	0.0200
Manning's n	0.24
2-year, 24-hour Rainfall, inches	3.3
Flow Velocity, feet per second	0.089
Travel Time, minutes	5.6
Tc, hours	0.09

TABLE C.12 Tc for Revised Rain Garden

Hydrograph Development—Worksheet 5a

Project: Front Yard Rain Garden
Outline one: Present | Developed
Location: East Coast
Frequency (yr): 2, 10, 100
By:
Checked by:
Date:
Date:

Subarea Name 1	Drainage Area sq mi 2	Time of Concentration hr 3	Travel Time thru Subarea hr 4	Downstream Subarea Names 5	Travel Time Summation to Outlet 6	24-hr Rainfall in 7	Runoff Curve Number 8	Runoff in 9	AmQ sq mi-in 10	Initial Abstraction in 11	Ia/P 12
Garage Roof	0.000028	0.10	0.00	—	0.00	3.3	98	3.1	0.000086	0.041	0.012
Front Yard	0.000108	0.10	0.00	—	0.00	3.3	40	0.0	0.000001	1.279	0.388
Garage Roof	0.000028	0.10	0.00	—	0.00	5.1	98	4.9	0.000136	0.041	0.008
Front Yard	0.000108	0.10	0.00	—	0.00	5.1	40	0.3	0.000028	1.279	0.251
Garage Roof	0.000028	0.10	0.00	—	0.00	7.2	98	7.0	0.000195	0.041	0.006
Front Yard	0.000108	0.10	0.00	—	0.00	7.2	40	0.9	0.000099	1.279	0.178

Table C.13 Revised Rain Garden Hydrograph Variables

C.4.3.1 Two-Year Storm
Ordinates are listed in Table C.14 with Ia/P used as 0.1 and 0.3. Peak inflow rate is 0.056 cfs at hour 12.2 from rain falling on the roof and front yard.

C.4.3.2 10-Year Storm
Ordinates are listed in Table C.15 with Ia/P used as 0.1 and 0.3. Peak inflow rate is 0.107 cfs at hour 12.2 from rain falling on the roof and front yard.

C.4.3.3 100-Year Storm
Ordinates are listed in Table C.16 with Ia/P used as 0.1. Peak inflow rate is 0.195 cfs at hour 12.2 from rain falling on the roof and front yard.

C.4.4 Storage Volume
Use a 30 × 20-ft size with 4:1 slopes. Depth-areas are shown in Table C.17. Volume is 0.5 × 1.0 × (30 × 20 + 22 × 12) or 432 ft³. Excavate 3-ft of soil. The first foot has amended soil, the next 9-in. are a 40 percent porosity gravel layer. Gravel volume is 30 × 20 × 0.75 × 0.4 or 180 ft³. Total storage is 612 ft³. Volume each 0.2 ft is shown in Table C.18 with total storage of 602 ft³.

C.4.5 Outflow Rate
Outflow is infiltration rate of 0.40 in/h. Outflow rate is a constant value of 30 × 20 × 0.40/12 or 20.0 ft³ per hour or 20.0/60/60 or 0.006 cfs.

C.4.6 Routing Curve
A routing curve for this rain garden is calculated in Table C.19. The routing curve does not need to be plotted because outflow into the soil is a constant 0.006 cfs.

C.4.7 Hydrograph Routing
C.4.7.1 Two-Year Storm
This storm's hydrograph is routed in Table C.20. Maximum water depth in the gravel is 0.39 ft at hour 13.4. The gravel is empty of water at hour 26.0, 2 hours after the rain stops.

C.4.7.2 10-Year Storm
This storm's hydrograph is routed in Tables C.21a and b. Maximum water depth on the surface is 0.08 ft at hour12.0. Water is off the surface at hour 19.6. The gravel is empty of water at hour 32.2, 8.2 hours after the rain stops.

C.4.7.3 100-Year Storm
This storm's hydrograph is routed in Tables C.22a to c. Maximum water depth on the rain garden's surface is 0.72 ft at hour 17.4. Water is on the surface from hours 12.2 to 35.8, a total of 23.6 hours. The gravel is empty of water at hour 45.2, 21.2 hours after the rain stops.

Hydrograph Development—Worksheet 5b

Project: Front Yard Rain Garden Location: East Coast By: Date:

Outline one: Present [Developed] Frequency (yr): 2 Checked by: Date:

Select and enter hydrograph times in hours from exhibit 5-

Discharge at selected hydrograph times cfs

Subarea Name	Subarea Tc hr	ΣTt to Outlet hr	Ia/P	AmQ sq mi-in	11.0	11.3	11.6	11.9	12.0	12.2	12.4	12.6	12.8	13.0	13.2	13.4
1	2	3	4	5	6	7	8	9	10	11	12	13	14	15	16	17
Garage Roof	0.10	0.00	0.10	0.000086	29	38	57	172	241	662	345	191	101	83	68	62
					0.002	0.003	0.005	0.015	0.021	0.057	0.030	0.016	0.009	0.007	0.006	0.005
Front Yard	0.10	0.00	0.30	0.000001	0	0	0	48	106	597	368	221	125	106	89	83
					0.000	0.000	0.000	0.000	0.000	0.001	0.000	0.000	0.000	0.000	0.000	0.000
					0.000	0.000	0.000	0.000	0.000	0.000	0.000	0.000	0.000	0.000	0.000	0.000
Total					0.002	0.003	0.005	0.015	0.021	0.058	0.030	0.017	0.009	0.007	0.006	0.005

Select and enter hydrograph times in hours from exhibit 5-

Subarea Name	Subarea Tc hr	ΣTt to Outlet hr	Ia/P	AmQ sq mi-in	13.6	13.8	14.0	14.3	14.6	15.0	16.0	17.0	18.0	20.0	22.0	26.0
					6	7	8	9	10	11	12	13	14	15	16	17
Garage Roof	0.10	0.00	0.10	0.000086	58	54	50	44	41	37	27	21	16	13	11	0
					0.005	0.005	0.004	0.004	0.004	0.003	0.002	0.002	0.001	0.001	0.001	0.000
Front Yard	0.10	0.00	0.30	0.000001	79	74	69	62	59	54	40	32	25	20	17	0
					0.000	0.000	0.000	0.000	0.000	0.000	0.000	0.000	0.000	0.000	0.000	0.000
					0.000	0.000	0.000	0.000	0.000	0.000	0.000	0.000	0.000	0.000	0.000	0.000
Total					0.005	0.005	0.004	0.004	0.004	0.003	0.002	0.002	0.001	0.001	0.001	0.000

TABLE C.14 2-Year Revised Rain Garden Hydrograph Ordinates

Hydrograph Development—Worksheet 5b

Project: Front Yard Rain Garden Location East Coast By: Date:

Outline one: Present | **Developed** Frequency (yr): 10 Checked by: Date:

Basic Watershed Data Used — Select and enter hydrograph times in hours from exhibit 5-

Discharge at selected hydrograph times (cfs)

Subarea Name	Subarea Tc hr	ΣTt to Outlet hr	Ia/P	AmQ sq mi-in	11.0	11.3	11.6	11.9	12.0	12.2	12.4	12.6	12.8	13.0	13.2	13.4
1	2	3	4	5	6	7	8	9	10	11	12	13	14	15	16	17
Garage Roof	0.10	0.00	0.10	0.000136	29	38	57	172	241	662	345	191	101	83	68	62
					0.004	0.005	0.008	0.023	0.033	0.090	0.047	0.026	0.014	0.011	0.009	0.008
Front Yard	0.10	0.00	0.30	0.000028	0	0	0	48	106	597	368	221	125	106	89	83
					0.000	0.000	0.000	0.001	0.003	0.017	0.010	0.006	0.004	0.003	0.002	0.002
					0.000	0.000	0.000	0.000	0.000	0.000	0.000	0.000	0.000	0.000	0.000	0.000
Total					0.004	0.005	0.008	0.025	0.036	0.107	0.057	0.032	0.017	0.014	0.012	0.011

Select and enter hydrograph times in hours from exhibit 5-

Subarea Name	Subarea Tc hr	ΣTt to Outlet hr	Ia/P	AmQ sq mi-in	13.6	13.8	14.0	14.3	14.6	15.0	16.0	17.0	18.0	20.0	22.0	26.0
1	2	3	4	5	6	7	8	9	10	11	12	13	14	15	16	17
Garage Roof	0.10	0.00	0.10	0.000136	58	54	50	44	41	37	27	21	16	13	11	0
					0.008	0.007	0.007	0.006	0.006	0.005	0.004	0.003	0.002	0.002	0.001	0.000
Front Yard	0.10	0.00	0.30	0.000028	79	74	69	62	59	54	40	32	25	20	17	0
					0.002	0.002	0.002	0.002	0.002	0.002	0.001	0.001	0.001	0.001	0.000	0.000
					0.000	0.000	0.000	0.000	0.000	0.000	0.000	0.000	0.000	0.000	0.000	0.000
Total					0.010	0.009	0.009	0.008	0.007	0.007	0.005	0.004	0.003	0.002	0.002	0.000

Table C.15 10-Year Revised Rain Garden Hydrograph Ordinates

Hydrograph Development—Worksheet 5b

Project: Front Yard Rain Garden
Outline one: Present [Developed]

Location: East Coast
Frequency (yr): 100

By:
Checked by:

Date:
Date:

Basic Watershed Data Used

Discharge at selected hydrograph times — cfs

Select and enter hydrograph times in hours from exhibit 5-

Subarea Name 1	Subarea Tc hr 2	ΣTt to Outlet hr 3	Ia/P 4	AmQ sq mi-in 5	11.0 / 6	11.3 / 7	11.6 / 8	11.9 / 9	12.0 / 10	12.2 / 11	12.4 / 12	12.6 / 13	12.8 / 14	13.0 / 15	13.2 / 16	13.4 / 17
Garage Roof	0.10	0.00	0.10	0.000195	29	38	57	172	241	662	345	191	101	83	68	62
					0.006	0.007	0.011	0.034	0.047	0.129	0.067	0.037	0.020	0.016	0.013	0.012
Front Yard	0.10	0.00	0.10	0.000099	29	38	57	172	241	662	345	191	101	83	68	62
					0.003	0.004	0.006	0.017	0.024	0.066	0.034	0.019	0.010	0.008	0.007	0.006
					0.000	0.000	0.000	0.000	0.000	0.000	0.000	0.000	0.000	0.000	0.000	0.000
Total					0.009	0.011	0.017	0.051	0.071	0.195	0.101	0.056	0.030	0.024	0.020	0.018

Select and enter hydrograph times in hours from exhibit 5-

Subarea Name 1	Subarea Tc hr 2	ΣTt to Outlet hr 3	Ia/P 4	AmQ sq mi-in 5	13.6 / 6	13.8 / 7	14.0 / 8	14.3 / 9	14.6 / 10	15.0 / 11	16.0 / 12	17.0 / 13	18.0 / 14	20.0 / 15	22.0 / 16	26.0 / 17
Garage Roof	0.10	0.00	0.10	0.000195	58	54	50	44	41	37	27	21	16	13	11	0
					0.011	0.011	0.010	0.009	0.008	0.007	0.005	0.004	0.003	0.003	0.002	0.000
Front Yard	0.10	0.00	0.10	0.000099	58	54	50	44	41	37	27	21	16	13	11	0
					0.006	0.005	0.005	0.004	0.004	0.004	0.003	0.002	0.002	0.001	0.001	0.000
					0.000	0.000	0.000	0.000	0.000	0.000	0.000	0.000	0.000	0.000	0.000	0.000
Total					0.017	0.016	0.015	0.013	0.012	0.011	0.008	0.006	0.005	0.004	0.003	0.000

TABLE C.16 100-Year Revised Rain Garden Hydrograph Ordinates

Depth ft	Length ft	Width ft	Area sq ft
−0.75	30.0	20.0	240
−0.60	30.0	20.0	240
−0.40	30.0	20.0	240
−0.20	30.0	20.0	240
0.00	30.0	20.0	240
0.00	22.0	12.0	264
0.20	23.6	13.6	321
0.40	25.2	15.2	383
0.60	26.8	16.8	450
0.80	28.4	18.4	523
1.00	30.0	20.0	600

TABLE C.17   Depth-Area Calculations for Revised Rain Garden

Depth ft	Area sq ft	Aver. Area sq ft	Δ Depth ft	Δ Volume cu ft	Tot. Vol. cu ft
−0.75	240				0
		240	0.15	36	
−0.60	240				36
		240	0.20	48	
−0.40	240				84
		240	0.20	48	
−0.20	240				132
		240	0.20	48	
0.00	240				180
		252	0.00	0	
0.00	264				180
		292.5	0.20	59	
0.20	321				239
		352	0.20	70	
0.40	383				309
		416.5	0.20	83	
0.60	450				392
		486.5	0.20	97	
0.80	523				490
		561.5	0.20	112	
1.00	600				602

TABLE C.18   Depth-Storage Calculations for Revised Rain Garden

Routing Period = 0.20 hr

$2S/\Delta T + O = (2S \text{ (cu ft)}/0.2 \text{ hr}) + O \text{ (cfs)} \times (1 \text{ hr}/60 \text{ min}) \times (1 \text{ min}/60 \text{ sec}) = 0.00278 *S + O$

Depth ft 1	S cu ft 2	O cfs 3	$2S/\Delta T + O$ cfs 4
−0.75	0	0.006	0.01
−0.60	36	0.006	0.11
−0.40	84	0.006	0.24
−0.20	132	0.006	0.37
0.00	180	0.006	0.51
0.20	239	0.006	0.67
0.40	309	0.006	0.87
0.60	392	0.006	1.10
0.80	490	0.006	1.37
1.00	602	0.006	1.68

TABLE C.19 Routing Curve Calculations for Revised Rain Garden

Line No. 1	Time hr 2	Inflow, I_2 cfs 3	$(I_1 + I_2) +$ cfs 4	$(2S_1/\Delta T - O_1) =$ cfs 5	$(2S_2/\Delta T + O_2)$ cfs 6	Outflow cfs 7	Depth ft 8
1	11.0	0.002	0.007	0.012	0.031	0.002	−0.73
2	11.2	0.003	0.005	0.027	0.032	0.003	
3	11.4	0.004	0.007	0.026	0.033	0.004	
4	11.6	0.005	0.009	0.025	0.034	0.005	
5	11.8	0.011	0.016	0.024	0.040	0.006	
6	12.0	0.021	0.032	0.029	0.061	0.006	
7	12.2	0.058	0.079	0.050	0.129	0.006	
8	12.4	0.030	0.088	0.118	0.206	0.006	
9	12.6	0.017	0.047	0.195	0.242	0.006	
10	12.8	0.009	0.026	0.231	0.257	0.006	
11	13.0	0.007	0.016	0.246	0.262	0.006	
12	13.2	0.006	0.013	0.251	0.264	0.006	
13	13.4	0.005	0.011	0.253	0.264	0.006	−0.36 peak
14	13.6	0.005	0.010	0.253	0.263	0.006	
15	13.8	0.005	0.010	0.252	0.262	0.006	
16	14.0	0.004	0.009	0.251	0.260	0.006	
17	14.2	0.004	0.008	0.249	0.257	0.006	
18	14.4	0.004	0.008	0.246	0.254	0.006	
19	14.6	0.004	0.008	0.243	0.251	0.006	
20	14.8	0.004	0.008	0.240	0.248	0.006	
21	15.0	0.003	0.007	0.237	0.244	0.006	

TABLE C.20 2-Year Routing for Revised Rain Garden

Line No. 1	Time hr 2	Inflow, I_2 cfs 3	$(I_1 + I_2)$ cfs 4	$+ (2S_1/\Delta T - 0_1) =$ cfs 5	$(2S_2/\Delta T + 0_2)$ cfs 6	Outflow cfs 7	Depth ft 8
22	15.2	0.003	0.006	0.233	0.239	0.006	−0.40
23	15.4	0.003	0.006	0.228	0.234	0.006	
24	15.6	0.003	0.006	0.223	0.229	0.006	
25	15.8	0.003	0.006	0.218	0.224	0.006	
26	16.0	0.002	0.005	0.213	0.218	0.006	
27	16.2	0.002	0.004	0.207	0.211	0.006	
28	16.4	0.002	0.004	0.200	0.204	0.006	
29	16.6	0.002	0.004	0.193	0.197	0.006	
30	16.8	0.002	0.004	0.186	0.190	0.006	
31	17.0	0.002	0.004	0.179	0.183	0.006	
32	17.2	0.002	0.004	0.172	0.176	0.006	
33	17.4	0.002	0.004	0.165	0.169	0.006	
34	17.6	0.002	0.004	0.158	0.162	0.006	
35	17.8	0.002	0.004	0.151	0.155	0.006	
36	18.0	0.001	0.003	0.144	0.147	0.006	
37	18.2	0.001	0.002	0.136	0.138	0.006	
38	18.4	0.001	0.002	0.127	0.129	0.006	
39	18.6	0.001	0.002	0.118	0.120	0.006	
40	18.8	0.001	0.002	0.109	0.111	0.006	−0.60
41	19.0	0.001	0.002	0.100	0.102	0.006	
42	19.2	0.001	0.002	0.091	0.093	0.006	
43	19.4	0.001	0.002	0.082	0.084	0.006	
44	19.6	0.001	0.002	0.073	0.075	0.006	
45	19.8	0.001	0.002	0.064	0.066	0.006	
46	20.0	0.001	0.002	0.055	0.057	0.006	
47	20.2	0.001	0.002	0.046	0.048	0.006	
48	20.4	0.001	0.002	0.037	0.039	0.006	
49	20.6	0.001	0.002	0.028	0.030	0.006	
50	20.8	0.001	0.002	0.019	0.021	0.006	
51	21.0	0.001	0.002	0.010	0.012	0.006	
52	21.2	0.001	0.002	0.001	0.003	0.002	
53	21.4	0.001	0.002	0.000	0.002	0.001	
54	21.6	0.001	0.002	0.000	0.002	0.001	
55	21.8	0.001	0.002	0.000	0.002	0.001	
56	22.0	0.001	0.002	0.000	0.002	0.001	
57	22.2	0.001	0.002	0.000	0.002	0.001	
58	22.4	0.001	0.002	0.000	0.002	0.001	
59	22.6	0.001	0.002	0.000	0.002	0.001	
60	22.8	0.001	0.002	0.000	0.002	0.001	
61	23.0	0.001	0.002	0.000	0.002	0.001	−0.74

TABLE C.20   2-Year Routing for Revised Rain Garden (*Continued*)

Line No. 1	Time hr 2	Inflow, I_2 cfs 3	$(I_1 + I_2) +$ cfs 4	$(2S_1/\Delta T - 0_1) =$ cfs 5	$(2S_2/\Delta T + 0_2)$ cfs 6	Outflow cfs 7	Depth ft 8
1	11.0	0.004	0.009	0.001	0.010	0.004	−0.73
2	11.2	0.005	0.009	0.002	0.011	0.006	
3	11.4	0.006	0.011	0.000	0.011	0.006	
4	11.6	0.008	0.014	0.000	0.014	0.006	
5	11.8	0.020	0.028	0.003	0.031	0.006	
6	12.0	0.036	0.056	0.020	0.076	0.006	
7	12.2	0.107	0.143	0.065	0.208	0.006	
8	12.4	0.057	0.164	0.197	0.361	0.006	−0.20
9	12.6	0.032	0.089	0.350	0.439	0.006	
10	12.8	0.017	0.049	0.428	0.477	0.006	
11	13.0	0.014	0.031	0.466	0.497	0.006	
12	13.2	0.012	0.026	0.486	0.512	0.006	0.00
13	13.4	0.011	0.023	0.501	0.524	0.006	
14	13.6	0.010	0.021	0.513	0.534	0.006	
15	13.8	0.009	0.019	0.523	0.542	0.006	
16	14.0	0.008	0.017	0.531	0.548	0.006	
17	14.2	0.008	0.016	0.537	0.553	0.006	
18	14.4	0.008	0.016	0.542	0.558	0.006	
19	14.6	0.007	0.015	0.547	0.562	0.006	
20	14.8	0.007	0.014	0.551	0.565	0.006	
21	15.0	0.007	0.014	0.554	0.568	0.006	
22	15.2	0.007	0.014	0.557	0.571	0.006	
23	15.4	0.006	0.013	0.560	0.572	0.006	
24	15.6	0.006	0.012	0.561	0.573	0.006	
25	15.8	0.005	0.011	0.562	0.574	0.006	0.08 peak
26	16.0	0.005	0.010	0.563	0.573	0.006	
27	16.2	0.005	0.010	0.562	0.572	0.006	
28	16.4	0.005	0.010	0.561	0.571	0.006	
29	16.6	0.005	0.010	0.560	0.570	0.006	
30	16.8	0.005	0.010	0.559	0.569	0.006	
31	17.0	0.004	0.009	0.558	0.567	0.006	
32	17.2	0.004	0.008	0.556	0.564	0.006	
33	17.4	0.004	0.007	0.553	0.560	0.006	
34	17.6	0.004	0.008	0.549	0.557	0.006	

TABLE C.21a 10-Year Routing for Revised Rain Garden

Line No. 1	Time hr 2	Inflow, I_2 cfs 3	$(I_1 + I_2) +$ cfs 4	$(2S_1/\Delta T - O_1) =$ cfs 5	$(2S_2/\Delta T + O_2)$ cfs 6	Outflow cfs 7	Depth ft 8
35	17.8	0.004	0.008	0.546	0.554	0.006	
36	18.0	0.003	0.007	0.543	0.549	0.006	
37	18.2	0.003	0.006	0.538	0.544	0.006	
38	18.4	0.003	0.006	0.533	0.539	0.006	
39	18.6	0.003	0.006	0.528	0.534	0.006	
40	18.8	0.003	0.006	0.523	0.529	0.006	
41	19.0	0.003	0.006	0.518	0.524	0.006	
42	19.2	0.003	0.006	0.513	0.519	0.006	
43	19.4	0.003	0.006	0.508	0.514	0.006	
44	19.6	0.002	0.005	0.503	0.508	0.006	0.00
45	19.8	0.002	0.004	0.497	0.501	0.006	
46	20.0	0.002	0.004	0.490	0.494	0.006	
47	20.2	0.002	0.004	0.483	0.487	0.006	
48	20.4	0.002	0.004	0.476	0.480	0.006	
49	20.6	0.002	0.004	0.469	0.473	0.006	
50	20.8	0.002	0.004	0.462	0.466	0.006	
51	21.0	0.002	0.004	0.455	0.459	0.006	
52	21.2	0.002	0.004	0.448	0.452	0.006	
53	21.4	0.002	0.004	0.441	0.445	0.006	
54	21.6	0.002	0.004	0.434	0.438	0.006	
55	21.8	0.002	0.004	0.427	0.431	0.006	
56	22.0	0.002	0.004	0.420	0.424	0.006	
57	22.2	0.002	0.004	0.413	0.417	0.006	
58	22.4	0.002	0.004	0.406	0.410	0.006	
59	22.6	0.002	0.004	0.399	0.403	0.006	
60	22.8	0.002	0.004	0.392	0.396	0.006	
61	23.0	0.002	0.004	0.385	0.389	0.006	−0.17

TABLE C.21a   10-Year Routing for Revised Rain Garden (*Continued*)

Line No. 1	Time hr 2	Inflow, I_2 cfs 3	$(I_1 + I_2) +$ cfs 4	$(2S_1/\Delta T - O_1) =$ cfs 5	$(2S_2/\Delta T + O_2)$ cfs 6	Outflow cfs 7	Depth ft 8
61	23.0	0.002	0.004	0.385	0.389	0.006	−0.17
62	23.2	0.003	0.005	0.378	0.383	0.006	
63	23.4	0.003	0.006	0.372	0.378	0.006	
64	23.6	0.003	0.006	0.367	0.373	0.006	
65	23.8	0.003	0.006	0.362	0.368	0.006	−0.20
66	24.0	0.003	0.006	0.357	0.363	0.006	
67	24.2	0.003	0.006	0.352	0.358	0.006	
68	24.4	0.003	0.006	0.347	0.353	0.006	
69	24.6	0.003	0.006	0.342	0.348	0.006	
70	24.8	0.003	0.006	0.337	0.343	0.006	
71	25.0	0.003	0.006	0.332	0.338	0.006	
72	25.2	0.003	0.006	0.327	0.333	0.006	
73	25.4	0.003	0.006	0.322	0.328	0.006	
74	25.6	0.003	0.006	0.317	0.323	0.006	
75	25.8	0.003	0.006	0.312	0.318	0.006	
76	26.0	0.002	0.005	0.307	0.312	0.006	
77	26.2	0.002	0.004	0.301	0.305	0.006	
78	26.4	0.002	0.004	0.294	0.298	0.006	
79	26.6	0.002	0.004	0.287	0.291	0.006	
80	26.8	0.002	0.004	0.280	0.284	0.006	
81	27.0	0.001	0.003	0.273	0.276	0.006	
82	27.2	0.001	0.002	0.265	0.267	0.006	
83	27.4	0.001	0.002	0.256	0.258	0.006	
84	27.6	0.001	0.002	0.247	0.249	0.006	
85	27.8	0.001	0.002	0.238	0.240	0.006	−0.40
86	28.0	0.000	0.001	0.229	0.230	0.006	
87	28.2	0.000	0.000	0.219	0.219	0.006	
88	28.4	0.000	0.000	0.208	0.208	0.006	
89	28.6	0.000	0.000	0.197	0.197	0.006	
90	28.8	0.000	0.000	0.186	0.186	0.006	
91	29.0	0.000	0.000	0.175	0.175	0.006	
92	29.2	0.000	0.000	0.164	0.164	0.006	
93	29.4	0.000	0.000	0.153	0.153	0.006	
94	29.6	0.000	0.000	0.142	0.142	0.006	

TABLE C.21b 10-Year Routing for Revised Rain Garden

Line No. 1	Time hr 2	Inflow, I_2 cfs 3	$(I_1 + I_2) + (2S_1/\Delta T - 0_1) =$ cfs 4	cfs 5	$(2S_2/\Delta T + 0_2)$ cfs 6	Outflow cfs 7	Depth ft 8
95	29.8	0.000	0.000	0.131	0.131	0.006	
96	30.0	0.000	0.000	0.120	0.120	0.006	
97	30.2	0.000	0.000	0.109	0.109	0.006	−0.60
98	30.4	0.000	0.000	0.098	0.098	0.006	
99	30.6	0.000	0.000	0.087	0.087	0.006	
100	30.8	0.000	0.000	0.076	0.076	0.006	
101	31.0	0.000	0.000	0.065	0.065	0.006	
102	31.2	0.000	0.000	0.054	0.054	0.006	
103	31.4	0.000	0.000	0.043	0.043	0.006	
104	31.6	0.000	0.000	0.032	0.032	0.006	
105	31.8	0.000	0.000	0.021	0.021	0.006	
106	32.0	0.000	0.000	0.010	0.010	0.005	
107	32.2	0.000	0.000	0.000	0.000	0.000	−0.75
108	32.4	0.000	0.000	0.000	0.000	0.000	
109	32.6	0.000	0.000	0.000	0.000	0.000	
110	32.8	0.000	0.000	0.000	0.000	0.000	
111	33.0	0.000	0.000	0.000	0.000	0.000	
112	33.2	0.000	0.000	0.000	0.000	0.000	
113	33.4	0.000	0.000	0.000	0.000	0.000	
114	33.6	0.000	0.000	0.000	0.000	0.000	
115	33.8	0.000	0.000	0.000	0.000	0.000	
116	34.0	0.000	0.000	0.000	0.000	0.000	
117	34.2	0.000	0.000	0.000	0.000	0.000	
118	34.4	0.000	0.000	0.000	0.000	0.000	
119	34.6	0.000	0.000	0.000	0.000	0.000	
120	34.8	0.000	0.000	0.000	0.000	0.000	
121	35.0	0.000	0.000	0.000	0.000	0.000	−0.75

TABLE C.21b 10-Year Routing for Revised Rain Garden (*Continued*)

Line No. 1	Time hr 2	Inflow, I_2 cfs 3	$(I_1 + I_2) +$ cfs 4	$(2S_1/\Delta T - 0_1) =$ cfs 5	$(2S_2/\Delta T + 0_2)$ cfs 6	Outflow cfs 7	Depth ft 8
1	11.0	0.009	0.017	0.001	0.018	0.006	−0.73
2	11.2	0.011	0.020	0.007	0.027	0.006	
3	11.4	0.013	0.024	0.016	0.040	0.006	
4	11.6	0.017	0.030	0.029	0.059	0.006	
5	11.8	0.042	0.059	0.048	0.107	0.006	
6	12.0	0.071	0.113	0.096	0.209	0.006	
7	12.2	0.195	0.266	0.198	0.464	0.006	−0.07
8	12.4	0.101	0.296	0.453	0.749	0.006	
9	12.6	0.056	0.157	0.738	0.895	0.006	
10	12.8	0.030	0.086	0.884	0.970	0.006	
11	13.0	0.024	0.054	0.959	1.013	0.006	
12	13.2	0.020	0.044	1.002	1.046	0.006	
13	13.4	0.018	0.038	1.035	1.073	0.006	
14	13.6	0.017	0.035	1.062	1.097	0.006	0.60
15	13.8	0.016	0.033	1.086	1.119	0.006	
16	14.0	0.015	0.031	1.108	1.139	0.006	
17	14.2	0.014	0.029	1.128	1.157	0.006	
18	14.4	0.013	0.027	1.146	1.173	0.006	
19	14.6	0.012	0.025	1.162	1.187	0.006	
20	14.8	0.012	0.024	1.176	1.200	0.006	
21	15.0	0.011	0.023	1.189	1.212	0.006	
22	15.2	0.010	0.021	1.201	1.222	0.006	
23	15.4	0.010	0.020	1.211	1.232	0.006	
24	15.6	0.009	0.019	1.221	1.240	0.006	
25	15.8	0.009	0.018	1.229	1.246	0.006	
26	16.0	0.008	0.017	1.235	1.252	0.006	
27	16.2	0.008	0.016	1.241	1.257	0.006	
28	16.4	0.007	0.015	1.246	1.261	0.006	
29	16.6	0.007	0.014	1.250	1.264	0.006	
30	16.8	0.006	0.013	1.253	1.266	0.006	
31	17.0	0.006	0.012	1.255	1.267	0.006	
32	17.2	0.006	0.012	1.256	1.268	0.006	
33	17.4	0.005	0.011	1.257	1.268	0.006	0.72 peak
34	17.6	0.005	0.010	1.257	1.267	0.006	

TABLE C.22a 100-Year Routing for Revised Rain Garden

Line No. 1	Time hr 2	Inflow, I_2 cfs 3	$(I_1 + I_2) +$ cfs 4	$(2S_1/\Delta T - 0_1) =$ cfs 5	$(2S_2/\Delta T + 0_2)$ cfs 6	Outflow cfs 7	Depth ft 8
35	17.8	0.005	0.010	1.256	1.266	0.006	
36	18.0	0.005	0.010	1.255	1.265	0.006	
37	18.2	0.005	0.010	1.254	1.264	0.006	
38	18.4	0.005	0.010	1.253	1.263	0.006	
39	18.6	0.004	0.009	1.252	1.261	0.006	
40	18.8	0.004	0.008	1.250	1.258	0.006	
41	19.0	0.004	0.008	1.247	1.255	0.006	
42	19.2	0.004	0.008	1.244	1.252	0.006	
43	19.4	0.004	0.008	1.241	1.249	0.006	
44	19.6	0.004	0.008	1.238	1.246	0.006	
45	19.8	0.004	0.008	1.235	1.243	0.006	
46	20.0	0.004	0.008	1.232	1.240	0.006	
47	20.2	0.004	0.008	1.229	1.237	0.006	
48	20.4	0.003	0.007	1.226	1.233	0.006	
49	20.6	0.003	0.006	1.222	1.228	0.006	
50	20.8	0.003	0.006	1.217	1.223	0.006	
51	21.0	0.003	0.006	1.212	1.218	0.006	
52	21.2	0.003	0.006	1.207	1.213	0.006	
53	21.4	0.003	0.006	1.202	1.208	0.006	
54	21.6	0.003	0.006	1.197	1.203	0.006	
55	21.8	0.003	0.006	1.192	1.198	0.006	
56	22.0	0.003	0.006	1.187	1.193	0.006	
57	22.2	0.003	0.006	1.182	1.188	0.006	
58	22.4	0.003	0.006	1.177	1.183	0.006	
59	22.6	0.003	0.006	1.172	1.178	0.006	
60	22.8	0.003	0.006	1.167	1.173	0.006	
61	23.0	0.003	0.006	1.162	1.168	0.006	0.65

TABLE C.22a   100-Year Routing for Revised Rain Garden (*Continued*)

Line No. 1	Time hr 2	Inflow, I_2 cfs 3	$(I_1 + I_2) +$ cfs 4	$(2S_1/\Delta T - O_1) =$ cfs 5	$(2S_2/\Delta T + O_2)$ cfs 6	Outflow cfs 7	Depth ft 8
61	23.0	0.003	0.006	1.162	1.168	0.006	0.65
62	23.2	0.003	0.006	1.157	1.163	0.006	
63	23.4	0.003	0.006	1.152	1.157	0.006	
64	23.6	0.003	0.005	1.146	1.152	0.006	
65	23.8	0.002	0.005	1.141	1.146	0.006	
66	24.0	0.002	0.005	1.135	1.140	0.006	
67	24.2	0.002	0.005	1.129	1.133	0.006	
68	24.4	0.002	0.004	1.122	1.126	0.006	
69	24.6	0.002	0.004	1.115	1.119	0.006	
70	24.8	0.001	0.003	1.108	1.111	0.006	
71	25.0	0.001	0.002	1.100	1.102	0.006	0.60
72	25.2	0.001	0.002	1.091	1.093	0.006	
73	25.4	0.001	0.001	1.082	1.083	0.006	
74	25.6	0.000	0.001	1.072	1.073	0.006	
75	25.8	0.000	0.001	1.062	1.063	0.006	
76	26.0	0.000	0.000	1.052	1.052	0.006	
77	26.2	0.000	0.000	1.041	1.041	0.006	
78	26.4	0.000	0.000	1.030	1.030	0.006	
79	26.6	0.000	0.000	1.019	1.019	0.006	
80	26.8	0.000	0.000	1.008	1.008	0.006	
81	27.0	0.000	0.000	0.997	0.997	0.006	
82	27.2	0.000	0.000	0.986	0.986	0.006	
83	27.4	0.000	0.000	0.975	0.975	0.006	
84	27.6	0.000	0.000	0.964	0.964	0.006	
85	27.8	0.000	0.000	0.953	0.953	0.006	
86	28.0	0.000	0.000	0.942	0.942	0.006	
87	28.2	0.000	0.000	0.931	0.931	0.006	
88	28.4	0.000	0.000	0.920	0.920	0.006	
89	28.6	0.000	0.000	0.909	0.909	0.006	
90	28.8	0.000	0.000	0.898	0.898	0.006	
91	29.0	0.000	0.000	0.887	0.887	0.006	
92	29.2	0.000	0.000	0.876	0.876	0.006	
93	29.4	0.000	0.000	0.865	0.865	0.006	0.40

TABLE **C.22b**   100-Year Routing for Revised Rain Garden

Line No. 1	Time hr 2	Inflow, I_2 cfs 3	$(I_1 + I_2) +$ cfs 4	$(2S_1/\Delta T - O_1) =$ cfs 5	$(2S_2/\Delta T + O_2)$ cfs 6	Outflow cfs 7	Depth ft 8
94	29.6	0.000	0.000	0.854	0.854	0.006	
95	29.8	0.000	0.000	0.843	0.843	0.006	
96	30.0	0.000	0.000	0.832	0.832	0.006	
97	30.2	0.000	0.000	0.821	0.821	0.006	
98	30.4	0.000	0.000	0.810	0.810	0.006	
99	30.6	0.000	0.000	0.799	0.799	0.006	
100	30.8	0.000	0.000	0.788	0.788	0.006	
101	31.0	0.000	0.000	0.777	0.777	0.006	
102	31.2	0.000	0.000	0.766	0.766	0.006	
103	31.4	0.000	0.000	0.755	0.755	0.006	
104	31.6	0.000	0.000	0.744	0.744	0.006	
105	31.8	0.000	0.000	0.733	0.733	0.006	
106	32.0	0.000	0.000	0.722	0.722	0.006	
107	32.2	0.000	0.000	0.711	0.711	0.006	
108	32.4	0.000	0.000	0.700	0.700	0.006	
109	32.6	0.000	0.000	0.689	0.689	0.006	
110	32.8	0.000	0.000	0.678	0.678	0.006	
111	33.0	0.000	0.000	0.667	0.667	0.006	0.20
112	33.2	0.000	0.000	0.656	0.656	0.006	
113	33.4	0.000	0.000	0.645	0.645	0.006	
114	33.6	0.000	0.000	0.634	0.634	0.006	
115	33.8	0.000	0.000	0.623	0.623	0.006	
116	34.0	0.000	0.000	0.612	0.612	0.006	
117	34.2	0.000	0.000	0.601	0.601	0.006	
118	34.4	0.000	0.000	0.590	0.590	0.006	
119	34.6	0.000	0.000	0.579	0.579	0.006	
120	34.8	0.000	0.000	0.568	0.568	0.006	
121	35.0	0.000	0.000	0.557	0.557	0.006	0.06

TABLE C.22b 100-Year Routing for Revised Rain Garden (*Continued*)

Line No. 1	Time hr 2	Inflow, I_2 cfs 3	$(I_1 + I_2) +$ cfs 4	$(2S_1/\Delta T - O_1) =$ cfs 5	$(2S_2/\Delta T + O_2)$ cfs 6	Outflow cfs 7	Depth ft 8
121	35.0	0.000	0.000	0.557	0.557	0.006	0.06
122	35.2	0.000	0.000	0.546	0.546	0.006	
123	35.4	0.000	0.000	0.535	0.535	0.006	
124	35.6	0.000	0.000	0.524	0.524	0.006	
125	35.8	0.000	0.000	0.513	0.513	0.006	0.00
126	36.0	0.000	0.000	0.502	0.502	0.006	
127	36.2	0.000	0.000	0.491	0.491	0.006	
128	36.4	0.000	0.000	0.480	0.480	0.006	
129	36.6	0.000	0.000	0.469	0.469	0.006	
130	36.8	0.000	0.000	0.458	0.458	0.006	
131	37.0	0.000	0.000	0.447	0.447	0.006	
132	37.2	0.000	0.000	0.436	0.436	0.006	
133	37.4	0.000	0.000	0.425	0.425	0.006	
134	37.6	0.000	0.000	0.414	0.414	0.006	
135	37.8	0.000	0.000	0.403	0.403	0.006	
136	38.0	0.000	0.000	0.392	0.392	0.006	
137	38.2	0.000	0.000	0.381	0.381	0.006	
138	38.4	0.000	0.000	0.370	0.370	0.006	−0.20
139	38.6	0.000	0.000	0.359	0.359	0.006	
140	38.8	0.000	0.000	0.348	0.348	0.006	
141	39.0	0.000	0.000	0.337	0.337	0.006	
142	39.2	0.000	0.000	0.326	0.326	0.006	
143	39.4	0.000	0.000	0.315	0.315	0.006	
144	39.6	0.000	0.000	0.304	0.304	0.006	
145	39.8	0.000	0.000	0.293	0.293	0.006	
146	40.0	0.000	0.000	0.282	0.282	0.006	
147	40.2	0.000	0.000	0.271	0.271	0.006	
148	40.4	0.000	0.000	0.260	0.260	0.006	
149	40.6	0.000	0.000	0.249	0.249	0.006	
150	40.8	0.000	0.000	0.238	0.238	0.006	−0.40
151	41.0	0.000	0.000	0.227	0.227	0.006	
152	41.2	0.000	0.000	0.216	0.216	0.006	
153	41.4	0.000	0.000	0.205	0.205	0.006	

TABLE C.22c   100-Year Routing for Revised Rain Garden

Line No. 1	Time hr 2	Inflow, I_2 cfs 3	$(I_1 + I_2) +$ cfs 4	$(2S_1/\Delta T - O_1) =$ cfs 5	$(2S_2/\Delta T + O_2)$ cfs 6	Outflow cfs 7	Depth ft 8
154	41.6	0.000	0.000	0.194	0.194	0.006	
155	41.8	0.000	0.000	0.183	0.183	0.006	
156	42.0	0.000	0.000	0.172	0.172	0.006	
157	42.2	0.000	0.000	0.161	0.161	0.006	
158	42.4	0.000	0.000	0.150	0.150	0.006	
159	42.6	0.000	0.000	0.139	0.139	0.006	
160	42.8	0.000	0.000	0.128	0.128	0.006	
161	43.0	0.000	0.000	0.117	0.117	0.006	
162	43.2	0.000	0.000	0.106	0.106	0.006	−0.60
163	43.4	0.000	0.000	0.095	0.095	0.006	
164	43.6	0.000	0.000	0.084	0.084	0.006	
165	43.8	0.000	0.000	0.073	0.073	0.006	
166	44.0	0.000	0.000	0.062	0.062	0.006	
167	44.2	0.000	0.000	0.051	0.051	0.006	
168	44.4	0.000	0.000	0.040	0.040	0.006	
169	44.6	0.000	0.000	0.029	0.029	0.006	
170	44.8	0.000	0.000	0.018	0.018	0.006	
171	45.0	0.000	0.000	0.007	0.007	0.004	
172	45.2	0.000	0.000	0.000	0.000	0.000	−0.75
173	45.4	0.000	0.000	0.000	0.000	0.000	
174	45.6	0.000	0.000	0.000	0.000	0.000	
175	45.8	0.000	0.000	0.000	0.000	0.000	
176	46.0	0.000	0.000	0.000	0.000	0.000	
177	46.2	0.000	0.000	0.000	0.000	0.000	
178	46.4	0.000	0.000	0.000	0.000	0.000	
179	46.6	0.000	0.000	0.000	0.000	0.000	
180	46.8	0.000	0.000	0.000	0.000	0.000	
181	47.0	0.000	0.000	0.000	0.000	0.000	−0.75

TABLE C.22c   100-Year Routing for Revised Rain Garden (*Continued*)

C.4.8 Summary

With a porous portland cement concrete street, sidewalk, and driveway, a front yard rain garden is reduced from 50×30 ft to 30×20 ft. It is underlain by 9 in of gravel rather than 1.5 ft. Also, the rain garden is empty on the surface 11.8 hours after the rain stops rather than one day as in the original rain garden problem. All street, driveway, and sidewalk runoff and pollutants are infiltrated into the pavements in the revised rain garden alternative and stored in the gravel and native soil. If a 30-ft by 20-ft rain garden is still too large, another alternative is to use a gravel layer 1.5 ft deep and reduce its size. Another alternative is to use a greenroof on the garage. Then the only runoff would be from the front yard itself plus reduced runoff from the roof.

Is this a new development or is it an existing neighborhood where the street, sidewalks, and driveways are being replaced due to age or condition? Who pays the cost of these?

A problem arises in water-rights states. Users are entitled to a historic and legal fair share of available water, both surface and groundwater. If all runoff is infiltrated, will some users lose a portion of their water? If the answer is yes, then historic runoff patterns must be preserved. In this case, the answer is to construct rain gardens for a more frequent rainfall that the 100-year, 24-hour event, do not use porous pavements, and include storm sewer systems so that historic runoff patterns are preserved with proper (historic) percentages of surface water and groundwater. The rain gardens will be smaller but if this is a new single-family residential development, then the storm sewer system's cost must be added to the other construction costs.

Note that in the revised rain garden example, the impervious garage roof and the front yard were treated as two separate subareas. Their combined hydrographs resulted in lower peak runoff rates.

APPENDIX D

Vegetated Swale (Bioswale)

D.1 Definition

Bioswales are landscaped elements to remove silt and pollutants from surface runoff. They consist of a swaled drainageway with gently sloped sides, filled with vegetation, compost, and/or riprap. Flow paths, along with a wide and shallow swale, are designed to maximize time water is in a swale, which aids trapping of pollutants and silt. Depending on land geometry available, a bioswale may have a meandering or almost straight alignment. For less permeable soils, bioswales can be underlain with perforated pipe surrounded by a geotextile and a gravel trench.

Use of bioretention media (amended soil) enhances infiltration, water retention, and nutrient and pollutant removal. Like bioretention cells and basins, bioswales encourage infiltration to retain runoff volume and use a variety of physical, biological, and chemical processes to reduce pollutant loadings. Native and other appropriate plants may be used as an alternative to grass. Use of amended soils (for native soils having a higher clay content) assist in preventing clogging of the soil and allowing increased infiltration rates to be maintained over a longer period of time.

D.2 Apartment Complex

Runoff from a small apartment complex's parking lot flows onto a 20-ft width of grass acting as a level spreader before it enters a vegetated swale as shown in Fig. D.1. Swale width is 10 ft, depth is 1.5 ft, side slopes are 5:1, and its surface is covered with native grasses and shrubs. It is made up of an amended soil similar to that of the bioswale in Chap. 5. Because soils are classified as Natural Resources Conservation Service (NRCS) Hydrologic Soil Group (HSG) C with an infiltration rate of 0.15 in/h, this swale is underlain with 1 ft of gravel. Runoff from apartment roofs are conveyed into rain gardens.

A local ordinance requires that a developed 100-year, 24-hour storm be reduced to a flow rate less than or equal to a peak rate from a 5-year, 24-hour storm with the undeveloped site, i.e., meadow in good condition. The reason for this is an inadequate storm-sewer system constructed 30 years ago. The outlet structure is the native soil. Route the 100-year, 24-hour design storm hydrograph through it.

D.2.1 Tributary Drainage Area

Tributary drainage area is made up of the parking lot, level spreader, and swale. Parking lot is 80×320 ft or 25,600 sq ft. The level spreader and swale are $320 \times (10 + 10 + 2 \times 1.5 \times 5 + 20) = 320 \times 55$ or 17,600 ft^2. Total tributary area is 43,200 ft^2 or 0.992 ac or 0.00155 mi^2.

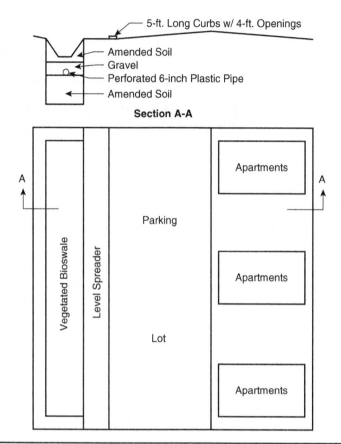

FIGURE D.1 Plan view of apartment complex.

D.2.2 Undeveloped Peak Flow Rate

Using a TR-55 (USDA, 1986) method, peak from a meadow for a 5-year, 24-hour storm is based on the formula: $Qp = q_u \, Am \, Q$

where Qp = peak flow rate (cfs)
q_u = unit peak flow rate (cfs/mi²/in. of runoff) (Chap 4 Exhibit in TR-55)
Am = drainage area (mi²)
Q = runoff (in)

Its length is $((320)^2 + (135)^2)^{0.5}$ or 347 ft. Time of concentration (Tc) was 47.7 min from Table D.1 or 0.80 hr. A spreadsheet for solving this equation is shown in Table D.2. The runoff curve number (CN) is 71 from TR-55 (USDA, 1986). The peak flow is estimated to be 0.48 cfs.

D.2.3 Time of Concentration

Tc for the developed condition is calculated using a spreadsheet incorporating the NRCS method of sheet flow, shallow concentrated flow, and channel flow as shown in Table D.3. Estimated Tc is 13.9 min or 0.23 hour. Use a value for Tc of 0.20 hour.

Time of Concentration Calculations

Nsh = 0.15 P2 = 3.2 in. Ngu = 0.016 Nch = 0.05 Sx = 0.0208 ft/ft

Subarea	Sheet Flow					Shallow Concentrated Flow						
	Length ft	El. Up ft	El. Dn. ft	Slope ft/ft	Travel Time min	Length ft	El. Up ft	El. Dn. ft	Slope ft/ft	Travel Time min	Length ft	El. Up ft
1	2	3	4	5	6	7	8	9	10	11	12	13
Meadow	347	110.00	108.40	0.00461	47.7	1	0.0		0.01000	0.0	1	0.0
	100	1.0		0.01000	0.0	1	0.0		0.01000	0.0	1	0.0
	100	1.0		0.01000	0.0	1	0.0		0.01000	0.0	1	0.0
	100	1.0		0.01000	0.0	1	0.0		0.01000	0.0	1	0.0
	100	1.0		0.01000	0.0	1	0.0		0.01000	0.0	1	0.0
	100	1.0		0.01000	0.0	1	0.0		0.01000	0.0	1	0.0
	100	1.0		0.01000	0.0	1	0.0		0.01000	0.0	1	0.0

Channel Flow									Gutter Flow			Tc
El. Dn. ft	Slope ft/ft	Area sq ft	Wet. Per. ft	R ft	Trav. Time min	Length ft	El. Up ft	El. Dn. ft	Slope ft/ft	Top Wid. ft	Trav. Time min	min
14	15	16	17	18	19	20	21	22	23	24	25	26
	0.01000	1.0	1.0	1.000	0.0	1	0.0		0.01000	1.0	0.0	47.7
	0.01000	1.0	1.0	1.000	0.0	1	0.0		0.01000	1.0	0.0	0.0
	0.01000	1.0	1.0	1.000	0.0	1	0.0		0.01000	1.0	0.0	0.0
	0.01000	1.0	1.0	1.000	0.0	1	0.0		0.01000	1.0	0.0	0.0
	0.01000	1.0	1.0	1.000	0.0	1	0.0		0.01000	1.0	0.0	0.0
	0.01000	1.0	1.0	1.000	0.0	1	0.0		0.01000	1.0	0.0	0.0
	0.01000	1.0	1.0	1.000	0.0	1	0.0		0.01000	1.0	0.0	0.0
	0.01000	1.0	1.0	1.000	0.0	1	0.0		0.01000	1.0	0.0	0.0

TABLE D.1 Tc for the Undeveloped Site

Worksheet 4: Graphical Peak Discharge Method				
Project: Apartment Complex	By:		Date:	
Location: Central Iowa	Checked:		Date:	
Outline one: Present Developed				

1. Data:
   Drainage area, Am 0.00155 square miles
   Runoff curve number, CN 71 dimensionless
   Time of concentration, Tc 0.8 hours
   Rainfall distribution type II (I, Ia, II, III)

	Storm #1	Storm #2	Storm #3	
2. Frequency	5			years
3. Rainfall, P (24 hours)	3.2			inches
4. Initial Abstraction	0.817			inches (Table 5-1 in TR-55)
5. Ia/P	0.255	#DIV/0!	#DIV/0!	dimensionless
6. Unit peak discharge, Qu Exhibit 4-II, TR-55	350			cfs/inch/square mile
7. Runoff, Q Figure 2-1, TR-55	0.9	0.0	0.0	inches
8. Peak discharge, Qp where Qp = Qu Am Q	0.48	0.0	0.0	cfs

TABLE D.2   5-Year Undeveloped Peak Flow Rate

D.2.4   Runoff Curve Number

A CN is developed as shown in Table D.4 using the NRCS method as developed in TR-55 (USDA,1986). CN is estimated to be 87 for the developed site.

D.2.5   100-Year Inflow Hydrograph

Development of a 100-year, 24-hour inflow hydrograph is shown in Table D.5 and Table D.6. Peak flow rate is 6.20 cfs at hour 12.2.

D.2.6   Storage Volume

The swale is underlain by 1 ft of gravel with a porosity of 0.40. Underground storage is $25 \times 300 \times 1 \times 0.40$ or 3,000 ft^3. Surface storage is equal to $1.5 \times 0.5 \times (25 \times 300 + 10 \times 285)$ or 7,760 ft^3. Total storage volume at the surface is 10,760 ft^3. Underground and surface areas and storage volumes every 0.2 ft are calculated in Table D.7 and Table D.8, respectively.

D.2.7   Depth-Outflow Calculations

Outflow from the bioswale is infiltration into the earth itself. Since its area is 25 ft by 300 ft and infiltration rate is 0.15 in/h, outflow rate is

$$Q = 0.15 \text{ in/h} \times 1 \text{ ft/12 in} \times 25 \text{ ft} \times 300 \text{ ft} \times 1 \text{ cfs-hr}/3{,}600 \text{ ft}^3 = 0.026 \text{ cfs}$$

Thus 0.026 cfs infiltrates into the earth beneath the vegetated swale.

Time of Concentration Calculations

Nsh = 0.011
0.240

P2 = 3.2 in.

Subarea	Sheet Flow								Shallow Concentrated Flow			
1	Length ft 2	El. Up ft 3	El. Dn. ft 4	Slope ft/ft 5	Travel Time min 6	Length ft 7	El. Up ft 8	El. Dn. ft 9	Slope ft/ft 10	Travel Time min 11	Length ft 12	El. Up ft 13
Parking	80	110.00	108.80	0.01500	1.1	1	0.0		0.01000	0.0	1	0.0
Spreader	20	108.80	108.40	0.02000	3.9	1	0.0		0.01000	0.0	1	0.0
Swale	8	1.0		0.12500	0.1	1	0.0		0.01000	0.0	320	106.9
	100	1.0		0.01000	0.0	1	0.0		0.01000	0.0	1	0.0
	100	1.0		0.01000	0.0	1	0.0		0.01000	0.0	1	0.0
	100	1.0		0.01000	0.0	1	0.0		0.01000	0.0	1	0.0
	100	1.0		0.01000	0.0	1	0.0		0.01000	0.0	1	0.0

Sx = 0.0208 ft/ft   Ngu = 0.016

Nch = 0.15

El. Dn. ft 14	Channel Flow						Gutter Flow					Tc
	Slope ft/ft 15	Area sq ft 16	Wet. Per. ft 17	R ft 18	Trav. Time min 19	Length ft 20	El. Up ft 21	El. Dn. ft 22	Slope ft/ft 23	Top Wid. ft 24	Trav. Time min 25	min 26
	0.01000	1.0	1.0	1.000	0.0	1	0.0		0.01000	4.0	0.0	1.2
	0.01000	1.0	1.0	1.000	0.0	1	0.0		0.01000	4.0	0.0	4.0
105.3	0.00500	25.0	30.2	0.828	8.6	1	0.0		0.01000	4.0	0.0	8.7
	0.01000	1.0	1.0	1.000	0.0	1	0.0		0.01000	4.0	0.0	0.0
	0.01000	1.0	1.0	1.000	0.0	1	0.0		0.01000	4.0	0.0	0.0
	0.01000	1.0	1.0	1.000	0.0	1	0.0		0.01000	4.0	0.0	0.0
	0.01000	1.0	1.0	1.000	0.0	1	0.0		0.01000	4.0	0.0	0.0

TABLE D.3 Tc for a Vegetated Swale

293

Worksheet 2: Runoff Curve Number and Runoff

Project: Vegetated Swale By: Date:

Location: Midwest Ckd: Date:

Outline one: Present [Developed]

1. Runoff Curve Number (CN)

Subarea Name 1	Soil Name and Hydrologic Group (App. A) 2	Cover Description (cover type, treatment, and hydrologic condition; percent imperviousness; unconnected/connected impervious area ratio) 3	Curve Number 4	Area acres or percent 5	Product of CN × Area 6
Parking	C	Concrete	98	59	5782
Spreader	C	Grass in good condition	74	15	1110
Swale	C	Meadow	71	26	1846
					0
					0
					0
					0
					0
					0
			Total =	[100]	[8738]

Weighted Curve Number = product/area = 8738 over 100 = 87.4

Use CN = 87

2. Runoff

	Storm #1	Storm #2	Storm #3
Frequency. . . . . .yr	100		
Rainfall, (P) 24-hour . . . . .in	6.5		
Runoff, Q. . . . . .in	5.0	0.1	0.1

TABLE D.4 Curve Number for a Parking Lot, Level Spreader, and Swale

Hydrograph Development—Worksheet 5a

Project: Vegetated Swale

Outline one: Present | Developed

Location: Midwest

Frequency (yr): 100

By:

Checked by:

Date:

Date:

Subarea Name 1	Drainage Area sq mi 2	Time of Concen. hr 3	Travel Time thru Subarea hr 4	Downstream Subarea Names 5	Travel Time Summation to Outlet 6	24-hr Rainfall in 7	Runoff Curve Number 8	Runoff in 9	AmQ sq mi-in 10	Initial Abstraction in 11	Ia/P 12
Swale	0.00155	0.20	0.00	—	0.00	6.5	87	5.0	0.00775	0.247	0.038
								#DIV/0!	#DIV/0!		#DIV/0!
								#DIV/0!	#DIV/0!		#DIV/0!

TABLE D.5 Hydrograph Variables for a Vegetated Swale

Hydrograph Development—Worksheet 5b

Project: Vegetated Swale
Outline one: Present [Developed]
Location: Midwest
Frequency (yr): 100
By: _____ Date: _____
Checked by: _____ Date: _____

Block 1

| | Basic Watershed Data Used | | | | Select and enter hydrograph times in hours from exhibit 5- | | | | | | | | | | | |
| | | | | | Discharge at selected hydrograph times cfs | | | | | | | | | | | |
Subarea Name 1	Subarea Tc hr 2	ΣTt to Outlet hr 3	Ia/P 4	AmQ sq mi-in 5	11.0 6	11.3 7	11.6 8	11.9 9	12.0 10	12.2 11	12.4 12	12.6 13	12.8 14	13.0 15	13.2 16	13.4 17
Swale	0.20	0.00	0.10	0.00775	23	31	47	209	403	800	250	128	86	70	61	54
					0.18	0.24	0.36	1.62	3.12	6.20	1.94	0.99	0.67	0.54	0.47	0.42
					0.0	0.0	0.0	0.0	0.0	0.0	0.0	0.0	0.0	0.0	0.0	0.0
					0.0	0.0	0.0	0.0	0.0	0.0	0.0	0.0	0.0	0.0	0.0	0.0
Total					0.18	0.24	0.36	1.62	3.12	6.20	1.94	0.99	0.67	0.54	0.47	0.42

Block 2

| | | | | | Select and enter hydrograph times in hours from exhibit 5- | | | | | | | | | | | |
Subarea Name 1	Subarea Tc hr 2	ΣTt to Outlet hr 3	Ia/P 4	AmQ sq mi-in 5	13.6 6	13.8 7	14.0 8	14.3 9	14.6 10	15.0 11	16.0 12	17.0 13	18.0 14	19.0 15	20.0 16	22.0 17
Swale	0.20	0.00	0.10	0.00775	49	44	40	35	33	30	24	20	18	16	13	12
					0.38	0.34	0.31	0.27	0.26	0.23	0.19	0.16	0.14	0.12	0.10	0.09
					0.0	0.0	0.0	0.0	0.0	0.0	0.0	0.0	0.0	0.0	0.0	0.0
					0.0	0.0	0.0	0.0	0.0	0.0	0.0	0.0	0.0	0.0	0.0	0.0
Total					0.38	0.34	0.31	0.27	0.26	0.23	0.19	0.16	0.14	0.12	0.10	0.09

Table D.6 100-Year Hydrograph Ordinates for a Vegetated Swale

Depth ft	Length ft	Width ft	Unit	Area sq ft
−1.0	300.0	25.0	0.4	3000
−0.8	300.0	25.0	0.4	3000
−0.6	300.0	25.0	0.4	3000
−0.4	300.0	25.0	0.4	3000
−0.2	300.0	25.0	0.4	3000
0.0	300.0	25.0	0.4	3000
0.0	285.0	10.0	1.0	2850
0.2	287.0	12.0	1.0	3444
0.4	289.0	14.0	1.0	4046
0.6	291.0	16.0	1.0	4656
0.8	293.0	18.0	1.0	5274
1.0	295.0	20.0	1.0	5900
1.2	297.0	22.0	1.0	6534
1.4	299.0	24.0	1.0	7176
1.5	300.0	25.0	1.0	7500

TABLE D.7   Depth-Area Calculations for Bioswale

Depth ft	Area sq ft	Aver. Area sq ft	Δ Depth ft	Δ Volume cu ft	Total Vol. cu ft
−1.0	3000				0
		3000	0.2	600	
−0.8	3000				600
		3000	0.2	600	
−0.6	3000				1200
		3000	0.2	600	
−0.4	3000				1800
		3000	0.2	600	
−0.2	3000				2400
		3000	0.2	600	
0.0	3000				3000
		1425	0.0	0	
0.0	2850				3000
		3147	0.2	629	
0.2	3444				3629
		3745	0.2	749	
0.4	4046				4378
		4351	0.2	870	

TABLE D.8   Depth-Storage Calculations for Bioswale

Depth ft	Area sq ft	Aver. Area sq ft	Δ Depth ft	Δ Volume cu ft	Total Vol. cu ft
0.6	4656				5249
		4965	0.2	993	
0.8	5274				6242
		5587	0.2	1117	
1.0	5900				7359
		6217	0.2	1243	
1.2	6534				8602
		6855	0.2	1371	
1.4	7176				9973
		7338	0.1	734	
1.5	7500				10707

TABLE D.8   Depth-Storage Calculations for Bioswale (*Continued*)

D.2.8   Routing Curve Calculations

The routing curve calculations are shown in Table D.9. The curve does not need to be plotted because the outflow rate is a constant 0.026 cfs into the earth.

Routing Period = 0.2 hr			
$2S/\Delta T + O = (2S \text{ (cu ft)}/0.2 \text{ hr}) + O \text{ (cfs)} \times (1 \text{ hr}/60 \text{ min}) \times$ $(1 \text{ min}/60 \text{ sec}) = 0.0028 *S + O$			
Depth ft 1	S cu ft 2	O cfs 3	$2S/\Delta T + O$ cfs 4
−1.0	0	0.026	0.03
−0.8	600	0.026	1.69
−0.6	1200	0.026	3.36
−0.4	1800	0.026	5.03
−0.2	2400	0.026	6.69
0.0	3000	0.026	8.36
0.2	3630	0.026	10.11
0.4	4380	0.026	12.19
0.6	5250	0.026	14.61
0.8	6240	0.026	17.36
1.0	7360	0.026	20.47
1.2	8600	0.026	23.91
1.4	9970	0.026	27.72
1.5	10710	0.026	29.78

TABLE D.9   Routing Curve Calculations for Bioswale

D.2.9   100-Year Hydrograph Routing

A 100-year, 24-hour storm is routed through the gravel and swale in Table D.10. Hydrograph routing was unsuccessful because there was insufficient storage in a 1.5 ft deep swale with a foot of gravel beneath it. This is because of the low infiltration rate of 0.15 in/h of the HSG C soil. A different solution is required.

D.2.10   Outlet Structure Summary

The gravel and swale dimensions listed above are insufficient to reduce a peak runoff rate from a 100-year, 24-hour storm in this apartment complex's developed condition to less than a 5-year, 24-hour storm in an undeveloped condition by using the native soil as the outlet structure. Some water would temporarily store on the level spreader and parking lot.

D.3   Alternative Outlet Structure

Redo the calculations by using a 6-in perforated plastic pipe surrounded by geofabric in the gravel beneath the swale. Connect this pipe to a 12-in reinforced concrete pipe (RCP) and match the soffits. Place a 3-in circular orifice plate between the 6-in and 12-in pipes to ensure that the flow rate into the existing storm sewer is less than the 5-year, 24-hour storm event with the site in its undeveloped condition. The inflow hydrograph and depth-storage calculations do not change. The change is the calculations for the depth-outflow curve. These are shown in Table D.11. The peak outflow rate is 0.39 cfs with 2.5 ft of head.

Line No. 1	Time hr 2	Inflow, I_2 cfs 3	I_1+I_2 cfs 4	$2S_1/\Delta T - O_1$ cfs 5	$2S_2/\Delta T + O_2$ cfs 6	Outflow cfs 7	Depth ft 8
1	11.0	0.18	0.29	0.17	0.46	0.026	−0.95
2	11.2	0.22	0.40	0.41	0.81	0.026	
3	11.4	0.28	0.50	0.76	1.26	0.026	
4	11.6	0.36	0.64	1.20	1.84	0.026	
5	11.8	1.00	1.36	1.79	3.15	0.026	−0.63
6	12.0	3.12	4.12	3.10	7.22	0.026	
7	12.2	6.20	9.32	7.17	16.49	0.026	0.72
8	12.4	1.94	8.14	16.44	24.58	0.026	
9	12.6	0.99	2.93	24.52	27.45	0.026	1.39
10	12.8	0.67	1.66	27.40	29.06	0.026	
11	13.0	0.54	1.21	29.01	30.22	0.026	1.50
12	13.2	0.47	1.01	30.17	31.18	0.026	maximum
13	13.4	0.42	0.89	31.13	32.02	0.026	depth
14	13.6	0.38	0.80	31.96	32.76	0.026	allowed

TABLE D.10   100-Year Routing for Vegetative Swale

Line No. 1	Time hr 2	Inflow, I_2 cfs 3	$I_1 + I_2$ cfs 4	$2S_1/\Delta T - O_1$ cfs 5	$2S_2/\Delta T + O_2$ cfs 6	Outflow cfs 7	Depth ft 8
15	13.8	0.34	0.72	32.71	33.43	0.026	
16	14.0	0.31	0.65	33.38	34.03	0.026	
17	14.2	0.30	0.61	33.98	34.59	0.026	
18	14.4	0.28	0.58	34.54	35.12	0.026	
19	14.6	0.26	0.54	35.06	35.60	0.026	
20	14.8	0.25	0.51	35.55	36.06	0.026	
21	15.0	0.24	0.49	36.01	36.50	0.026	
22	15.2	0.23	0.47	36.45	36.92	0.026	
23	15.4	0.22	0.45	36.87	37.32	0.026	
24	15.6	0.21	0.43	37.26	37.69	0.026	
25	15.8	0.20	0.41	37.64	38.05	0.026	
26	16.0	0.19	0.39	38.00	38.39	0.026	
27	16.2	0.18	0.37	38.34	38.71	0.026	
28	16.4	0.18	0.36	38.66	39.02	0.026	
29	16.6	0.17	0.35	38.97	39.32	0.026	
30	16.8	0.17	0.34	39.27	39.61	0.026	
31	17.0	0.16	0.33	39.55	39.88	0.026	
32	17.2	0.16	0.32	39.83	40.14	0.026	
33	17.4	0.15	0.31	40.09	40.40	0.026	
34	17.6	0.15	0.30	40.35	40.65	0.026	
35	17.8	0.14	0.29	40.60	40.89	0.026	
36	18.0	0.14	0.28	40.84	41.12	0.026	
37	18.2	0.14	0.28	41.07	41.35	0.026	
38	18.4	0.14	0.27	41.29	41.57	0.026	
39	18.6	0.13	0.27	41.52	41.79	0.026	
40	18.8	0.13	0.27	41.73	42.00	0.026	
41	19.0	0.12	0.25	41.95	42.20	0.026	
42	19.2	0.12	0.24	42.15	42.39	0.026	
43	19.4	0.11	0.23	42.34	42.57	0.026	
44	19.6	0.11	0.22	42.52	42.74	0.026	
45	19.8	0.10	0.21	42.69	42.90	0.026	
46	20.0	0.10	0.20	42.85	43.06	0.026	

TABLE D.10　100-year Routing for Vegetative Swale (*Continued*)

Outflow Curve For a Vertical Circular Orifice Flowing Partially Full and Full								

Head, H, listed in Col. 4, assumes that an orifice is acting with a free discharge. If an orifice is submerged, then head in Col. 4 must reflect the second definition of head. Insert a number in Col. 2 and revise it until head plus downstream water surface elevation equals elevation in Col. 1.

$Q = C a (2g)^{0.5} (H)^{0.5}$ Obtain Cŷ from Table 14-7   Obtain Ca from Table 14-6

Using a 3-inch circular orifice

D = 0.25 feet in diameter

$Q = 0.6 * Ca * D^2 * (64.32)^{0.5} * (H)^{0.5} = 4.812\ Ca * D^2 * H^{0.5}$

If C is other than 0.6, change equation in spreadsheet grid space G20.

Elevation ft 1	Depth ft 2	Cŷ 3	H ft 4	H^0.5 5	Ca 6	Q cfs 7	Q soil cfs 8	Total Q cfs 9
−1.0	0.00	0.500	0.000	0.000	0.0000	0.000	0.026	0.026
−0.8	0.20	0.500	0.100	0.316	0.7854	0.075	0.026	0.101
−0.6	0.40	0.500	0.275	0.524	0.7854	0.124	0.026	0.150
−0.4	0.60	0.500	0.475	0.689	0.7854	0.163	0.026	0.189
−0.2	0.80	0.500	0.675	0.822	0.7854	0.194	0.026	0.220
0.0	1.00	0.500	0.875	0.935	0.7854	0.221	0.026	0.247
0.2	1.20	0.500	1.075	1.037	0.7854	0.245	0.026	0.271
0.4	1.40	0.500	1.275	1.129	0.7854	0.267	0.026	0.293
0.6	1.60	0.500	1.475	1.214	0.7854	0.287	0.026	0.313
0.8	1.80	0.500	1.675	1.294	0.7854	0.306	0.026	0.332
1.0	2.00	0.500	1.875	1.369	0.7854	0.323	0.026	0.349
1.2	2.20	0.500	2.075	1.440	0.7854	0.340	0.026	0.366
1.4	2.40	0.500	2.275	1.508	0.7854	0.356	0.026	0.382
1.5	2.50	0.500	2.375	1.541	0.7854	0.364	0.026	0.390

TABLE D.11   Elevation-Outflow Curve for a 3.0-inch Orifice

D.3.1   Routing Curve Calculations

The routing curve calculations are shown in Table D.12 and are plotted in Fig. D.2.

D.3.2   100-Year Routing Calculations

A 100-year, 24-hour storm is routed through the gravel, swale, and underground pipe in Table D.13a and b. Maximum swale depth is 1.37 ft at hour 13.6. Peak inflow rate of 6.20 cfs is reduced to a peak outflow rate of 0.38 cfs. This is less than the existing flow rate of 0.48 cfs required by a local ordinance when the site was a pasture. The bioswale has water on the surface from hour 12.1 to hour 24.0, a total of 11.9 hours. The gravel is completely drained at hour 29.4, 5.4 hours after the rain has stopped.

Routing Period = 0.2 hr			
2S/ΔT + O = (2S (cu ft)/0.1 hr) + O (cfs) × (1 hr/60 min) × (1 min/60 sec) = 0.0027778 *S + O			
Depth ft 1	S cu ft 2	O cfs 3	2S/ΔT + O cfs 4
−1.0	0	0.03	0.03
−0.8	600	0.10	1.77
−0.6	1200	0.15	3.48
−0.4	1800	0.19	5.19
−0.2	2400	0.22	6.89
0.0	3000	0.25	8.58
0.2	3630	0.27	10.35
0.4	4380	0.29	12.46
0.6	5250	0.31	14.89
0.8	6240	0.33	17.66
1.0	7360	0.35	20.79
1.2	8600	0.37	24.26
1.4	9970	0.38	28.07
1.5	10710	0.39	30.14

TABLE D.12   Routing Curve Calculations for a New Bioswale Outlet

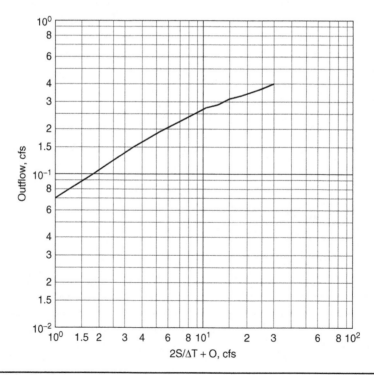

FIGURE D.2   Routing curve for vegetated bioswale.

Line No. 1	Time hr 2	Inflow, I_2 cfs 3	$I_1 + I_2$ cfs 4	$2S_1/\Delta T - O_1$ cfs 5	$2S_2/\Delta T + O_2$ cfs 6	Outflow cfs 7	Depth ft 8
1	11.0	0.18	0.35	0.71	1.00	0.06	−0.92
2	11.2	0.22	0.40	0.88	1.28	0.08	
3	11.4	0.28	0.50	1.12	1.62	0.09	
4	11.6	0.36	0.64	1.44	2.08	0.11	
5	11.8	1.00	1.36	1.86	3.22	0.14	
6	12.0	3.12	4.12	2.94	7.06	0.22	−0.18
7	12.2	6.20	9.32	6.62	15.94	0.31	
8	12.4	1.94	8.14	15.32	23.46	0.35	
9	12.6	0.99	2.93	22.76	25.69	0.36	1.25
10	12.8	0.67	1.66	24.97	26.63	0.37	
11	13.0	0.54	1.21	25.89	27.10	0.37	
12	13.2	0.47	1.01	26.36	27.37	0.38	
13	13.4	0.42	0.89	26.61	27.50	0.38	
14	13.6	0.38	0.80	26.74	27.54	0.38	1.37 peak
15	13.8	0.34	0.72	26.78	27.50	0.38	
16	14.0	0.31	0.65	26.74	27.39	0.38	
17	14.2	0.30	0.61	26.63	27.24	0.38	
18	14.4	0.28	0.58	26.48	27.06	0.37	
19	14.6	0.26	0.54	26.32	26.86	0.37	
20	14.8	0.25	0.51	26.12	26.63	0.37	
21	15.0	0.24	0.49	25.89	26.38	0.37	
22	15.2	0.23	0.47	25.64	26.11	0.37	
23	15.4	0.22	0.45	25.37	25.82	0.37	
24	15.6	0.21	0.43	25.08	25.51	0.36	
25	15.8	0.20	0.41	24.79	25.20	0.36	
26	16.0	0.19	0.39	24.48	24.87	0.36	
27	16.2	0.18	0.37	24.15	24.52	0.36	
28	16.4	0.18	0.36	23.80	24.16	0.36	1.20
29	16.6	0.17	0.35	23.44	23.79	0.36	
30	16.8	0.17	0.34	23.07	23.41	0.36	
31	17.0	0.16	0.33	22.69	23.02	0.36	
32	17.2	0.16	0.32	22.30	22.62	0.36	
33	17.4	0.15	0.31	21.90	22.21	0.36	
34	17.6	0.15	0.30	21.49	21.79	0.35	
35	17.8	0.14	0.29	21.09	21.38	0.35	

TABLE D.13a 100-Year Routing for Revised Vegetative Swale Outlets

Line No. 1	Time hr 2	Inflow, I_2 cfs 3	$I_1 + I_2$ cfs 4	$2S_1/\Delta T - O_1$ cfs 5	$2S_2/\Delta T + O_2$ cfs 6	Outflow cfs 7	Depth ft 8
36	18.0	0.14	0.28	20.68	20.96	0.35	1.00
37	18.2	0.14	0.28	20.26	20.54	0.35	
38	18.4	0.14	0.28	19.84	20.12	0.35	
39	18.6	0.13	0.27	19.42	19.69	0.34	
40	18.8	0.13	0.26	19.01	19.27	0.34	
41	19.0	0.12	0.25	18.59	18.84	0.34	
42	19.2	0.12	0.24	18.16	18.40	0.34	
43	19.4	0.11	0.23	17.72	17.95	0.33	
44	19.6	0.11	0.22	17.29	17.51	0.33	0.80
45	19.8	0.10	0.21	16.85	17.06	0.33	
46	20.0	0.10	0.20	16.40	16.60	0.31	
47	20.2	0.10	0.20	15.98	16.18	0.31	
48	20.4	0.10	0.20	15.56	15.76	0.31	
49	20.6	0.10	0.20	15.14	15.34	0.31	
50	20.8	0.10	0.20	14.72	14.92	0.31	0.60
51	21.0	0.10	0.20	14.30	14.50	0.31	
52	21.2	0.10	0.20	13.88	14.08	0.30	
53	21.4	0.10	0.20	13.48	13.68	0.30	
54	21.6	0.10	0.20	13.08	13.28	0.29	
55	21.8	0.10	0.20	12.70	12.90	0.29	
56	22.0	0.09	0.19	12.32	12.51	0.29	0.40
57	22.2	0.09	0.18	11.93	12.11	0.28	
58	22.4	0.09	0.18	11.55	11.73	0.28	
59	22.6	0.09	0.18	11.17	11.35	0.28	
60	22.8	0.09	0.18	10.79	10.97	0.27	
61	23.0	0.09	0.18	10.43	10.61	0.27	0.22

TABLE **D.13a** 100-Year Routing for Revised Vegetative Swale Outlets (*Continued*)

D.3.3 Summary

The revised outlet structure used a 3-in circular orifice in a steel plate between a 6-in perforated plastic pipe and a 12-in RCP that then outlets into an existing storm sewer. Maximum flow was reduced from 6.20 cfs to 0.38 cfs, less than the required maximum outflow of 0.48 cfs. Maximum depth in the swale was 1.37 ft in the 1.5-ft deep swale. Rock berms were placed every 50-ft in the swale to maximize storage within the swale. The swale was empty on the surface at hour 24.0. The 1-ft depth of gravel was completely drained at hour 29.4, 5.4 hours after the rain stopped.

Line No. 1	Time hr 2	Inflow, I_2 cfs 3	$I_1 + I_2$ cfs 4	$2S_1/\Delta T - O_1$ cfs 5	$2S_2/\Delta T + O_2$ cfs 6	Outflow cfs 7	Depth ft 8
61	23.0	0.09	0.18	10.43	10.61	0.27	0.22
62	23.2	0.08	0.17	10.07	10.24	0.26	
63	23.4	0.08	0.16	9.72	9.88	0.26	
64	23.6	0.07	0.15	9.36	9.51	0.25	
65	23.8	0.06	0.13	9.01	9.14	0.25	
66	24.0	0.06	0.12	8.64	8.76	0.25	0.00
67	24.2	0.05	0.11	8.27	8.38	0.24	
68	24.4	0.04	0.09	7.90	7.99	0.24	
69	24.6	0.04	0.08	7.52	7.60	0.23	
70	24.8	0.03	0.07	7.14	7.21	0.23	
71	25.0	0.02	0.05	6.76	6.81	0.22	−0.20
72	25.2	0.02	0.04	6.37	6.41	0.22	
73	25.4	0.01	0.03	5.98	6.01	0.21	
74	25.6	0.10	0.11	5.59	5.70	0.21	
75	25.8	0.01	0.11	5.29	5.40	0.20	
76	26.0	0.00	0.01	5.00	5.01	0.20	−0.40
77	26.2	0.00	0.00	4.62	4.62	0.19	
78	26.4	0.00	0.00	4.24	4.24	0.19	
79	26.6	0.00	0.00	3.87	3.87	0.18	
80	26.8	0.00	0.00	3.51	3.51	0.18	−0.60
81	27.0	0.00	0.00	3.16	3.16	0.17	
82	27.2	0.00	0.00	2.82	2.82	0.17	
83	27.4	0.00	0.00	2.49	2.49	0.16	
84	27.6	0.00	0.00	2.17	2.17	0.16	
85	27.8	0.00	0.00	1.86	1.86	0.15	−0.80
86	28.0	0.00	0.00	1.56	1.56	0.15	
87	28.2	0.00	0.00	1.27	1.27	0.14	
88	28.4	0.00	0.00	0.99	0.99	0.14	
89	28.6	0.00	0.00	0.72	0.72	0.13	
90	28.8	0.00	0.00	0.46	0.46	0.13	
91	29.0	0.00	0.00	0.21	0.21	0.10	
92	29.2	0.00	0.00	0.02	0.02	0.01	
93	29.4	0.00	0.00	0.00	0.00	0.00	−1.00
94	29.6	0.00	0.00	0.00	0.00	0.00	

TABLE D.13b   100-Year Routing for Revised Vegetative Swale Outlets

Line No. 1	Time hr 2	Inflow, I_2 cfs 3	$I_1 + I_2$ cfs 4	$2S_1/\Delta T - 0_1$ cfs 5	$2S_2/\Delta T + 0_2$ cfs 6	Outflow cfs 7	Depth ft 8
95	29.8	0.00	0.00	0.00	0.00	0.00	
96	30.0	0.00	0.00	0.00	0.00	0.00	
97	30.2	0.00	0.00	0.00	0.00	0.00	
98	30.4	0.00	0.00	0.00	0.00	0.00	
99	30.6	0.00	0.00	0.00	0.00	0.00	
100	30.8	0.00	0.00	0.00	0.00	0.00	
101	31.0	0.00	0.00	0.00	0.00	0.00	
102	31.2	0.00	0.00	0.00	0.00	0.00	
103	31.4	0.00	0.00	0.00	0.00	0.00	
104	31.6	0.00	0.00	0.00	0.00	0.00	
105	31.8	0.00	0.00	0.00	0.00	0.00	
106	32.0	0.00	0.00	0.00	0.00	0.00	
107	32.2	0.00	0.00	0.00	0.00	0.00	
108	32.4	0.00	0.00	0.00	0.00	0.00	
109	32.6	0.00	0.00	0.00	0.00	0.00	
110	32.8	0.00	0.00	0.00	0.00	0.00	
111	33.0	0.00	0.00	0.00	0.00	0.00	
112	33.2	0.00	0.00	0.00	0.00	0.00	
113	33.4	0.00	0.00	0.00	0.00	0.00	
114	33.6	0.00	0.00	0.00	0.00	0.00	
115	33.8	0.00	0.00	0.00	0.00	0.00	
116	34.0	0.00	0.00	0.00	0.00	0.00	
117	34.2	0.00	0.00	0.00	0.00	0.00	
118	34.4	0.00	0.00	0.00	0.00	0.00	
119	34.6	0.00	0.00	0.00	0.00	0.00	
120	34.8	0.00	0.00	0.00	0.00	0.00	
121	35.0	0.00	0.00	0.00	0.00	0.00	−1.00

TABLE D.13b 100-Year Routing for Revised Vegetative Swale Outlets (*Continued*)

The combination of the swale, gravel, perforated pipe, and 3-in vertical orifice were sufficient to allow them and the native soil to reduce the 100-year, 24-hour outflow from the developed site to less than the outflow during a 5-year, 24-hour storm from the undeveloped site.

APPENDIX E

Parking Lots

E.1　Introduction

Parking lots are included in all land uses. They are used in schools or shopping malls or industrial plants. There is more to their design than just determining how many parking places are needed to meet local ordinances or developing good traffic flow into and throughout them.

Two other considerations are important: safety and drainage. When people are in cars, they are drivers and passengers. When people get out of cars, they become pedestrians. How is the parking lot laid out so that interactions between drivers and pedestrians are minimized? Do cars and pedestrians use the same aisles, or are there pedestrian walkways between two rows of parked cars? At the ends of aisles, are there marked walkways and signs at roadways? Are walkways raised so pedestrians do not need to wade through water? What measures are taken in the parking lot so pedestrians and cars are visible to each other and cars cannot be driven fast?

The second is drainage. How often have we driven down into a parking lot, parked the car, and then walked up a ramp or some stairs to enter a building? The designer has developed a detention basin. How many outlets are there and how deep does water get during a 100-year storm? In the United States, the 100-year, 24-hour event drops from 4 to 15 in of rain on a site. People do not like to wade through even a few inches of water. Compact cars get their engines and insides of the car ruined if water is deeper than 7 in. Remember, there are two chances in three that a 100-year storm *will* occur in the next 100 years and one chance in four that it will occur in the next 30 years.

Does the parking lot slope toward buildings? It is deeper there. Are walkways elevated to keep people above water flowing to the building trying to find an outlet? Water runs downhill under gravity's influence. Does the building shade the parking lot during winter so snow and ice do not melt? How many drainage outlets are there in the parking lot so runoff does not get deep in any one location? Where are outlets located in relation to where people will be walking?

Finally, how is the parking lot drained? Are there inlets to storm sewers: in the interior, the periphery, both? If so, how deep is water in a gutter or at an interior inlet so it has enough head to enter the inlet? Are these used as outlets from temporarily detained runoff on the surface? If so, how deep does the water get? Or do the inlets lead to a series of large underground pipes?

Is the surface concrete or asphalt or porous concrete or asphalt underlain with a gravel layer as an underground detention basin? Or does runoff from it flow into surrounding rain gardens? In large parking lots, does runoff flow into a series of interior bioswales underlain with gravel designed to contain the 100-year, 24-hour runoff that are planted with grass, shrubs, and trees?

These are all detention basins as part of a low impact development (LID). Three of these are designed in this chapter. If these three examples are not designed for a 100-year storm, then the additional runoff must be conveyed in storm sewer systems. How and where are some of the pollutants in this additional runoff captured?

About 40 years ago, I was teaching a class on cost-effective storm sewer design in Chicago. One way is to locate the first inlet as far down the street as possible. For our subdivision, we could move the first inlets down to a manhole location. We saved the cost of two inlets, one manhole, and 240 ft of pipe. We then extrapolated that cost savings to the greater Chicago area. Our estimate about 40 years ago was just over $2 billion. How many billion dollars could we save in storm sewer construction and maintenance costs by eliminating them or by reducing sizes needed by using LID concepts not only on parking lots but on all future developments? Then we could institute the coving concept in future neighborhood developments to save more billions of dollars for construction and maintenance of steets and utilities.

E.2   Original Problem Description

Assume an office building has a concrete parking lot. The lot's size is 400 by 300 ft with the 400-ft dimension parallel to its building, an area of 120,000 ft^2 or 2.75 ac. It slopes away from the building at 0.4 percent and is surrounded on the other sides by a 7-in high earth berm covered with grass as shown in Fig. E.1. Dirt and grass are cheaper than curbs and gutters. A 1.0-ft gravel base is placed beneath the concrete pavement. Soils under it are Natural Resources Conservation Service (NRCS) hydrologic soil group (HSG) B.

Keep peak flows as low as possible so area storm sewers can be smaller when constructing a new parking lot. During a 100-year event, however, ponded water should be no deeper than 7 in. This number is based on measurements of compact cars in parking lots. Purposes of a parking lot are to temporarily store cars, temporarily store water, and ensure safety of people as they switch from drivers to pedestrians. Outlet structures are four grate inlets at 100-ft centers along the lot's lower edge that lead to 6-in-diameter vertical pipes as shown in Fig. E.1. These pipes are connected to a 12-in corrugated steel pipe (CSP) that leads to an existing storm sewer.

E.2.1   100-Year Inflow Hydrograph

The runoff curve number (CN) for pavement is 98. The parking lot inflow hydrograph is developed on three worksheets. as shown in Tables E.1 to E.3. Tc from Table E.1 is 5.7 minutes or 0.10 hour. Hydrograph variables are listed in Table E.2 and hydrograph ordinates are estimated in Table E.3. Peak-inflow rate is 26.5 cfs at hour 12.1. ΔT was selected as 0.1 hour because of the small area.

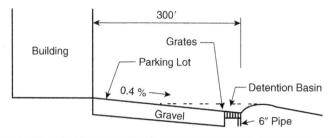

Figure E.1   Basic parking lot.

Worksheet 2				
Overland (Sheet)	**1**	**2**	**3**	**Total**
Pathway Length, ft	300	0.0	0.0	
Upstream Elevation, ft	101.2	0.1	0.1	
Downstream Elevation, ft	100.0	0.0	0.0	
Pathway Slope, ft/ft	0.00400	10.00000	10.00000	
Manning's n	0.011	0.011	0.011	
2-yr, 24-hr Rainfall, in	3.0	3.0	3.0	
Flow Velocity, fps	0.87	2.54	2.54	
Travel Time, min	5.7	0.0	0.0	5.7
Shallow Concentrated	**1**	**2**	**3**	
Pathway Length, ft	0.1	0.1	0.1	
Upstream Elevation, ft	0.1	0.1	0.1	
Downstream Elevation, ft	0.0	0.0	0.0	
Pathway Slope, ft/ft	1.00000	1.00000	1.00000	
Equation Coefficient	16.1	20.3	20.3	
Flow Velocity, fps	16.10	20.30	20.30	
Travel Time, min	0.0	0.0	0.0	0.0
Channel	**1**	**2**	**3**	
Pathway Length, ft	0.1	0.1	0.1	
Upstream Elevation, ft	0.1	0.1	0.1	
Downstream Elevation, ft	0.0	0.0	0.0	
Pathway Slope, ft/ft	1.00000	1.00000	1.00000	
Bottom Width, ft	1.0	1.0	1.0	
Flow Depth, ft	1.0	1.0	1.0	
Side Slope, H:V	1.0	1.0	1.0	
Area, sq ft	2.00	2.00	2.00	
Wetted Perimeter, ft	3.83	3.83	3.83	
Manning's n	0.035	0.035	0.035	
Flow Velocity, fps	27.61	27.61	27.61	
Travel Time, min	0.0	0.0	0.0	0.0
Gutter	**1**	**2**	**3**	
Pathway Length, ft	0.1	0.1	0.1	
Upstream Elevation, ft	0.1	0.1	0.1	
Downstream Elevation, ft	0.0	0.0	0.0	
Pathway Slope, ft/ft	1.0	1.0	1.0	
Street Cross Slope, ft/ft	0.02	0.02	0.02	
Flow Depth, ft	0.25	0.25	0.25	
Flow Top Width, ft	12.5	12.5	12.5	
Manning's n	0.016	0.016	0.016	
Flow Velocity, fps	27.78	27.78	27.78	
Travel Time, min	0.0	0.0	0.0	0.0
Total Travel Time, min	5.7	0.0	0.0	5.7

TABLE E.1 Time of Concentration for Parking Lot

Worksheet 5a

Project: Parking Lot

Outline one: Present | **Developed**

Location: Midwest

Frequency (yr): 100

By:

Checked by:

Date:

Date:

Subarea Name 1	Drainage Area sq mi 2	Time of Concen. hr 3	Travel Time thru Subarea hr 4	Downstream Subarea Names 5	Travel Time Summation to Outlet 6	24-hr Rainfall in 7	Runoff Curve Number 8	Runoff in 9	AmQ sq mi-in 10	Initial Abstraction in 11	Ia/P 12
Parking Lot	0.00430	0.1	0	—	0	6.2	98	6.1	0.02623	0.041	0.0066
	From Worksheet 2				From Worksheet 1					From Table 5-1	

TABLE E.2 Hydrograph Variables for Parking Lot

Project: Parking Lot   Location: Midwest   By: Date:

Outline one: Present | Developed   Frequency (yr): 100   Checked by: Date:

Basic Watershed Data Used					Select and enter hydrograph times in hours from exhibit 5-											
Subarea Name	Subarea Tc hr	ΣTt to Outlet hr	Ia/P	AmQ sq mi-in	Discharge at selected hydrograph times cfs											
1	2	3	4	5	11.0	11.3	11.6	11.9	12.0	12.1	12.2	12.3	12.4	12.5	12.6	12.7
					6	7	8	9	10	11	12	13	14	15	16	17
PL	0.10	0.00	0.10	0.02623	24	34	53	334	647	1010	628	217	147	123	104	86
					0.63	0.89	1.39	8.76	16.97	26.49	16.47	5.69	3.86	3.23	2.73	2.26
					12.8	13.0	13.2	13.4	13.6	13.8	14.0	14.3	14.6	15.0	15.5	16.0
PL	0.10	0.00	0.10	0.02623	76	66	57	51	46	42	38	34	32	29	26	23
					1.99	1.73	1.50	1.34	1.21	1.10	1.00	0.89	0.84	0.76	0.68	0.60
					16.5	17.0	17.5	18.0	19.0	20.0	22.0	26.0				
PL	0.10	0.00	0.10	0.02623	21	20	19	18	15	13	12	0				
					0.55	0.52	0.50	0.47	0.39	0.34	0.31	0.00				

TABLE E.3   100-Year Hydrograph Ordinates for Original Parking Lot

Depth ft	Width ft	Area ft²	Length ft	Volume ft³
0.0	0	0.00	400	0
0.1	25	1.25	400	500
0.2	50	5.00	400	2000
0.3	75	11.25	400	4500
0.4	100	20.00	400	8000
0.5	125	31.25	400	12500
0.6	150	45.00	400	18000

TABLE E.4   Depth-Storage Calculations for a Parking Lot

E.2.2   Depth-Storage Calculations

These are shown in Table E.4. Volume at each 0.1-ft is a triangular cross-sectional area times a 400-ft length. Total storage is 18,000 ft³ as water overtops a 7-in-high grass berm.

E.2.3   Depth-Outflow Calculations

Each 6-in-diameter vertical pipe acts as an orifice. By placing an orifice's crest 1.5 ft below the parking lot's surface, outflow is increased before runoff ponds on the surface. This reduces time and depth water is temporarily ponded on it. Its discharge coefficient has a nominal value of 0.6. Area of each orifice is 0.196 ft². Acceleration caused by gravity is 32.16 fps/sec. Thus, the orifice equation becomes:

$$Q = Ca(2gH)^{0.5} = 0.6 \times 0.196 \times (2 \times 32.16)^{1/2} \times H^{1/2} = 0.943\ H^{1/2} \qquad (E.1)$$

Calculations are shown in Table E.5. Maximum outflow rate from all four inlets is 5.48 cfs.

Q = C a (2g)*0.5 (H)^0.5				
Orifice = 6 inches in diameter				
Using four 6-inch horizontal circular orifices 18 inches below the parking lot				
Q = 4*0.6*0.7854*(D)^2*(64.32)^0.5 (H)^0.5 = 15.117 *D^2*(H)^0.5				
If Co is not 0.6, change equation in spreadsheet grid spaces A12 and E12.				
Elevation ft 1	Depth ft 2	H ft 3	H^0.5 4	Qo cfs 5
98.5	0.00	0.00	0.000	0.00
100.0	1.50	1.50	1.225	4.63
100.1	1.60	1.60	1.265	4.78
100.2	1.70	1.70	1.304	4.93
100.3	1.80	1.80	1.342	5.07
100.4	1.90	1.90	1.378	5.21
100.5	2.00	2.00	1.414	5.34
100.6	2.10	2.10	1.449	5.48

TABLE E.5   Depth-Outflow Calculations for a Parking Lot

Volume in Cubic Feet			
Routing Period, $\Delta T = 0.1$ hr			
$2S/\Delta T + O = (2S\,(\text{ft}^3)/\Delta T\,(\text{hr}) + O\,(\text{cfs}) \times (1\,\text{hr}/60\,\text{min}) \times (1\,\text{min}/60\,\text{sec}) = 0.00556\ {}^*S + O$			
Elev. ft 1	S ft^3 2	O cfs 3	$2S/\Delta T + O$ cfs 4
98.5	0	0.00	0.0
100.0	0	4.63	4.6
100.1	500	4.78	7.6
100.2	2000	4.93	16.0
100.3	4500	5.07	30.1
100.4	8000	5.21	49.7
100.5	12500	5.34	74.8
100.6	18000	5.48	105.5

TABLE E.6   Routing Curve Calculations for Parking Lot

E.2.4   Routing Curve Calculations

Calculations for a routing curve are shown in Table E.6. Storage below 1.5-ft of head is volume within the four grates, so this storage is assumed to be zero. Other storage and outflow values are obtained from Tables E.4 and E.5, respectively. ΔT is assumed to be 0.1 hour because the drainage area is small. Value of $2S/\Delta T + O$ in cfs for ΔT of 0.1 hour is

$$2S/\Delta T + O = 2S/(0.1 \times 60 \times 60) + O = (S/180) + O \qquad (E.2)$$

The last two columns of Table E.6 are plotted on log-log paper as shown in Fig. E.2.

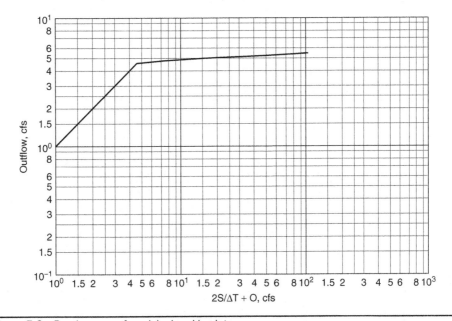

FIGURE E.2   Routing curve for original parking lot.

E.2.5   100-Year Hydrograph Routing

The 100-year, 24-hour storm hydrograph is placed in Col. 3 of the routing form. See Table E.7. As long as the inflow rate is less than or equal to 4.63 cfs, inflow equals outflow rate because there is no storage volume inside the inlets. Maximum depth obtained on this parking lot is interpreted from the depth-outflow curve. The outflow hydrograph's peak is 5.5 cfs. Thus, peak outflow rate from this parking lot has been reduced from 26.5 cfs to just 5.5 cfs simply by temporarily ponding water on half of it furthest from the building. However, since Murphy always shows up, a 100-year event will occur when the parking lot is full of cars. Some office workers will need to walk in a few inches of water. Water is ponded on the lot from hour 11.8 to hour 13.9, a total of 2.1 hours.

Line 1	Time 2	Inflow I_2 3	$I_1 + I_2$ 4	$2S_1/\Delta T - O$ 5	$2S_2/\Delta T + O$ 6	Outflow O_2 7	Elevation
1	11.0	0.63	1.17	0.43	1.60	0.63	98.7
2	11.1	0.71	1.34	0.34	1.68	0.71	
3	11.2	0.80	1.51	0.26	1.77	0.80	
4	11.3	0.89	1.69	0.17	1.86	0.89	
5	11.4	1.03	1.92	0.08	2.00	1.03	
6	11.5	1.20	2.23	−0.06	2.17	1.20	
7	11.6	1.39	2.59	−0.23	2.36	1.39	
8	11.7	2.00	3.39	−0.42	2.97	2.00	
9	11.8	4.50	6.50	−1.03	5.47	4.50	100.0
10	11.9	8.76	13.26	−3.53	9.73	4.90	
11	12.0	16.97	25.73	−0.07	25.66	5.01	
12	12.1	26.49	43.46	15.64	59.10	5.12	
13	12.2	16.47	42.96	48.86	91.82	5.42	
14	12.3	5.69	22.16	80.98	103.14	5.47	100.6 peak
15	12.4	3.86	9.55	92.20	101.75	5.46	
16	12.5	3.23	7.09	90.83	97.92	5.44	
17	12.6	2.73	5.96	87.04	93.00	5.42	
18	12.7	2.26	4.99	82.16	87.15	5.39	
19	12.8	1.99	4.25	76.37	80.62	5.35	
20	12.9	1.85	3.84	69.92	73.76	5.31	100.5
21	13.0	1.73	3.58	63.15	66.73	5.26	
22	13.1	1.62	3.35	56.20	59.55	5.22	
23	13.2	1.50	3.12	49.11	52.23	5.18	
24	13.3	1.42	2.92	41.87	44.79	5.14	
25	13.4	1.34	2.76	34.51	37.27	5.10	
26	13.5	1.27	2.61	27.08	29.69	5.05	100.3

TABLE E.7   100-Year Routing for the Original Parking Lot

Line 1	Time 2	Inflow I_2 3	$I_1 + I_2$ 4	$2S_1/\Delta T - 0$ 5	$2S_2/\Delta T + 0$ 6	Outflow 0_2 7	Elevation
27	13.6	1.21	2.48	19.58	22.06	4.94	
28	13.7	1.15	2.36	12.19	14.55	4.82	
29	13.8	1.10	2.25	4.90	7.15	4.76	
30	13.9	1.05	2.15	−2.37	−0.22	4.65	100.0
31	14.0	1.00	2.05	−1.05	1.00	1.00	
32	14.1	0.96	1.96	−1.00	0.96	0.96	
33	14.2	0.92	1.88	−0.96	0.92	0.92	
34	14.3	0.89	1.81	−0.92	0.89	0.89	
35	14.4	0.87	1.76	−0.89	0.87	0.87	
36	14.5	0.85	1.72	−0.87	0.85	0.85	
37	14.6	0.84	1.69	−0.85	0.84	0.84	
38	14.7	0.82	1.66	−0.84	0.82	0.82	
39	14.8	0.80	1.62	−0.82	0.80	0.80	
40	14.9	0.78	1.58	−0.80	0.78	0.78	
41	15.0	0.76	1.54	−0.78	0.76	0.76	
42	15.1	0.74	1.50	−0.76	0.74	0.74	
43	15.2	0.72	1.46	−0.74	0.72	0.72	
44	15.3	0.70	1.42	−0.72	0.70	0.70	
45	15.4	0.69	1.39	−0.70	0.69	0.69	
46	15.5	0.68	1.37	−0.69	0.68	0.68	
47	15.6	0.66	1.34	−0.68	0.66	0.66	
48	15.7	0.64	1.30	−0.66	0.64	0.64	
49	15.8	0.62	1.26	−0.64	0.62	0.62	
50	15.9	0.61	1.23	−0.62	0.61	0.61	
51	16.0	0.60	1.21	−0.61	0.60	0.60	
52	16.1	0.59	1.19	−0.60	0.59	0.59	
53	16.2	0.58	1.17	−0.59	0.58	0.58	
54	16.3	0.57	1.15	−0.58	0.57	0.57	
55	16.4	0.56	1.13	−0.57	0.56	0.56	
56	16.5	0.55	1.11	−0.56	0.55	0.55	
57	16.6	0.54	1.09	−0.55	0.54	0.54	
58	16.7	0.54	1.08	−0.54	0.54	0.54	
59	16.8	0.53	1.07	−0.54	0.53	0.53	
60	16.9	0.53	1.06	−0.53	0.53	0.53	
61	17.0	0.52	1.05	−0.53	0.52	0.52	98.7

TABLE E.7　100-Year Routing for the Original Parking Lot (*Continued*)

E.2.6 Parking Lot Cost

Cost of the parking lot is site preparation, subgrade, concrete paving, inlets, 6-in CSP, and 12-in CSP. Area of the parking lot is $300 \times 400/9$ or 13,333 yd^2. Site preparation is $6,000. Cost of excavation is $300 \times 400 \times 1/27 \times 4.00$ or $17,800. Gravel subgrade is $300 \times 400 \times 1.0/27 \times 20.00 or $88,900. Concrete pavement is $13,333 \times 25.00 or $333,300. Eight feet of 6-in CSP is $8 \times 13.00 or $100. 450 ft of 12-in CSP is $450 \times 25.00 or $11,200. Four grate inlets are $4 \times $2,500$ or $10,000. Total cost is $457,300.

E.2.7 Summary

Peak inflow rate from a 100-year, 24-hour storm event of 26.5 cfs is reduced to a peak outflow rate of 5.5 cfs. Impact of this reduced flow rate for storm sewers along this street could mean that existing storm sewers designed for a 5-year storm could be able to convey flow from a larger storm event. However, no improvement is made in water quality. Total cost is $457,300.

Calculations are not shown, but runoff from the 2-, 5-, 10-, 25-, and 50-year storm events were routed to estimate lengths of time water is physically ponded on the parking lot. For a 100-year storm, this was 2.1 hours. Water did not pond during a 2-year event. Allowing for all storms during a 100 year period, water would temporarily pond to some depth for a total of about 3,200 hours. Since there are $24 \times 365.25 \times 100$ equals 876,600 hours in 100 years, water is ponded to some depth about $100 \times 3,200/876,600$ or 0.36 of 1 percent of the time.

Therefore, half of the parking lot away from the building will be only wet or contain a few inches of water one-third of one percent of the time. Half of the parking lot closest to the building will be dry or only wet 100 percent of the time. However, since Murphy always shows up, this 0.36 percent will occur when people need to wade through water. Some people will still complain on those days because they do not want to shoes, socks, stockings, and/or feet wet.

E.3 Modified Parking Lot Drainage

Five underground pipes (see Fig. E.3) are used to further decrease flow exiting the lot. The inflow hydrograph to them is the outflow hydrograph from the original parking lot. The area of a 60-in CSP is 19.64 ft^2. Volume of the five pipes is $5 \times 19.64 \times 400$ or 39,270 ft^3. Multiplying the area of a partially filled CSP at each half foot of depth by 2,000 ft yields a depth-storage relationship for these pipes. These calculations are shown in Table E.8.

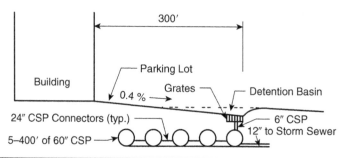

Figure E.3 Modified parking lot.

| Elevation/Depth–Storage Calculations | | | | | |
| Incremental Area Method | | | | | |
Depth ft 1	Area ft² 2	Incre. Area ft² 3	Length ft 4	Δ Volume ft³ 5	Tot. Vol. ft³ 6
0.00	0.00				0
		1.02	2000	2040	
0.50	1.02				2040
		1.78	2000	3560	
1.00	2.80				5600
		2.16	2000	4320	
1.50	4.96				9920
		2.38	2000	4760	
2.00	7.34				14680
		2.49	2000	4980	
2.50	9.83				19660
		2.47	2000	4940	
3.00	12.30				24600
		2.38	2000	4760	
3.50	14.68				29360
		2.17	2000	4340	
4.00	16.85				33700
		1.78	2000	3560	
4.50	18.63				37260
		1.01	2000	2020	
5.00	19.64				39280

TABLE E.8   Depth-Storage Calculations for 5–60-inch CSPs

E.3.1   Depth-Outflow Calculations

Outflow from these pipes to a storm sewer is a 12-in CSP with a 6-in orifice in a plate at its entrance. Flow through the 6-in vertical circular orifice at each half foot of depth yields a depth-ouflow relationship. Peak outflow rate is 2.06 cfs as shown in Table E.9.

E.3.2   Routing Curve Calculations

Combining Table E.8 and Table E.9 yields the routing curve. These calculations are shown in Table E.10 and are plotted in Fig. E.4.

E.3.3   100-Year Routing Calculations

Using this routing curve with the outflow hydrograph from the inlets yields a peak flow through the orifice into an existing storm sewer system. These calculations are

Depth-Outflow Curve for a Vertical Circular Orifice Flowing Partially Full and Full						
The head, H, listed in col. 4 assumes that the orifice is acting with a free discharge. If the orifice is submerged, then the head in col. 4 must reflect the second definition of head. Insert a number in col. 2 and revise it until the head plus the downstream water surface elevation equals the elevation in col. 1.						
Q = C a (2g)*0.5 (H)^0.5 Obtain Cŷ from Table 15.7 Obtain Ca from Table 15.6						
Using a 6-inch vertical circular orifice						
D = 0.50 feet in diameter If Co is other than 0.6, change equation in F21.						
Q = 0.6 * Ca * D^2 * (64.32)^0.5 * (H)^0.5 = 4.812 Ca * D^2 * H^0.5						
Elevation ft 1	Depth ft 2	Cŷ 3	H ft 4	H^0.5 5	Ca 6	Q cfs 7
0.0	0.0	0.000	0.000	0.000	0.0000	0.00
0.5	0.5	0.500	0.250	0.500	0.7854	0.47
1.0	1.0	0.500	0.750	0.866	0.7854	0.82
1.5	1.5	0.500	1.250	1.118	0.7854	1.06
2.0	2.0	0.500	1.750	1.323	0.7854	1.25
2.5	2.5	0.500	2.250	1.500	0.7854	1.42
3.0	3.0	0.500	2.750	1.658	0.7854	1.57
3.5	3.5	0.500	3.250	1.803	0.7854	1.70
4.0	4.0	0.500	3.750	1.936	0.7854	1.83
4.5	4.5	0.500	4.250	2.062	0.7854	1.95
5.0	5.0	0.500	4.750	2.179	0.7854	2.06

TABLE E.9   Depth-Outflow Calculations for Revised Parking Lot

shown in Tables E.11a to c. The peak outflow into the storm sewer is 1.8 cfs, a reduction of an additional 3.7 cfs.

Adding five 60-in CSPs reduces the peak outflow rate to just 1.80 cfs with a maximum depth of 3.86 ft in the 5-ft diameter pipes. At hour 24, the water is still 0.45 ft deep in the pipes. The pipes are empty at about hour 25.9. How much money could be saved on storm sewer systems if all parking lots in the United States would have similar 93 percent decreases in outflow rates?

E.3.4   Cost

Adding the underground pipe is more expensive. Site preparation is $6,000. Cost of excavation is $300 \times 400 \times 1/27 \times 4.00$ or $17,800. Gravel subgrade is $300 \times 400 \times 1/27 \times$ $20.00 or $88,900. Concrete pavement is $13,333 \times 25.00 or $333,300. Excavation and backfill costs $10.00 \times 45 \times 400 \times 7/27$ or $46,700. CSP costs about $2.10/in diameter. Thus, cost of the 60-in CSPs is $2.10 \times 60 \times 2,000$ or $252,000. Cost of the 24-in CSPs is $2.10 \times 24 \times 60$ or $3,000. Cost of the 12-in CSP is $2.10 \times 12 \times 24$ or $600. Cost of the 6-in CSP is $2.1 \times 6 \times 8$ or $100. Grate inlets are $2,500 each, so four cost $10,000. Total cost is $758,400.

Volume in Cubic Feet			
Routing Period, $\Delta T = 0.1$ hr			
$2S/\Delta T + O = (2S\ (ft^3)/\Delta T\ (hr) + O\ (cfs) \times (1\ hr/60\ min) \times (1\ min/60\ sec) = 1/180\ *S + O$			
Depth ft 1	Storage ft³ 2	Outflow cfs 3	$2S/\Delta T + O$ cfs 4
0.00	0	0.00	0.0
0.50	2040	0.47	11.8
1.00	5600	0.82	31.9
1.50	9920	1.06	56.2
2.00	14680	1.25	82.8
2.50	19660	1.42	110.6
3.00	24600	1.57	138.2
3.50	29360	1.70	164.8
4.00	33700	1.83	189.1
4.50	37260	1.95	209.0
5.00	39280	2.06	220.3

TABLE **E.10**   Routing Curve for Revised Parking Lot

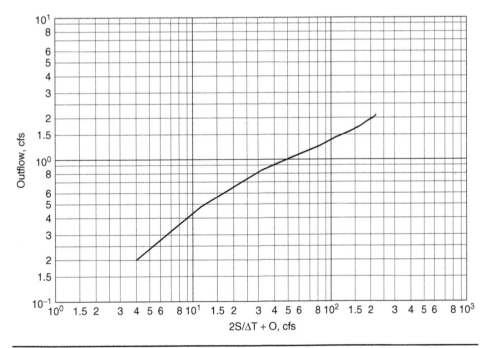

FIGURE **E.4**   Routing curve for revised parking lot.

Line 1	Time 2	Inflow I_2 3	$I_1 + I_2$ 4	$2S_1/\Delta T - 0$ 5	$2S_2/\Delta T + 0$ 6	Outflow 0_2 7	Depth ft 8
1	11.0	0.63	1.20	2.80	4.00	0.20	0.20
2	11.1	0.71	1.34	3.60	4.94	0.25	
3	11.2	0.80	1.51	4.44	5.95	0.26	
4	11.3	0.89	1.69	5.43	7.12	0.31	
5	11.4	1.03	1.92	6.50	8.42	0.38	
6	11.5	1.20	2.23	7.66	9.89	0.42	
7	11.6	1.39	2.59	9.05	11.64	0.46	0.50
8	11.7	2.00	3.39	10.72	14.11	0.55	
9	11.8	4.50	6.50	13.01	19.51	0.63	
10	11.9	4.90	9.40	18.25	27.65	0.78	0.90
11	12.0	5.01	9.91	26.09	36.00	0.88	
12	12.1	5.12	10.13	34.24	44.37	0.94	
13	12.2	5.42	10.54	42.49	53.03	1.01	1.50
14	12.3	5.47	10.89	51.01	61.90	1.12	
15	12.4	5.46	10.93	59.66	70.59	1.20	
16	12.5	5.44	10.90	68.19	79.09	1.25	2.00
17	12.6	5.42	10.86	76.59	87.45	1.30	
18	12.7	5.39	10.81	84.85	95.66	1.35	
19	12.8	5.35	10.74	92.96	103.70	1.40	
20	12.9	5.31	10.66	100.90	111.56	1.45	2.50
21	13.0	5.26	10.57	108.66	119.23	1.49	
22	13.1	5.22	10.48	116.25	126.73	1.53	
23	13.2	5.18	10.40	123.67	134.07	1.57	
24	13.3	5.14	10.32	130.93	141.25	1.61	3.00
25	13.4	5.10	10.24	138.03	148.27	1.64	
26	13.5	5.05	10.15	144.99	155.14	1.66	
27	13.6	4.94	9.99	151.82	161.81	1.68	3.50
28	13.7	4.82	9.76	158.45	168.21	1.71	
29	13.8	4.76	9.58	164.79	174.37	1.74	
30	13.9	4.65	9.41	170.89	180.30	1.77	
31	14.0	1.00	5.65	176.76	182.41	1.80	3.86 peak
32	14.1	0.96	1.96	178.81	180.77	1.79	
33	14.2	0.92	1.88	177.19	179.07	1.78	
34	14.3	0.89	1.81	175.51	177.32	1.77	
35	14.4	0.87	1.76	173.78	175.54	1.76	

TABLE E.11A   100-Year Hydrograph Routing for Revised Parking Lot

Line 1	Time 2	Inflow I_2 3	$I_1 + I_2$ 4	$2S_1/\Delta T - 0$ 5	$2S_2/\Delta T + 0$ 6	Outflow 0_2 7	Depth ft 8
36	14.5	0.85	1.72	172.02	173.74	1.75	
37	14.6	0.84	1.69	170.24	171.93	1.74	
38	14.7	0.82	1.66	168.45	170.11	1.73	
39	14.8	0.80	1.62	166.65	168.27	1.72	
40	14.9	0.78	1.58	164.83	166.41	1.71	
41	15.0	0.76	1.54	162.99	164.53	1.70	3.50
42	15.1	0.74	1.50	161.13	162.63	1.69	
43	15.2	0.72	1.46	159.25	160.71	1.68	
44	15.3	0.70	1.42	157.35	158.77	1.67	
45	15.4	0.69	1.39	155.43	156.82	1.67	
46	15.5	0.68	1.37	153.48	154.85	1.66	
47	15.6	0.66	1.34	151.53	152.87	1.66	
48	15.7	0.64	1.30	149.55	150.85	1.65	
49	15.8	0.62	1.26	147.55	148.81	1.65	
50	15.9	0.61	1.23	145.51	146.74	1.64	
51	16.0	0.60	1.21	143.46	144.67	1.63	
52	16.1	0.59	1.19	141.41	142.60	1.62	
53	16.2	0.58	1.17	139.36	140.53	1.61	
54	16.3	0.57	1.15	137.31	138.46	1.60	3.00
55	16.4	0.56	1.13	135.26	136.39	1.59	
56	16.5	0.55	1.11	133.21	134.32	1.57	
57	16.6	0.54	1.09	131.18	132.27	1.56	
58	16.7	0.54	1.08	129.15	130.23	1.55	
59	16.8	0.53	1.07	127.13	128.20	1.54	
60	16.9	0.53	1.06	125.12	126.18	1.53	
61	17.0	0.52	1.05	123.12	124.17	1.52	2.90

TABLE E.11a   100-Year Hydrograph Routing for Revised Parking Lot (*Continued*)

E.3.5   Summary of Revised Parking Lot Drainage

Peak inflow rate from a 100-year, 24-hour storm event of 26.5 cfs is reduced to a peak outflow of 1.80 cfs. Impact of this reduced flow rate for storm sewers along this street could mean that existing storm sewers designed for a 5-year storm could be able to convey flow from a much larger storm event. However, there is still no improvement to runoff water quality.

Total cost to develop and drain the parking lot is $758,400. Spending another $301,100 to reduce the outflow by an additional 3.7 cfs is a decision to be made based on local circumstances. Half of the parking lot further away from the building will still have some water temporarily ponded on it for up to 2 hours whenever a greater than 2-year storm event occurs.

Line 1	Time 2	Inflow I_2 3	$I_1 + I_2$ 4	$2S_1/\Delta T - 0$ 5	$2S_2/\Delta T + 0$ 6	Outflow O_2 7	Depth ft 8
61	17.0	0.52	1.05	123.12	124.17	1.52	2.90
62	17.1	0.51	1.03	121.13	122.16	1.51	
63	17.2	0.50	1.01	119.15	120.16	1.49	
64	17.3	0.49	0.99	117.17	118.16	1.48	
65	17.4	0.48	0.97	115.21	116.18	1.46	
66	17.5	0.47	0.95	113.25	114.20	1.45	
67	17.6	0.46	0.93	111.30	112.23	1.44	
68	17.7	0.46	0.92	109.36	110.27	1.42	2.50
69	17.8	0.45	0.91	107.43	108.33	1.41	
70	17.9	0.45	0.90	105.51	106.41	1.40	
71	18.0	0.44	0.89	103.61	104.49	1.39	
72	18.1	0.44	0.88	101.72	102.59	1.38	
73	18.2	0.43	0.87	99.84	100.70	1.37	
74	18.3	0.43	0.86	97.97	98.82	1.36	
75	18.4	0.42	0.85	96.11	96.96	1.34	
76	18.5	0.42	0.84	94.27	95.10	1.33	
77	18.6	0.41	0.83	92.44	93.26	1.32	
78	18.7	0.41	0.82	90.62	91.43	1.31	
79	18.8	0.40	0.81	88.81	89.62	1.30	
80	18.9	0.40	0.80	87.02	87.81	1.29	
81	19.0	0.39	0.79	85.23	86.02	1.28	
82	19.1	0.39	0.78	83.46	84.24	1.27	
83	19.2	0.38	0.77	81.70	82.47	1.26	2.00
84	19.3	0.38	0.76	79.96	80.71	1.24	
85	19.4	0.37	0.74	78.22	78.96	1.23	
86	19.5	0.36	0.72	76.50	77.22	1.22	
87	19.6	0.35	0.70	74.78	75.48	1.21	
88	19.7	0.34	0.68	73.07	73.75	1.19	
89	19.8	0.33	0.66	71.37	72.03	1.18	
90	19.9	0.32	0.64	69.66	70.30	1.17	
91	20.0	0.31	0.62	67.97	68.59	1.15	

TABLE E.11b   100-Year Hydrograph Routing for Revised Parking Lot

Line 1	Time 2	Inflow I_2 3	$I_1 + I_2$ 4	$2S_1/\Delta T - 0$ 5	$2S_2/\Delta T + 0$ 6	Outflow 0_2 7	Depth ft 8
92	20.1	0.30	0.60	66.28	66.88	1.14	
93	20.2	0.29	0.58	64.60	65.18	1.13	
94	20.3	0.28	0.56	62.92	63.48	1.12	
95	20.4	0.27	0.54	61.24	61.78	1.10	
96	20.5	0.26	0.52	59.58	60.10	1.09	
97	20.6	0.25	0.50	57.92	58.42	1.08	
98	20.7	0.24	0.48	56.26	56.74	1.07	1.50
99	20.8	0.23	0.46	54.61	55.07	1.05	
100	20.9	0.22	0.44	52.96	53.40	1.04	
101	21.0	0.21	0.42	51.33	51.75	1.02	
102	21.1	0.20	0.40	49.71	50.11	1.01	
103	21.2	0.19	0.38	48.10	48.48	0.99	
104	21.3	0.18	0.36	46.50	46.86	0.97	
105	21.4	0.17	0.34	44.91	45.25	0.96	
106	21.5	0.16	0.32	43.33	43.65	0.94	
107	21.6	0.15	0.30	41.76	42.06	0.93	
108	21.7	0.14	0.28	40.20	40.48	0.91	
109	21.8	0.13	0.26	38.66	38.92	0.90	
110	21.9	0.12	0.24	37.12	37.36	0.88	
111	22.0	0.11	0.22	35.60	35.82	0.87	
112	22.1	0.09	0.20	34.08	34.28	0.85	
113	22.2	0.08	0.18	32.58	32.76	0.84	
114	22.3	0.07	0.16	31.09	31.25	0.82	1.00
115	22.4	0.06	0.14	29.61	29.75	0.80	
116	22.5	0.05	0.12	28.16	28.28	0.77	
117	22.6	0.04	0.10	26.74	26.84	0.75	
118	22.7	0.03	0.08	25.34	25.42	0.72	
119	22.8	0.02	0.06	23.97	24.03	0.70	
120	22.9	0.01	0.04	22.63	22.67	0.68	
121	23.0	0.00	0.02	21.32	21.34	0.65	0.74

TABLE **E.11b**   100-Year Hydrograph Routing for Revised Parking Lot (*Continued*)

Line 1	Time 2	Inflow I_2 3	$I_1 + I_2$ 4	$2S_1/\Delta T - 0$ 5	$2S_2/\Delta T + 0$ 6	Outflow 0_2 7	Depth ft 8
121	23.0	0.00	0.02	21.32	21.34	0.65	0.74
122	23.1	0.00	0.00	20.04	20.04	0.63	
123	23.2	0.00	0.00	18.78	18.78	0.61	
124	23.3	0.00	0.00	17.56	17.56	0.59	
125	23.4	0.00	0.00	16.38	16.38	0.57	
126	23.5	0.00	0.00	15.24	15.24	0.55	
127	23.6	0.00	0.00	14.14	14.14	0.53	
128	23.7	0.00	0.00	13.08	13.08	0.51	
129	23.8	0.00	0.00	12.06	12.06	0.49	0.50
130	23.9	0.00	0.00	11.08	11.08	0.47	
131	24.0	0.00	0.00	10.14	10.14	0.45	
132	24.1	0.00	0.00	9.24	9.24	0.43	
133	24.2	0.00	0.00	8.38	8.38	0.41	
134	24.3	0.00	0.00	7.56	7.56	0.39	
135	24.4	0.00	0.00	6.78	6.78	0.37	
136	24.5	0.00	0.00	6.04	6.04	0.35	
137	24.6	0.00	0.00	5.34	5.34	0.33	
138	24.7	0.00	0.00	4.68	4.68	0.31	
139	24.8	0.00	0.00	4.06	4.06	0.29	
140	24.9	0.00	0.00	3.48	3.48	0.27	
141	25.0	0.00	0.00	2.94	2.94	0.25	
142	25.1	0.00	0.00	2.44	2.44	0.23	
143	25.2	0.00	0.00	1.98	1.98	0.21	
144	25.3	0.00	0.00	1.56	1.56	0.19	
145	25.4	0.00	0.00	1.18	1.18	0.17	
146	25.5	0.00	0.00	0.84	0.84	0.15	
147	25.6	0.00	0.00	0.54	0.54	0.13	
148	25.7	0.00	0.00	0.28	0.28	0.11	
149	25.8	0.00	0.00	0.06	0.06	0.03	
150	25.9	0.00	0.00	0.00	0.00	0.00	0.00
151	26.0	0.00	0.00	0.00	0.00	0.00	
152	26.1	0.00	0.00	0.00	0.00	0.00	
153	26.2	0.00	0.00	0.00	0.00	0.00	
154	26.3	0.00	0.00	0.00	0.00	0.00	
155	26.4	0.00	0.00	0.00	0.00	0.00	

TABLE E.11c   100-Year Hydrograph Routing for Revised Parking Lot

Line 1	Time 2	Inflow I_2 3	$I_1 + I_2$ 4	$2S_1/\Delta T - 0$ 5	$2S_2/\Delta T + 0$ 6	Outflow 0_2 7	Depth ft 8
156	26.5	0.00	0.00	0.00	0.00	0.00	
157	26.6	0.00	0.00	0.00	0.00	0.00	
158	26.7	0.00	0.00	0.00	0.00	0.00	
159	26.8	0.00	0.00	0.00	0.00	0.00	
160	26.9	0.00	0.00	0.00	0.00	0.00	
161	27.0	0.00	0.00	0.00	0.00	0.00	
162	27.1	0.00	0.00	0.00	0.00	0.00	
163	27.2	0.00	0.00	0.00	0.00	0.00	
164	27.3	0.00	0.00	0.00	0.00	0.00	
165	27.4	0.00	0.00	0.00	0.00	0.00	
166	27.5	0.00	0.00	0.00	0.00	0.00	
167	27.6	0.00	0.00	0.00	0.00	0.00	
168	27.7	0.00	0.00	0.00	0.00	0.00	
169	27.8	0.00	0.00	0.00	0.00	0.00	
170	27.9	0.00	0.00	0.00	0.00	0.00	
171	28.0	0.00	0.00	0.00	0.00	0.00	0.00

TABLE E.11c 100-Year Hydrograph Routing for Revised Parking Lot (Continued)

E.4 Porous Pavement Parking Lot

The porous portland cement pavement shown in Fig. E.5 is underlain with 12 in of gravel with a porosity of 40 percent. The pavement and gravel will temporarily store the 100-year, 24-hour storm event. There will be no surface runoff, and all rain will eventually drain into the native soils which are NRCS HSG B. There will be no runoff to an existing storm sewer system. Almost all pollutants will be captured onsite. Any urban trash deposited by the wind, drivers, and passengers will be picked up by maintenance workers as they sweep the parking lot. Also, all portions will only get wet. There is no water ponded on the surface at any time. Thus, there will be no reason for anyone to complain about getting his/her feet wet.

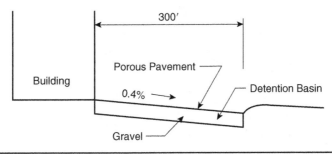

FIGURE E.5 Porous pavement parking lot.

E.4.1 100-Year Inflow Hydrograph

The inflow hydrograph is the same as shown in Table E.3 since we have not changed any of the physical variables. Only two items have changed. One is the detention basin, the pore spaces in the gravel, which acts not only as the subgrade material but the underground detention reservoir as well. The second is the outlet structure, the earth itself.

E.4.2 Depth-Storage Calculations

The depth-storage calculations are listed in Table E.12. Total runoff volume is $300 \times 400 \times 6.1/12$ or $61,000$ ft³. Total storage volume is $300 \times 400 \times 1 \times 0.4$ or $48,000$ ft³.

E.4.3 Depth-Outflow Calculations

Outflow is infiltration into the earth itself. Minimum infiltration rate for this B-type soil is 0.30 in/hr. Flow rate is:

$$Q = 300 \times 400 \times 0.30/12/3,600 = 0.83 \text{ cfs}$$

Elevation/Depth-Storage Calculations Incremental Area Method					
Depth ft 1	Area ft² 2	Incre. Area ft² 3	Δ Depth ft 4	Δ Volume ft³ 5	Tot. Vol. ft³ 6
0.00	48000				0
		48000	0.10	4800	
0.10	48000				4800
		48000	0.10	4800	
0.20	48000				9600
		48000	0.10	4800	
0.30	48000				14400
		48000	0.10	4800	
0.40	48000				19200
		48000	0.10	4800	
0.50	48000				24000
		48000	0.10	4800	
0.60	48000				28800
		48000	0.10	4800	
0.70	48000				33600
		48000	0.10	4800	
0.80	48000				38400
		48000	0.10	4800	
0.90	48000				43200
		48000	0.10	4800	
1.00	48000				48000

TABLE E.12 Depth-Storage Calculations for 12-in of Gravel

Volume in Cubic Feet			
Routing Period, $\Delta T = 0.1$ hr			
$2S/\Delta T + O = (2S \ (ft^3)/\Delta T \ (hr) + O \ (cfs) \times (1 \ hr/60 \ min) \times (1 \ min/60 \ sec) = 0.00556 *S + O$			
Elev. ft 1	S ft^3 2	O cfs 3	$2S/\Delta T + O$ cfs 4
0.0	0	0.83	0.8
0.1	4800	0.83	27.5
0.2	9600	0.83	54.2
0.3	14400	0.83	80.8
0.4	19200	0.83	107.5
0.5	24000	0.83	134.2
0.6	28800	0.83	160.8
0.7	33600	0.83	187.5
0.8	38400	0.83	214.2
0.9	43200	0.83	240.8
1.0	48000	0.83	267.5

TABLE E.13   Routing Curve Calculations for Gravel

E.4.4   Routing Curve Calculations

Routing curve calculations are shown in Table E.13. The curve does not need to be plotted because outflow is a constant 0.83 cfs.

E.4.5   100-Year Hydrograph Routing

The 100-year, 24-hour storm hydrograph is routed in Tables E.14a to c. The peak inflow rate of 26.5 cfs is reduced to a peak outflow rate of 0.83 cfs into the soil. Maximum depth in the gravel is 0.70 ft at hour 14.7. When the rain stops at hour 24.0, depth in the gravel is still 0.42 ft. The gravel is completely drained 9.0 hours after the rain stops.

E.4.6   Cost

Cost to construct the parking lot includes site preparation, excavation, gravel, and porous concrete pavement. Cost of site preparation is $6,000. Cost of excavation is $300 \times 400 \times 1/27 \times 4.00$ or $17,800. Cost of gravel is $300 \times 400 \times 1/27 \times 20.00$ or $88,900. Cost of porous pavement is $300 \times 400/9 \times 25.00$ or $333,300. Total cost is $446,000.

E.4.7   Summary of Porous Pavement Alternative

Peak inflow rate from a 100-year, 24-hour storm event of 26.5 cfs is reduced to a peak outflow rate of 0.83 cfs for a total cost of $446,000. Maximum depth in the gravel is 0.70 ft and is totally drained in a total of 1.4 days from the beginning of rainfall. All runoff is infiltrated so there is no surface runoff to an existing storm sewer. This allows the storm sewer to convey additional flow—if the inlets are large enough. Also, since there is no surface runoff, almost all pollutants are captured onsite. Any remaining pollutants are swept up by maintenance personnel. Office workers never have to contend with temporarily ponded water.

Line 1	Time 2	Inflow I_2 3	$I_1 + I_2$ 4	$2S_1/\Delta T - 0$ 5	$2S_2/\Delta T + 0$ 6	Outflow 0_2 7	Depth ft 8
1	11.0	0.63	1.20	0.00	1.20	0.63	0.00
2	11.1	0.71	1.34	−0.06	1.28	0.71	
3	11.2	0.80	1.51	−0.14	1.37	0.80	
4	11.3	0.89	1.69	−0.23	1.46	0.83	
5	11.4	1.03	1.92	−0.20	1.72	0.83	
6	11.5	1.20	2.23	0.06	2.29	0.83	
7	11.6	1.39	2.59	0.63	3.22	0.83	
8	11.7	2.00	3.39	1.56	4.95	0.83	
9	11.8	4.50	6.50	3.29	9.79	0.83	
10	11.9	8.76	13.26	8.13	21.39	0.83	0.08
11	12.0	16.97	25.73	19.73	45.46	0.83	
12	12.1	26.49	43.46	43.80	87.26	0.83	
13	12.2	16.47	42.96	85.60	128.56	0.83	
14	12.3	5.69	22.16	126.90	149.06	0.83	
15	12.4	3.86	9.55	147.40	156.95	0.83	
16	12.5	3.23	7.09	155.29	162.38	0.83	0.61
17	12.6	2.73	5.96	160.72	166.68	0.83	
18	12.7	2.26	4.99	165.02	170.01	0.83	
19	12.8	1.99	4.25	168.35	172.60	0.83	
20	12.9	1.85	3.84	170.94	174.78	0.83	
21	13.0	1.73	3.58	173.12	176.70	0.83	
22	13.1	1.62	3.35	175.04	178.39	0.83	
23	13.2	1.50	3.12	176.73	179.85	0.83	
24	13.3	1.42	2.92	178.19	181.11	0.83	
25	13.4	1.34	2.76	179.45	182.21	0.83	
26	13.5	1.27	2.61	180.55	183.16	0.83	
27	13.6	1.21	2.48	181.50	183.98	0.83	
28	13.7	1.15	2.36	182.32	184.68	0.83	
29	13.8	1.10	2.25	183.02	185.27	0.83	
30	13.9	1.05	2.15	183.61	185.76	0.83	
31	14.0	1.00	2.05	184.10	186.15	0.83	

TABLE E.14a   100-Year Hydrograph Routing for Porous Parking Lot

Line 1	Time 2	Inflow I_2 3	$I_1 + I_2$ 4	$2S_1/\Delta T - 0$ 5	$2S_2/\Delta T + 0$ 6	Outflow 0_2 7	Depth ft 8
32	14.1	0.96	1.96	184.49	186.45	0.83	
33	14.2	0.92	1.88	184.79	186.67	0.83	
34	14.3	0.89	1.81	185.01	186.82	0.83	
35	14.4	0.87	1.76	185.16	186.92	0.83	
36	14.5	0.85	1.72	185.26	186.98	0.83	
37	14.6	0.84	1.69	185.32	187.01	0.83	
38	14.7	0.82	1.66	185.35	187.01	0.83	0.70 peak
39	14.8	0.80	1.62	185.35	186.97	0.83	
40	14.9	0.78	1.58	185.31	186.89	0.83	
41	15.0	0.76	1.54	185.23	186.77	0.83	
42	15.1	0.74	1.50	185.11	186.61	0.83	
43	15.2	0.72	1.46	184.95	186.41	0.83	
44	15.3	0.70	1.42	184.75	186.17	0.83	
45	15.4	0.69	1.39	184.51	185.90	0.83	
46	15.5	0.68	1.37	184.24	185.61	0.83	
47	15.6	0.66	1.34	183.95	185.29	0.83	
48	15.7	0.64	1.30	183.63	184.93	0.83	
49	15.8	0.62	1.26	183.27	184.53	0.83	
50	15.9	0.61	1.23	182.87	184.10	0.83	
51	16.0	0.60	1.21	182.44	183.65	0.83	
52	16.1	0.59	1.19	181.99	183.18	0.83	
53	16.2	0.58	1.17	181.52	182.69	0.83	
54	16.3	0.57	1.15	181.03	182.18	0.83	
55	16.4	0.56	1.13	180.52	181.65	0.83	
56	16.5	0.55	1.11	179.99	181.10	0.83	
57	16.6	0.54	1.09	179.44	180.53	0.83	
58	16.7	0.54	1.08	178.87	179.95	0.83	
59	16.8	0.53	1.07	178.29	179.36	0.83	
60	16.9	0.53	1.06	177.70	178.76	0.83	
61	17.0	0.52	1.05	177.10	178.15	0.83	0.66

TABLE E.14a   100-Year Hydrograph Routing for Porous Parking Lot (*Continued*)

Line 1	Time 2	Inflow I_2 3	$I_1 + I_2$ 4	$2S_1/\Delta T - 0$ 5	$2S_2/\Delta T + 0$ 6	Outflow 0_2 7	Depth ft 8
61	17.0	0.52	1.05	177.10	178.15	0.83	0.66
62	17.1	0.52	1.04	176.49	177.53	0.83	
63	17.2	0.51	1.03	175.87	176.90	0.83	
64	17.3	0.51	1.02	175.24	176.26	0.83	
65	17.4	0.50	1.01	174.60	175.61	0.83	
66	17.5	0.50	1.00	173.95	174.95	0.83	
67	17.6	0.49	0.99	173.29	174.28	0.83	
68	17.7	0.49	0.98	172.62	173.60	0.83	
69	17.8	0.48	0.97	171.94	172.91	0.83	
70	17.9	0.48	0.96	171.25	172.21	0.83	
71	18.0	0.47	0.95	170.55	171.50	0.83	
72	18.1	0.47	0.94	169.84	170.78	0.83	
73	18.2	0.46	0.93	169.12	170.05	0.83	
74	18.3	0.46	0.92	168.39	169.31	0.83	
75	18.4	0.45	0.91	167.65	168.56	0.83	
76	18.5	0.44	0.89	166.90	167.79	0.83	
77	18.6	0.43	0.87	166.13	167.00	0.83	
78	18.7	0.42	0.85	165.34	166.19	0.83	
79	18.8	0.41	0.83	164.53	165.36	0.83	
80	18.9	0.40	0.81	163.70	164.51	0.83	
81	19.0	0.39	0.79	162.85	163.64	0.83	
82	19.1	0.39	0.78	161.98	162.76	0.83	
83	19.2	0.38	0.77	161.10	161.86	0.83	
84	19.3	0.38	0.76	160.20	160.96	0.83	0.60
85	19.4	0.37	0.75	159.30	160.06	0.83	
86	19.5	0.37	0.74	158.40	159.14	0.83	
87	19.6	0.36	0.73	157.48	158.21	0.83	
88	19.7	0.36	0.72	156.55	157.27	0.83	
89	19.8	0.35	0.71	155.61	156.31	0.83	
90	19.9	0.35	0.70	154.65	155.35	0.83	
91	20.0	0.34	0.69	153.69	154.38	0.83	

TABLE E.14b   100-Year Hydrograph Routing for Porous Parking Lot

Line 1	Time 2	Inflow I_2 3	$I_1 + I_2$ 4	$2S_1/\Delta T - 0$ 5	$2S_2/\Delta T + 0$ 6	Outflow 0_2 7	Depth ft 8
92	20.1	0.34	0.68	152.72	153.40	0.83	
93	20.2	0.34	0.68	151.74	152.42	0.83	
94	20.3	0.34	0.68	150.76	151.43	0.83	
95	20.4	0.34	0.67	149.77	150.45	0.83	
96	20.5	0.33	0.67	148.79	149.46	0.83	
97	20.6	0.33	0.67	147.80	148.47	0.83	
98	20.7	0.33	0.66	146.81	147.47	0.83	
99	20.8	0.33	0.66	145.81	146.47	0.83	
100	20.9	0.33	0.66	144.81	145.47	0.83	
101	21.0	0.33	0.66	143.81	144.47	0.83	
102	21.1	0.33	0.65	142.81	143.46	0.83	
103	21.2	0.32	0.65	141.80	142.45	0.83	
104	21.3	0.32	0.65	140.79	141.43	0.83	
105	21.4	0.32	0.64	139.77	140.42	0.83	
106	21.5	0.32	0.64	138.76	139.40	0.83	
107	21.6	0.32	0.64	137.74	138.38	0.83	
108	21.7	0.32	0.63	136.72	137.35	0.83	
109	21.8	0.32	0.63	135.69	136.32	0.83	
110	21.9	0.31	0.63	134.66	135.29	0.83	
111	22.0	0.31	0.63	133.63	134.26	0.83	0.50
112	22.1	0.31	0.62	132.60	133.22	0.83	
113	22.2	0.31	0.62	131.56	132.18	0.83	
114	22.3	0.31	0.62	130.52	131.13	0.83	
115	22.4	0.31	0.61	129.47	130.09	0.83	
116	22.5	0.30	0.61	128.43	129.04	0.83	
117	22.6	0.30	0.61	127.38	127.99	0.83	
118	22.7	0.30	0.60	126.33	126.93	0.83	
119	22.8	0.30	0.60	125.27	125.87	0.83	
120	22.9	0.30	0.60	124.21	124.81	0.83	
121	23.0	0.30	0.60	123.15	123.75	0.83	0.46

TABLE E.14b   100-Year Hydrograph Routing for Porous Parking Lot (*Continued*)

Line 1	Time 2	Inflow I_2 3	$I_1 + I_2$ 4	$2S_1/\Delta T - 0$ 5	$2S_2/\Delta T + 0$ 6	Outflow 0_2 7	Depth ft 8
121	23.0	0.30	0.60	123.15	123.75	0.83	0.46
122	23.1	0.29	0.59	122.09	122.68	0.83	
123	23.2	0.28	0.57	121.02	121.59	0.83	
124	23.3	0.27	0.55	119.93	120.48	0.83	
125	23.4	0.26	0.53	118.82	119.35	0.83	
126	23.5	0.25	0.51	117.69	118.20	0.83	
127	23.6	0.24	0.49	116.54	117.03	0.83	
128	23.7	0.23	0.47	115.37	115.84	0.83	
129	23.8	0.22	0.45	114.18	114.63	0.83	
130	23.9	0.21	0.43	112.97	113.40	0.83	
131	24.0	0.20	0.41	111.74	112.15	0.83	
132	24.1	0.19	0.39	110.49	110.88	0.83	
133	24.2	0.18	0.37	109.22	109.59	0.83	
134	24.3	0.17	0.35	107.93	108.28	0.83	
135	24.4	0.16	0.33	106.62	106.95	0.83	0.40
136	24.5	0.15	0.31	105.29	105.60	0.83	
137	24.6	0.14	0.29	103.94	104.23	0.83	
138	24.7	0.13	0.27	102.57	102.84	0.83	
139	24.8	0.12	0.25	101.18	101.43	0.83	
140	24.9	0.11	0.23	99.77	100.00	0.83	
141	25.0	0.10	0.21	98.34	98.55	0.83	
142	25.1	0.09	0.19	96.89	97.08	0.83	
143	25.2	0.08	0.17	95.42	95.59	0.83	
144	25.3	0.07	0.15	93.93	94.08	0.83	
145	25.4	0.06	0.13	92.42	92.55	0.83	
146	25.5	0.05	0.11	90.89	91.00	0.83	
147	25.6	0.04	0.09	89.34	89.43	0.83	
148	25.7	0.03	0.07	87.77	87.84	0.83	
149	25.8	0.02	0.05	86.18	86.23	0.83	
150	25.9	0.01	0.03	84.57	84.60	0.83	
151	26.0	0.00	0.01	82.94	82.95	0.83	
152	26.1	0.00	0.00	81.29	81.29	0.83	0.30
153	26.2	0.00	0.00	79.63	79.63	0.83	
154	26.3	0.00	0.00	77.97	77.97	0.83	
155	26.4	0.00	0.00	76.31	76.31	0.83	
156	26.5	0.00	0.00	74.65	74.65	0.83	
157	26.6	0.00	0.00	72.99	72.99	0.83	
158	26.7	0.00	0.00	71.33	71.33	0.83	

TABLE E.14c   100-Year Hydrograph Routing for Porous Parking Lot

Line 1	Time 2	Inflow I_2 3	$I_1 + I_2$ 4	$2S_1/\Delta T - 0$ 5	$2S_2/\Delta T + 0$ 6	Outflow 0_2 7	Depth ft 8
159	26.8	0.00	0.00	69.67	69.67	0.83	
160	26.9	0.00	0.00	68.01	68.01	0.83	
161	27.0	0.00	0.00	66.35	66.35	0.83	
162	27.1	0.00	0.00	64.69	64.69	0.83	
163	27.2	0.00	0.00	63.03	63.03	0.83	
164	27.3	0.00	0.00	61.37	61.37	0.83	
165	27.4	0.00	0.00	59.71	59.71	0.83	
166	27.5	0.00	0.00	58.05	58.05	0.83	
167	27.6	0.00	0.00	56.39	56.39	0.83	
168	27.7	0.00	0.00	54.73	54.73	0.83	0.20
169	27.8	0.00	0.00	53.07	53.07	0.83	
170	27.9	0.00	0.00	51.41	51.41	0.83	
171	28.0	0.00	0.00	49.75	49.75	0.83	
172	28.1	0.00	0.00	48.09	48.09	0.83	
173	28.2	0.00	0.00	46.43	46.43	0.83	
174	28.3	0.00	0.00	44.77	44.77	0.83	
175	28.4	0.00	0.00	43.11	43.11	0.83	
176	28.5	0.00	0.00	41.45	41.45	0.83	
177	28.6	0.00	0.00	39.79	39.79	0.83	
178	28.7	0.00	0.00	38.13	38.13	0.83	
179	28.8	0.00	0.00	36.47	36.47	0.83	
180	28.9	0.00	0.00	34.81	34.81	0.83	
181	29.0	0.00	0.00	33.15	33.15	0.83	0.12

TABLE E.14c   100-Year Hydrograph Routing for Porous Parking Lot (*Continued*)

E.5   Summary of the Three Alternatives

Three alternatives for draining a parking lot were presented and results summarized in Table E.15. The best alternative is 3, a porous pavement parking lot for the following reasons.

1. Concrete cost is assumed to be the same, whether it is normal or porous concrete. Its cost is $11,300 less than Alternative 1.

2. It has no surface runoff so no storm sewer is needed, or if a storm sewer system already exists, it can convey a greater storm event—if the inlets are large enough.

3. All pollutants are captured and retained onsite and meets local ordinances.

4. There is no temporary ponding on the surface so office workers do not need to wade through water whenever a greater than 2-year storm occurs.

Alternative 1, normal concrete with grate inlets, is ranked second since it costs $11,300 more. It still has surface runoff, does nothing to enhance stormwater runoff

Alter.	Description	Inflow Q cfs	Outflow Q cfs	Quality Enhancement	Surface Ponding	Cost $	Difference $	Rank
1	Normal Concrete Underlain with 1 ft of gravel Grate Inlets Outlet to Storm Sewer	26.5	5.47	No	Yes	457,300	11,300	Second
2	Normal Concrete Underlain with 1 ft of gravel Grate Inlets Underground Pipes Outlet to Storm Sewer	26.5	1.80	No	Yes	758,400	312,400	Third
3	Porous Concrete Underlain with 1 ft of Gravel No Outlet to Storm Sewer	26.5	0.83	Yes	No	446,000	0	First

TABLE E.15 Summary of the Three Alternatives

quality, and office workers need to wade through some water whenever a greater than 2-year storm occurs.

Alternative 2, normal concrete with grate inlets and five rows of larger underground pipes, is ranked third because it costs $312,400 more, still has some surface runoff to a storm sewer, does nothing to enhance stormwater runoff quality, and office workers need to wade through some water whenever a greater than 2-year storm occurs.

Alternative 3, porous concrete underlain with gravel, is ranked first because it has no surface runoff, retains all pollutants onsite, nobody gets their feet wet, and costs slightly less.

Underground storage in these oversized pipes adds another 40,000 ft³ of storage to this system. The outflow hydrograph from the parking lot becomes the inflow hydrograph to the 60-in pipes. Overall peak reduction into an existing storm sewer is just 7 percent of this parking lot's peak inflow rate.

The porous concrete pavement reduces the peak runoff rate on the surface to just 3 percent into the earth. There is no surface or pollutant runoff. All is retained on-site.

Using these types of storage in an urbanized area substantially reduces conduit sizes needed in urban areas. This practice also has implications for storm sewer systems. If all parking lots had been designed using surface and/or underground storage, billions of dollars would have been saved in public storm sewer costs—or at least existing systems could contain a rarer storm event—if the inlets were large and long enough.

Single-Family Neighborhood

F.1 Introduction

Figure F.1 is an enlarged portion (1 in = 250 ft) of a 7.5-minute United States Geological Survey (USGS) quadrangle map in central Iowa published in 1972. Figure F.2 shows a highway on its southern boundary and the site's major channels and ridgelines. Then, the land was near the city of Des Moines and used as a meadow. Now, this suburb of Des Moines has expanded its boundaries and this 45.5-ac site is located within it and zoned for 0.38-ac lots. Soils are hydrologic soil group (HSG) B. Rainfall amounts for the 2-, 10-, and 100-year, 24-hour storms are 2.7, 4.6, and 6.6 in, respectively.

F.2 Traditional Subdivision

A developer had laid out a traditional single-family subdivision with rectangular lots and a grid street pattern. Most natural channels were to be eliminated during grading and replaced with storm sewers designed for a 5-year event. The only open channel left was located in the northern end of the parcel and shown as a park (see Fig. F.3). Streets are 30-ft wide within 60-ft right-of-ways (ROWs). Four-feet wide sidewalks were placed on both sides of streets. A total of 6,480 ft of streets contained 8.9 ac. The park had 3.6 ac. The other 33.0 ac had ninety-three 70 × 220 ft lots.

A storm sewer system would empty into the remaining portion of open channel. Since the land would be developed, runoff from a 5-year storm event would be increased. So would runoff from the 10- and 100-year storm events. But nothing would be done to reduce these flows to those occurring when the land was in agricultural uses. Also, nothing was included in this traditional design to reduce the pollutant load exiting the site at its northern end.

F.3 Curvilinear Streets

The developer and engineering firm revised the original plan using curvilinear streets and preserved the open channels onsite to comply with current city rules concerning runoff quantity and quality. The revised layout is shown in Fig. F.4. Lots are graded so runoff flows over grass into manmade bioswales adjacent to streets, at lots' rears, or directly into existing channels. This increases Tc, reduces flow rates, and traps pollutants. Streets' centerlines act as drainage divides.

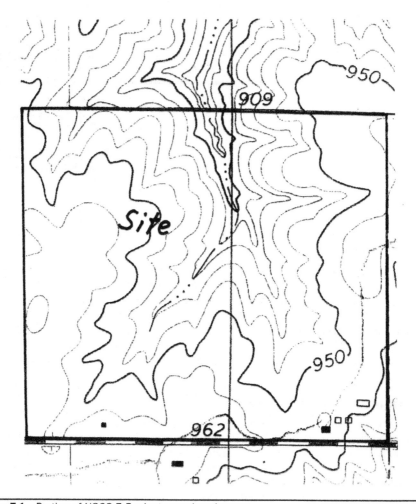

FIGURE F.1   Portion of USGS 7.5 minute quadrangle map.

Most lot runoff is to their front or rear depending on existing land slopes. It then flows in the channel to the site's north end, exiting the site through an outlet under a new street. The street becomes a detention basin's berm. Six culverts lead runoff into the basin. Small existing swales in the site's NW and SE corners are relocated to their lots' rears to allow for better lot placement.

Street length totals 5,250 ft and contains 6.0 ac in 50-ft ROWs plus 0.5 ac in a parking lot for a total of 6.5 ac. Two parks contain 2.4 ac and natural swales' ROWs occupy 2.2 ac. This totals 11.1 ac leaving 34.4 ac for 97 lots with an average size of 15,400 ft². Four more lots are created and open space is increased from 3.6 to 4.6 ac. Street length and all utilities are reduced from 6,480 ft to 5,250 ft, a reduction of 1,230 ft or 23.4 percent. Costs of streets and utilities are reduced by a similar amount. The 5-year storm sewers are eliminated. With use of porous concrete curb and gutter and front yard swales, the 100-year runoff is conveyed in manmade and natural channels.

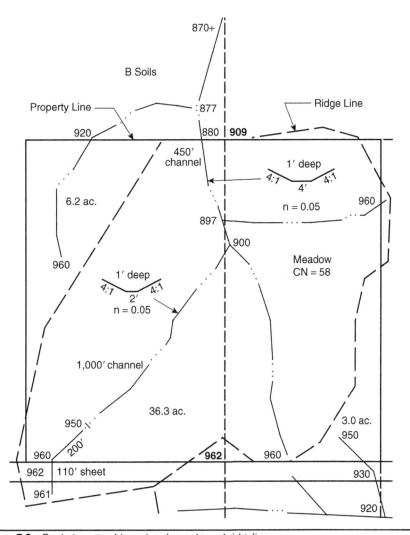

FIGURE F.2   Basic topographic major channels and ridgelines.

Coving the front yard setbacks gives the feel of estate-sized lots. Sidewalks are 7 ft wide and placed on only one side of a street, allowing a couple and a cyclist to pass each other without interference. Their curved alignments are independent of streets. The curvalinear streets act as traffic calming devices and reduction of intersections from eight to five, makes this a safer neighborhood in which to live. Residents and their visitors can easily walk to the parks and other portions of the neighborhood via sidewalks and pathways.

City ordinances requires flows for all storms including the 100-year event be reduced to at least preexisting meadow conditions and more, if possible, to further reduce downstream flooding and requires that pollutants be reduced using approved best management practices (BMPs).

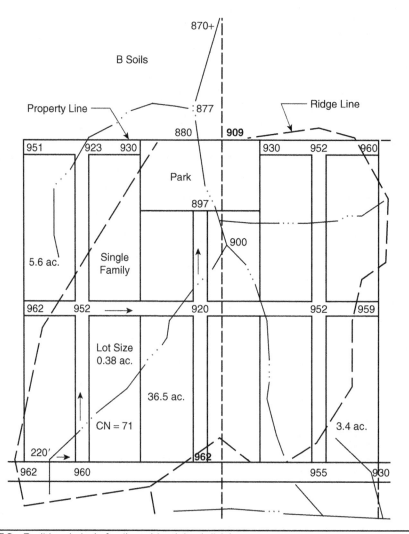

FIGURE F.3 Traditional single-family residential subdivision.

F.4 Questions Needing to Be Answered

The ordinances are met using the street across the channel at the north end as a detention basin's berm. Do we include or exclude that portion in the northwestern portion of the site that joins the main channel downstream of the property line? In or out?

1. If "out," how do we convince the city that increased flow rates and pollutant load will not worsen conditions downstream? Figure F.4 shows it to be included.

2. If "in," then how do we divert flows into the detention basin upstream of the street's berm?

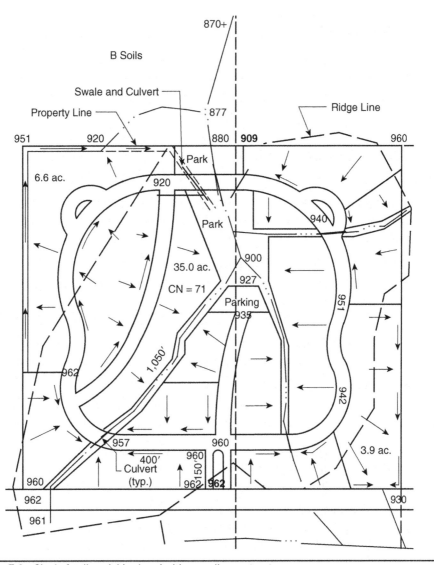

870+

B Soils

Swale and Culvert

Property Line

Ridge Line

951   920   877   880 | **909**   960

Park

6.6 ac.   920

Park

35.0 ac.   900

CN = 71   927

Parking
935

1,050'

951

962   942

957   960   3.9 ac.
400'   960
Culvert   150'
(typ.)   962 | **962**

960   930

962

961

Figure F.4   Single-family neighborhood with curvalinear streets.

3. The same questions need to be asked concerning that portion in the southeastern portion of the site? Figure F.4 shows it to be excluded.

4. Can we do more at this site than just reducing flows to the agricultural rates?

5. If the answer is yes, how much more could we reduce these runoff rates?

6. What can we do to improve the runoff's water quality?

The developer and engineering team answer these questions and are approved by city staff. With them answered, the firm begins making calculations and studies needed to obtain approvals from city, state, and federal agencies involved in the approval process.

F.5   Answers to These Questions

My answers are based on two facts: (1) ordinances and state and federal laws require runoff rates be reduced and runoff water quality enhanced, and (2) reducing or eliminating flooding and trapping pollutants on-site to improve water quality in our creeks is the right thing to do. This is accomplished by constructing greenroofs on homes and garages with net runoff conveyed to front- and rear-yard rain gardens; porous concrete streets, sidewalks, driveways, and patios underlain with a layer of gravel; and decks underlain with gravel.

F.5.1   Site's Southeastern Portion

The 3.9 ac in the site's SE portion (see Fig. F.4) flow to an existing culvert under the highway. This portion is regraded so runoff flows from an interior street to the lots' rears and down a manmade swale in the lots' rears to the existing culvert. The swale's depth is enough so that the developed 100-year, 24-hour storm's peak remaining runoff rate from grass is contained within the bioswale.

F.5.2   Site's Northwestern Portion

The 6.6 ac in the site's NW portion (see Fig. F.4) flows to the lots' rears, in a manmade bioswale to a new culvert, and into the detention basin. The swale's depth is enough that the developed 100-year, 24-hour storm's peak runoff rate from grass is contained in the swale.

F.5.3   Site's Main Portion

All lots within the western portion between the two internal streets drain from rear to front, then north in manmade bioswales to a new culvert, and then into the detention basin. Lots in the northwest portion drain from front to rear into a manmade bioswale, then through a culvert into the detention basin. Lots on the south side drain from rear to front into manmade bioswales and then into existing swales. The swales' depths are enough so the developed 100-year, 24-hour storm's peak remaining runoff rate from grass is contained within them. Except for the NE corner, the other lots all drain to the rear into existing swales, and then into the detention basin.

The additional lot length allows rain gardens underlain with gravel to temporarily store all runoff from roofs and gardens during the 2-, 10-, and 100-year, 24-hour storm event from all portions of the site.

Calculations in tables for all of the following were made but are not included. These types of calculations were made and included as tables in each of the first five appendices. Only the results of these calculations are listed in the following paragraphs.

F.5.4   Peak Flow Rates and Runoff Water Quality

The detention basin in the site's northern portion has the volume to do two things: (1) reduce peak flow rates below those when the site was a meadow and (2) trap and capture many pollutants in the runoff. The site's porous concrete portions trap all runoff and pollutants. Remaining runoff from roofs and grass eventually flows into the basin.

F.6   Runoff Curve Numbers

The site was originally a meadow and had a runoff curve number (CN) of 58. As a traditional subdivision with lots averaging 0.35 ac and a B soil, CN is 71. As a single-family neighborhood with curvalinear streets, coved yards, streets, driveways, sidewalks, patios of porous concrete, and wood decks, CN is 61. These values were obtained from

TR-55 (USDA, 1986). Note CNs' reduction from 71 to 61, traditional to newer layout based on use of on-site BMPs.

Net grass areas contributing runoff to the basin were estimated as follows. Eighty-nine lots drain into the basin with an average size of 15,400 ft^2 or 0.353 ac. With front yard setbacks ranging from 30 to 60 ft, average front yards contain $50 \times (45 + 3)$ ft or 2,400 ft^2. Driveways and sidewalks are 20×45 and 7×50 or 1,250 ft^2. Home and garage are $40 \times 50 + 24 \times 32$ or 2,770 ft^2. A patio and deck are $20 \times 20 + 12 \times 50$ or 1,000 ft^2. Total non-contributing area on each lot is 5,020 ft^2. Contributing grass area per lot is 15,400 − 5,050 or 10,350 ft^2 or 0.239 ac. Total contributing lot area is 0.239×89 or 21.2 ac. Total area contributing to the detention basin of lots and open space is 21.2 + 4.6 or 25.8 ac or 0.04031 mi^2—not 45.5 ac—based on use of on-site BMPs.

F.7 Times of Concentration

Time of concentration (Tc) for the three alternatives are the undeveloped site, traditional subdivision, and newer neighborhood concept. Values of Tc for these three are estimated as 27.6, 44.8, and 32.8 minutes, respectively. The difference is in time of sheet flow. They are based on assumptions I made for the lengths and slopes of sheet, shallow concentrated, and channel flows from Figs. F.2, F.3, and F.4. Your assumptions could be different from mine. Tc estimation is a subjective exercise.

F.8 Peak Flow Rate Reductions

Peak flows for the undeveloped site are 3.0, 30.6, and 76.1 cfs for the 2-, 10-, and 100-year, 24-hour storms. Peak flows for the site as a traditional subdivision are 14.8, 50.4, and 98.8 cfs for these storms. No flow reductions are made because a 5-year storm sewer system was constructed.

Peak flow rates for the newer neighborhood concept are 2.4, 19.5, and 44.5 cfs for the three storms and are lower than subdivision flows due to BMPs used. Runoff volumes are 0.26, 1.14, and 2.42 AF, and volumes needed to reduce flows and improve water quality to meet current requirements are 0.25, 1.11, and 2.37 AF.

F.9 Hydrograph Development

Hydrographs were developed using Worksheets 5a and 5b from TR-55 (USDA, 1986). Peak flows for the newer neighborhood flood hydrographs are 2.39, 20.09, and 51.55 cfs. These flow rates are slightly higher than those in section F.8 due to the methods used in TR-55 (USDA, 1986).

F.10 Basin Elevation-Storage Calculations

These calculations are based on Fig. F.1. Areas and volumes were determined and plotted in Fig. F.5. Maximum storage volume is 9.1 AF at elevation 920.

F.11 Outflow Structure

To reduce peak flows and improve water quality, the structure in Fig. F.6 was developed. The street was set at elevation 920. A 6-ft $\times$ 6-ft reinforced concrete riser was designed with various inlets to it.

The 911.0 pond elevation has an area of 0.46 acre. Its ratio to the drainage area is $100 \times 0.46/41.6$ or about 1.1 percent. This is sufficient for a pond in the Midwest to

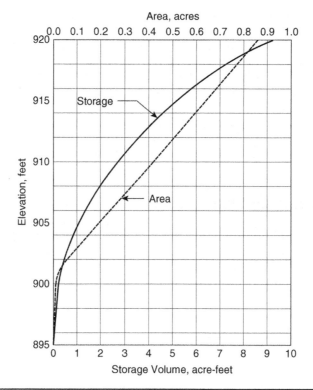

FIGURE F.5   Area and storage volume for 41.6-acre site.

balance rainfall and evaporation. It is maintained by using a reverse sloped pipe with inlet and outlet at 908.0 and 911.0. This traps trash and debris in the pond and is used as an extended detention outlet for a 2-year quality storm and is small enough to detain its runoff volume for an average of 2 days.

A gated riser pipe drains to allow removal of sediment every 20 years. A reinforced concrete pipe (RCP) outlet is used with elevations at 892.0 and 881.0. An orifice at 912.5 is a 10-year storm's outlet. The 914.0 riser's top is partially open for 100-year storm runoff to exit the pond with freeboard below the street.

F.12   Revised Elevation-Storage Calculations

A pond reduces storage volumes to temporarily capture the 2-year water quality event plus helps to reduce outflows from the 10- and 100-year storm to or below flow rates required by the local ordinance. These calculations were made and plotted in Fig. F.7. Maximum storage volume at elevation 920.0 is reduced to 5.84 AF.

F.13   Outlet Structure Hydraulics

A small plastic pipe is the extended detention outlet for a 2-year water quality storm. A vertical rectangular orifice outlets the 10-year storm. An orifice in the riser's top outlets flows from the 100-year storm.

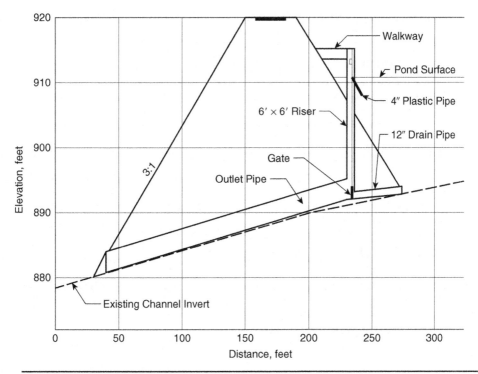

FIGURE F.6   Outlet pipe and riser for newer neighborhood concept.

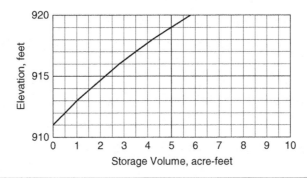

FIGURE F.7   Revised storage volumes for 41.6-acre site.

F.13.1   Outlet Pipe Calculations

Calculations for a 24-in RCP outlet pipe were made. The maximum water elevation inside the riser for total outflow during the 100-year storm event must be below 911.0.

F.13.2   Orifice for the 2-Year Water Quality Event

Calculations for a 4-in diameter orifice were made. Runoff from this storm must be captured and temporarily stored below the next outlet's crest elevation. Total runoff volume during the 2-year storm is $0.26 \times 25.8/12$ or 0.56 AF. Storage at elevation 912.5 is 0.75 AF. Peak flow at elevation 920 is 1.22 cfs.

F.13.3 Orifice for the 10-Year Storm Event

Calculations for a 6-in wide by 12-in high rectangular orifice were made. This orifice reduces flows from the 5- and 10-year event. Its crest elevation is at elevation 912.5. Flow through the lower 4-in orifice also contributes flow to the outlet pipe during these storm events. Peak flow at 920 is 6.59 cfs.

F.13.4 Orifice for the 100-Year Storm Event

Calculations for a horizontal 12-in circular orifice were made. This orifice reduces flows during the 100-year event. Its crest elevation is 914.0. Flow through the two lower orifices also contribute flow to the outlet pipe during the 100-year event. Peak flow at 920 is 10.00 cfs.

F.13.5 Total Flow through the Outlet Pipe

Total outflow through the outlet pipe is the sum of the flows through each of the inlets to the outlet structure for each foot of depth. Total outflow through the three openings in the riser structure at elevation 920.0 is 17.81 cfs.

F.13.6 Routing Curve Calculations

The routing curve calculations were made and plotted in Fig. F.8. Peak value of $2S/\Delta T + O$ at elevation 920 is 718.6.

F.14 Hydrograph Routing through Fig. F.4

F.14.1 Routing of the 2-Year, 24-Hour Hydrograph

Its peak inflow of 2.39 cfs at hour 12.6 is reduced to a peak outflow of 0.27 cfs at hour 23.0, a delay of 10.4 hours. Maximum pond elevation is 911.76, below the 912.5 crest of the next higher outlet. This peak outflow is also well below the allowable 2.7 cfs estimated for the undeveloped site. At hour 33.0, the flow rate is 0.16 cfs and the temporary pond is at 911.55. The pond continues to drain and is empty at hour 100, 4.2 days after the rain begins.

F.14.2 Routing of the 10-Year, 24-Hour Hydrograph

Peak inflow rate of 20.09 cfs at hour 12.4 is reduced to a peak outflow rate of 2.77 cfs at hour 14.2, a delay of 1.8 hours. Maximum temporary pond elevation is 913.34, below the crest of the next higher outlet at elevation 914.0. This peak outflow rate is also well below the undeveloped rate of 30.6 cfs estimated for the undeveloped site.

F.14.3 Routing of the 100-Year, 24-Hour Hydrograph

Peak inflow rate of 51.55 cfs at hour 12.4 is reduced to a peak outflow rate of 10.60 cfs at hour 13.2, a delay of just 0.8 hour. Maximum pond elevation is 915.5, 4.5 ft below the street elevation of 920.0. The temporary pool subsides to 912.1 at hour 35.0. This peak outflow rate is well below the undeveloped rate of 76.1 cfs estimated for the undeveloped site.

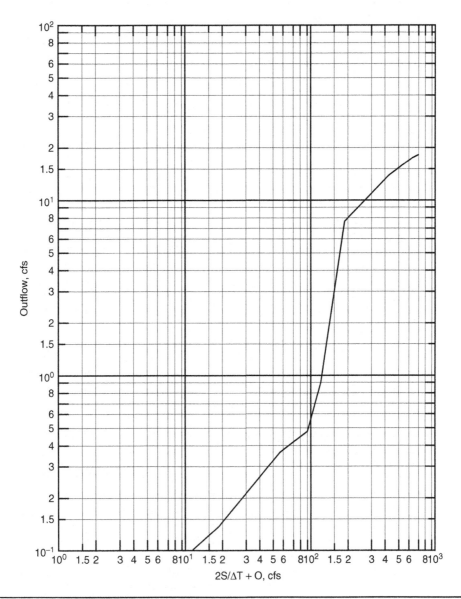

FIGURE F.8   Routing curve for neighborhood concept.

F.15   Summary of Appendix F

A 45.5-acre parcel of land in southcentral Iowa was selected to compare what a meadow would look like if it were developed as:

1. A traditional single-family subdivision with its grid pattern of streets, rectangular lots, and uniform placement of houses, or

2. A newer neighborhood concept with curvalinear streets, varying front yard setbacks, homes at angles to each other, and various BMPs in line with the concepts of low impact development (LID) and the triple bolttom line (TBL) (people, planet, profit).

The traditional subdivision has 6,480 ft of streets in 60-ft ROWs with 4-ft sidewalks on both sides of streets, eight intersections, complete regrading of the site and elimination of most existing on-site swales, ninety-three 70 × 220 ft (15,400 ft²) lots, a 3.6-acre park, a 5-year storm sewer system, no provision for runoff from rarer event storms, and no way to keep pollutants on site.

The newer neighborhood concept contains 5,250 ft of streets in 50-ft ROWs with 7-ft wide sidewalks on one side of streets, five intersections, regrading of a portion of the site, use of existing on-site natural swales and small bioswales, 97 homes on lots with an average size of 15,400 ft², parks and swales totaling 4.6 ac, elimination of storm sewers, swales containing the 100-year, 24-hour storm runoff, use of greenroofs, porous concrete, bioswales, a detention basin, and capture and retention of most pollutants on-site.

Reduction of 1,230 ft of street means reduction of 1,230 ft of all utilities, and elimination of a storm sewer system means savings in construction and long-term maintenance cost of streets and utilities of several hundred thousand dollars. All swales are in dedicated ROWs so maintenance of them and the detention basin are done by city staff.

The newer neighborhood concept design meets current ordinances and state and federal laws and regulations concerning flooding and water quality enhancement. Peak-flow rates for the 2-, 10-, and 100-year, 24-hour storm events are reduced from 89 to 79 percent and most pollutants are captured and retained on site. Residents and their visitors can use sidewalks and pathways along the swales to wander throughout the neighborhood. The above results yield a much more attractive, safer, economical, and environmentally sustainable location in which to live and play. This single-family neighborhood development again demonstrates that it and other land uses can blend in drainage facilities as attractive BMPs as part of an LID and TBL.

Because the outflows from the detention basin are so much lower (0.27, 2.77, and 10.60 cfs for the 2-, 10-, and 100-year storms) than the allowable flow in the site's undeveloped condition (3.0, 30.6, and 76.1 cfs for these three storms), some BMPs might be omitted from the newer neighborhood concept to increase inflows to the detention basin and to save some costs. For example, the roofs could be constructed with shingles rather than as greenroofs.

Also, because 4.5 ft of freeboard remained during the 100-year storm, the sizes of the outlet pipe and riser could be reduced plus decreasing the sizes of the 10- and 100-year outlets to reduce somewhat the amount of remaining freeboard. This reduces construction cost. Remember two items:

1. Storage volumes in the proposed detention basin were obtained without moving one shovel of dirt. Fill for the street serving as its berm was obtained from other portions of the site to develop street and lot grading.

2. Four or six 6 × 6-in vertical concrete protusions must be constructed around the perimeter of the riser to counteract the coriolis force which wants to make the water rotate inside the riser as it is falling to the bottom.

APPENDIX G

Office Park

G.1 Original Problem Description

A 30-ac parcel in St. Louis, Missouri, is to be developed as an office complex. Figure G.1 shows its existing topography. Figure G.2 is a proposed plan for three 2-story office buildings, parking lot, and grassy areas containing a detention basin. There are existing streets on all sides. Parking requirements are one space for each 300 ft^2 of gross building space. Soil is Natural Resources Conservation Service (NRCS) hydrologic soil group (HSG) B with a minimum infiltration rate of 0.35 in/h.

How many subareas should be created in order to develop good inflow hydrographs? Where are time of concentration (Tc) pathways? Grass is located between buildings with concrete sidewalks. Is there a curb between parking lot and grassy areas? End grassy areas contain picnic tables, horseshoe pits, basketball hoops, and a running track for office workers to enjoy during lunch times in good weather. There are front and side entrances to buildings and trucks unload materials at their rears.

What is the runoff volume in relationship to volume of temporary storage available in the basin during the 2-, 10-, and 100-year, 24-hour rainfall events? Develop inflow hydrographs, size an outflow pipe, and route the hydrographs through the detention basin. Assume a 150-ft-long outlet pipe has a 2.0 percent slope and is in inlet control with low tailwater (TW).

G.1.1 Rainfall

A 100-year, 24-hour rainfall in St. Louis is 7.8 in. With an infiltration rate of 0.35 in/h, runoff occurs only during hours 11 through 14 and is 3.82 in. in pervious areas. The 2- and 10-year rains are 3.2 and 5.3 in, respectively, from App. B in TR-55 (USDA, 1986).

G.1.2 Times of Concentration

How many subareas should be used and where are Tc pathways? It could be a single area and Tc estimated using the longest in time path. Or buildings, parking lot, and grassy areas could be subareas. A clue is to follow paths water takes as it flows to the basin from elevations shown on Fig. G.2. Water from the outer offices flows across its parking, thru grassy areas, to the basin. Or water from them flows across its parking, along a curb between parking and grass, to the basin. Water from office B and grass flows across its parking to the basin.

The deciding factor is: is there a curb or no curb along the parking lot's lower end? Tc is longer with lower peaks and pollutants trapped in grass if no curb is used.

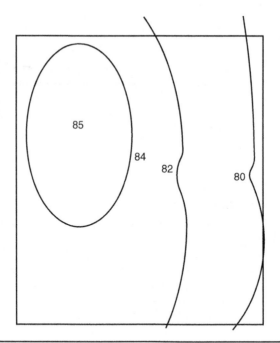

Figure G.1 Existing topography of 30-ac parcel.

The area B roof is sloped from its rear to the parking lot. Its Tc is 21.4 minutes. The A and C roofs are sloped from their rears near area B to their fronts at the site's edges. Its Tc is 37.6 minutes. Areas A and C have a total Tc of 119.1 minutes or 2.0 hours. Area B has a total Tc of 87.4 minutes or 1.5 hours.

G.1.3   Runoff Curve Numbers

The detention basin is assumed to be full of water and has a runoff curve number (CN) of 100. This assumption was made to put a more severe test on it because, during a 100-year event's first portion and large portions of more frequent events, it is not full and some rainfall will infiltrate, resulting in a lower CN. A and C have a CN of 89 and B has a CN of 93.

G.1.4   Inflow Hydrographs

G.1.4.1   2-Year Storm

Peak inflow rate is 21.3 cfs at hour 13.2. Tc of 2.0 hours for subareas A and C and 1.5 hours for area B. Ia/P used was 0.1.

G.1.4.2   10-Year Storm

Peak inflow rate is 39.6 cfs at hour 13.2. Tc of 2.0 hours for subareas A and C and 1.5 hours for area B. Ia/P used was 0.1.

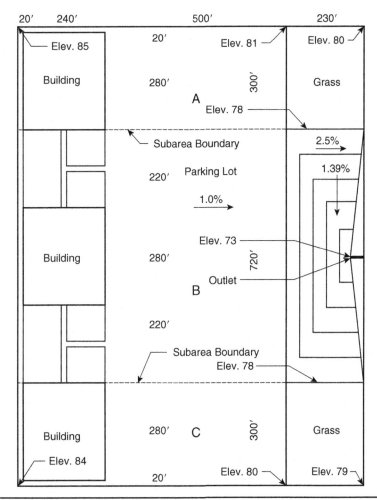

FIGURE G.2   Proposed plan view of office park.

G.1.4.3   100-Year Storm

Peak inflow rate is 61.8 cfs at hour 13.2. Tc of 2.0 hours for subareas A and C and 1.5 hours for area B. Ia/P used was 0.1.

G.1.5   Depth-Storage Relationship

Frustum of a cone method was used because it yields slightly more accurate values that are smaller than the average end area method. Total storage volume at a depth of 5-ft is 6.4 AF.

G.1.6   Depth-Outflow Relationship

Runoff volume from a 100-year storm is 30 × 7.0/12 or 17.5 AF. Total storage volume is 6.4 AF. Ratio of these two values is 6.4/17.5 or 0.37. From Fig. 6-1 in TR-55 (USDA, 1986),

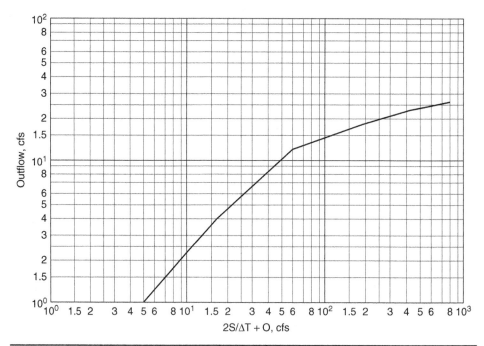

Figure G.3 Hydrograph routing curve for office park.

runoff peak is reduced to 61.8 × 0.335 or 20.9 cfs. Allowable headwater (AHW) depth for the outlet pipe is 5 ft.

A 150-ft pipe on a 2.0 percent slope has inlet control and TW is low. Size the pipe so its headwater depth is less than 5-ft. A 21-in reinforced concrete pipe (RCP) works with an outflow rate of 26 cfs at a HW depth of 5.0 ft under inlet control.

G.1.7 Routing Curve

A routing curve combines a depth-storage and depth-outflow curves. The result is plotted in Fig. G.3.

G.1.8 2-Year Hydrograph Routing

Peak outflow is 13 cfs at a depth of 2.1 ft, 2.9 ft of freeboard. Flow is reduced 39 percent, delayed for 0.0 hour, and inflow equals outflow at hour 19.2. There is no water-quality enhancement.

G.1.9 10-Year Hydrograph Routing

Peak outflow is 20.1 cfs at a depth of 3.3 ft with 1.7 ft of freeboard. Flow is reduced 49 percent and delayed for 1.0 hour. Outflow is almost equal to inflow at hour 20.4.

G.1.10 100-Year Hydrograph Routing

Peak outflow rate is 24.9 cfs at a depth of 4.2 ft with 0.8 ft of freeboard. Peak flow was delayed for 1.2 hour. Depth is 0.9 ft at hour 23.0. Thus, rather than sizing this culvert for 61.8 cfs, it need only be designed for 24.9 cfs. An outlet is designed for a peak flowing *through* it and not a peak flow *approaching* it. If there is little or no storage volume available upstream of an outlet, then a peak flow rate approaching and flowing through an outlet structure will be the same.

G.1.11   Cost

A 150-ft, 21-in RCP costs $46.00/lineal foot. One end section is $400.00 and connection to a storm sewer costs $1,000. Total cost is $46.00 × 150 + 400 + 1,000 or $8,300.

G.1.12   Summary

Peak-flow rate has been reduced from 61.8 cfs to 24.9 cfs and was delayed 1.2 hours. Runoff from two rooftops and parking lots is detained and infiltrated in grassy areas before entering the detention basin. Runoff from the center building and its parking lot is detained in the detention basin. Basin bottom and sides are covered with grass. Few pollutants are removed from a basin that is usually empty with a relatively large outlet. The grassy areas contain picnic tables and some recreational equipment so office workers can eat their lunches, converse, and engage in some exercise during their lunch hours on days when the weather is pleasant.

G.2   Revised Problem Description

A revised 30-acre office park in St. Louis has eight 2-story office buildings with green roofs as shown in Fig. G.4. Each is 320 by 240 ft and covers 1.76 ac. Parking is replaced by 2-story garages under the buildings. 10-ft-wide porous concrete sidewalks underlain with gravel surround the development using 1.1 ac. Building entrances are located in the middle of their sides. Garage entrances and exits are near the corners. They give employees access to their offices with shelter for them and their cars. Garages have one parking space for each 300 ft^2 of gross office space. Greenroofs total 14.1 ac covered with plants, shrubs, and grass. Some is covered space for meetings, lunches, and after-hours activities. Six inches of amended soil retains 2 in of rain. A drainage layer has storage for 100 percent of the 5.8 in of rain during a 100-year, 24-hour storm.

Rain garden areas total 5.6 ac and accept eventual runoff from roofs and rain on these gardens during this storm event. They are 1 ft lower than surrounding areas. Amended soil in them is composed of 50 percent sand, 25 percent organic matter, and 25 percent wood chips.

A central area of 8.1 ac contain grass, shrubs and trees on amended soil. Wide porous concrete pathways wander through the area to benches, picnic tables, and gazebos. A separate porous asphalt jogging trail wanders through the area. Other recreational activities are located in the lower right-hand corner of this open area including tennis, basketball, and badminton courts, horseshoe pits, and basketball hoops that serve as a 2.2-ac detention basin.

These interior areas drain to the lower 2.2-ac BMP. Basketball and tennis areas use porous asphalt with grass on the others. Surface storage and a layer of gravel is sufficient to temporarily detain all runoff from a 100-year storm. Goals are to infiltrate all rainfall up to a 100-year, 24-hour event, capture all pollutants, and replenish groundwater. This alleviates runoff into streets and existing storm sewer systems. Local stormwater fees should be partially or totally forgiven.

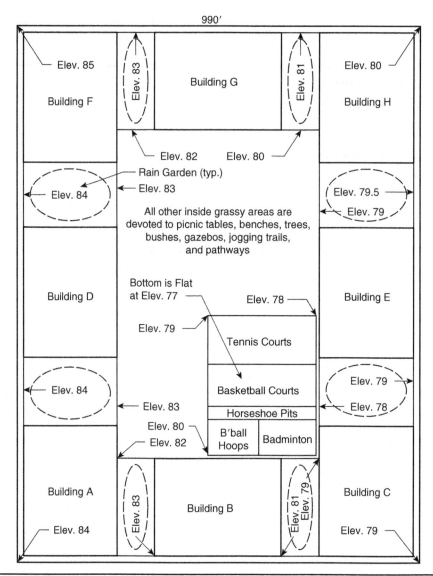

Figure G.4 Plan view of revised office park.

G.3 Rooftops

G.3.1 Description

They are extensive greenroofs with 6 in of amended soils. One layer is a drainage layer of 12 in of gravel with a porosity of 40 percent. Storage volume is $0.40 \times 1 \times 1 \times 12 = 4.8$ in/ft^2 of roof area. Thus, all rainfall is retained or temporarily stored with some rain used for plant growth. Runoff drains into 3-in by 3-in square downspouts that lead to adjacent rain gardens.

G.3.2   Times of Concentration

Downspouts are located in the middle of buildings B, D, E, and G's ends. These roofs have low center walls so water drains in two directions. Longest Tc path is from a building's side in the middle in a diagonal line along half its length to its end at its center. It is equal to $((160)^2+(120)^2)^{0.5}$ or 200 ft. Tc is estimated as 17.2 minutes or 0.30 hour.

Downspouts are located at the edge in the buildings' middles of A, C, F, and H's ends with no center walls. Longest Tc path is from a building's corner in a diagonal line along its length to the building's other end at its center. This length is equal to $((320)^2+(120)^2)^{0.5}$ or 342 ft. Tc is estimated as 32.7 minutes or 0.50 hour.

G.3.3   Inflow Hydrographs on Roofs A, C, F, H

When rain begins, water flows through the soil, into gravel, and to a downspout in 0.5 hour, 5.8 in of rain flows into the gravel, then drains into the rain garden that receives this runoff.

G.3.3.1   2-Year Storm
Peak inflow rate to the gravel is 1.75 cfs at hour 12.4. Peak outflow rate in the downspout is much less.

G.3.3.2   10-Year Storm
Peak inflow rate to the gravel is 4.81 cfs at hour 12.4. Peak flow in the downspout is less.

G.3.3.3   100-Year Storm
Peak inflow rate to the gravel is 8.45 cfs at hour 12.4. Peak flow in the downspout is less.

G.3.4   Inflow Hydrographs on Roofs B, D, E, G

When rain begins, water infiltrates through the soil, into gravel, and to a downspout in 0.3 hour. 5.8 in of rain flows into the gravel, then into the rain gardens that receive this runoff.

G.3.4.1   2-Year Storm
Peak inflow rate to the gravel is 1.12 cfs at hour 12.2. Peak flow in the downspout is less.

G.3.4.2   10-Year Storm
Peak inflow rate to the gravel is 3.07 cfs at hour 12.2. Peak flow in the downspout is less.

G.3.4.3   100-Year Storm
Peak inflow rate to the gravel is 5.40 cfs at hour 12.2. Peak flow in the downspout is less.

G.3.5   Depth-Storage Calculations
G.3.5.1   Calculations on Roofs A, C, F, H
The foot of gravel has a porosity of 40 percent. Their pore spaces can temporarily store 1.76 AF of water in the roof's gravel layer. Storage values are determined every 0.2 ft to yield a better estimate of how deep water will be temporarily stored in this drainage layer.

G.3.5.2   Calculations on Roofs B, D, E, G
The foot of gravel has a porosity of 40 percent. Their pore spaces can temporarily store 0.88 AF of water in the roof's gravel layer with storage values determined every 0.2 ft to yield a better estimate of how deep water will be temporarily stored in this drainage layer.

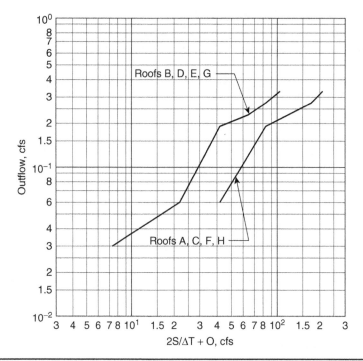

FIGURE **G.5** Rounting curves for greenroofs.

G.3.6 Depth-Outflow Calculations

A 3-in by 3-in horizontal orifice (downspout) has values determined every 0.2 ft. The downspout outflows 0.34 cfs with a head of 1.0 ft.

G.3.7 Routing Curve Calculations

The routing curves for calculations for roofs A, C, F, and H are plotted in Fig. G.5. They are also plotted in Fig. G.5 for roofs B, D, E, and G.

G.3.8 Hydrograph Routing for Roofs A, C, F, H

G.3.8.1 2-Year Storm

Its peak of 1.75 cfs at hour 12.4 is reduced to 0.09 cfs at hour 15.8. It becomes a portion of an inflow hydrograph to the larger rain gardens. Peak depth in the gravel is 0.26 ft.

G.3.8.2 10-Year Storm

Its peak of 4.81 cfs at hour 12.4 is reduced to only 0.18 cfs at hour 17.4. It becomes a portion of an inflow hydrograph to the larger rain gardens. Peak depth in the gravel is 0.38 ft.

G.3.8.3 100-Year Storm

Its peak of 8.45 cfs at hour 12.4 is reduced to 0.24 cfs at hour 19.4. It becomes a portion of an inflow hydrograph to the larger rain gardens. Peak depth in the gravel is 0.64 ft. The other 0.36 ft of storage is for less use of water by plants, partial plugging of gravel, and a greater than 100-year storm.

G.3.9 Hydrograph Routing for Roofs B, D, E, G

G.3.9.1 2-Year Storm
Its peak of 1.12 cfs at hour 12.2 is reduced to only 0.04 cfs at hour 16.4. It becomes a portion of an inflow hydrograph to each rain garden. Peak depth in the gravel is 0.13 ft.

G.3.9.2 10-Year Storm
Its peak of 3.07 cfs at hour 12.2 is reduced to only 0.08 cfs at hour 18.0. It becomes a portion of an inflow hydrograph to each rain garden. Peak depth in the gravel is 0.25 ft.

G.3.9.3 100-Year Storm
Its peak of 5.40 cfs at hour 12.2 is reduced to only 0.21 cfs at hour 15.0. It becomes a portion of an inflow hydrograph to each rain garden. Peak depth in the gravel is 0.50 ft. The other 0.50 ft of storage is for less use of water by plants, partial plugging of gravel, and a greater than 100-year storm.

G.4 Larger Rain Gardens

G.4.1 Description
Rain gardens have oval shapes. They have rain falling on them plus greenroof runoff, are 1-ft deep, with 5:1 side slopes. Three-feet of soil is removed and replaced with 0.5-ft of amended soil, 1.5-ft of gravel, and 1.0-ft of amended soil consisting of a sand, organic matter, and wood chips mixture.

G.4.2 Inflow Hydrographs

G.4.2.1 2-Year Storm
Assume Tc is 6 minutes, the shortest time period used in TR-55 (USDA, 1986). With a B-type soil, its NRCS CN is 61 from Table 2-2a in TR-55 (USDA, 1986). Ia/P used was 0.30 to yield larger ordinates for a greater test. Peak-inflow rate is 0.41 cfs at hour 12.2 from rain falling on the garden itself.

G.4.2.2 10-Year Storm
Assume Tc is 6 minutes, the shortest time period used in TR-55 (USDA, 1986). With a B-type soil, its NRCS CN is 61 from Table 2-2a in TR-55 (USDA, 1986). Ia/P used was 0.10 to yield larger ordinates for a greater test. Peak inflow rate is 2.16 at hour 12.2 cfs from rain falling on the garden itself.

G.4.2.3 100-Year Storm
Assume Tc is 6 minutes, the shortest time period used in TR-55 (USDA, 1986). With a B-type soil, its NRCS CN is 61 from Table 2-2a in TR-55 (USDA, 1986). Ia/P used was 0.10 to yield larger ordinates for a greater test. Peak inflow rate is 4.59 cfs at hour 12.2 from rain falling on the garden itself.

G.4.3 Depth-Storage Calculations
A rain garden is 100 × 220 ft. Area at each depth is equal to area of a circle plus area of a rectangle: Area = L × W + 0.7854 × Diam.^{2}. The 1.0-ft of gravel's area is 160 ft by 240 ft. Total storage volume is 31,500 ft^3. Runoff volume to each garden is 160 × 240 × 3.29/12 = 10,530 ft^3 plus runoff from 1.5 roofs during a 100-year, 24-hour storm.

G.4.4   Depth-Outflow Calculations

Outflow from each larger rain garden is infiltration into the earth itself. Gravel area is 160 ft by 240 ft and infiltration rate is 0.35 in/hr, outflow rate is:

$$Q = 0.35 \text{ in/hr} \times 1 \text{ ft/12 in} \times 160 \text{ ft} \times 240 \text{ ft} \times 1 \text{ cfs-hr}/3{,}600 \text{ ft}^3 = 0.31 \text{ cfs}$$

Thus 0.31 ft^3 every second infiltrates into the earth beneath these rain gardens.

G.4.5   Routing Curve Calculations

There is no need to plot this curve because outflow is a constant 0.31 cfs.

G.4.6   Hydrograph Routing

G.4.6.1   2-Year Storm

Inflow hydrographs from a rain garden's area and outflow hydrograph ordinates from greenroofs A, C, F, and H plus greenroofs D and E are combined and this combined hydrograph is then routed. Results indicate that maximum depth is only 0.02 ft in the gravel. Runoff in the rain garden is infiltrated into the amended soil, so no runoff ponds on the rain garden's surface.

G.4.6.2   10-Year Storm

Inflow hydrographs from a rain garden's area and outflow hydrograph ordinates from greenroofs A, C, F, and H plus greenroof D and E are combined and this combined hydrograph is then routed. Results indicate that maximum depth is only 0.25 deep in the gravel. Runoff in the rain garden is infiltrated into the amended soil, so no runoff ponds on the rain garden's surface.

G.4.6.3   100-Year Storm

Inflow hydrographs from a rain garden's area and outflow hydrograph ordinates from greenroofs A, C, F, and H plus greenroofs D and E are combined and this combined hydrograph is then routed. Results indicate that maximum depth in a rain garden is 0.42 ft at hour 36.8 and is empty on the surface at hour 60. Runoff continues to drain from the gravel for about 1.2 more days. Thus, all rainfall up to and including the 100-year, 24-hour storm event from six buildings and the larger rain gardens infiltrates. Detaining runoff on roofs reduces peak rates and delays its arrival into the rain gardens until hours after the peak flow into the rain gardens themselves occurs.

G.5   Smaller Rain Gardens

G.5.1   Description

Rain gardens adjacent to buildings B and G have oval shapes with rain falling on them and runoff from half a greenroof. They are a foot deep, 5:1 side slopes, and underlain with 1.0-ft of gravel. Remove 3-ft of earth and replace with 1.0-ft of amended soil, 1.0-ft of gravel, and 1.0-ft of amended soil consisting of 50 percent sand, 25 percent organic matter, and 25 percent wood chips. These store all runoff up to and including a 100-year, 24-hour storm.

G.5.2 Inflow Hydrographs

G.5.2.1 2-Year Storm

Assume Tc is 6 minutes, the shortest time period used in TR-55 (USDA, 1986). With a B-type soil, its NRCS CN is 61 from Table 2-2a in TR-55 (USDA, 1986). Ia/P used was 0.3 to yield larger ordinates for a greater test. Peak-inflow rate is 0.25 cfs at hour 12.2 from rain falling on the garden itself.

G.5.2.2 10-Year Storm

Assume Tc is 6 minutes, the shortest time period used in TR-55 (USDA, 1986). With a B-type soil, its NRCS CN is 61 from Table 2-2a in TR-55 (USDA, 1986). Ia/P used was 0.1 to yield larger ordinates for a greater test. Peak-inflow rate is 1.29 cfs at hour 12.2 from rain falling on the garden itself.

G.5.2.3 100-Year Storm

Assume Tc is 6 minutes, the shortest time period used in TR-55 (USDA, 1986). With a B-type soil, its NRCS CN is 61 from Table 2-2a in TR-55 (USDA, 1986). Ia/P used was 0.1 to yield larger ordinates for a greater test. Peak inflow rate is 2.75 cfs at hour 12.2 from rain falling on the garden itself.

G.5.3 Depth-Storage Calculations

Each garden is an oval. Top width of each garden is 76 ft by 220 ft. Area at each depth is equal to a circle's area plus a rectangle's area. Its equation is: Area = 0.7854 × Diam.^{2} + L × W. The 1.0 ft of gravel underneath has an area of 96 × 240 ft. Total runoff volume into each garden is 96 × 240 × 3.3/12 equals 6,340 ft^3 plus roof runoff. Thus, the smaller rain gardens can store all runoff from a 100-year event on itself and from half of a greenroof.

G.5.4 Depth-Outflow Calculations

Outflow from each smaller rain garden is infiltration into the earth itself. Since each area is 96 × 240 ft and infiltration rate is 0.35 in/hr, outflow rate is:

$$Q = 0.35 \text{ in/hr} \times 1 \text{ ft} / 12 \text{ in} \times 96 \text{ ft} \times 240 \text{ ft} \times 1 \text{ cfs-hr}/3{,}600 \text{ ft}^3 = 0.19 \text{ cfs}$$

Thus 0.19 cfs infiltrates into the earth beneath these rain gardens.

G.5.5 Routing Curve Calculations

The routing curve is not plotted because infiltration into the soil is constant at 0.19 cfs.

G.5.6 Hydrograph Routing

G.5.6.1 2-Year Storm

The combined hydrograph from roof and grass is routed. Results indicate that maximum depth is only 0.03 ft in the gravel. Runoff in the rain garden is infiltrated into the soil, so no runoff ponds on the surface.

G.5.6.2 10-Year Storm

The combined hydrograph from roof and grass is routed. Results indicate that maximum depth is 0.31 ft deep in the gravel. Runoff in the rain garden is infiltrated into the soil, so no runoff ponds on the surface.

G.5.6.3   100-Year Storm

The combined hydrograph from roof and grass is routed. Peak inflow rate is 2.87 cfs at hour 12.2 and is reduced to a peak outflow rate of 0.19 cfs at hour 26.4. Results indicate that maximum depth in the gravel is 0.90 ft. Gravel is empty at hour 51.6. Surface storage is for partial plugging of the gravel and occurrence of a larger than 100-year storm event.

For both the larger and smaller rain gardens, all runoff from roofs, rain gardens, and grass surrounding the rain gardens infiltrates into the soil. No runoff or pollutants leave the site for all storms up to and including the 100-year, 24-hour event.

G.6   Interior Courtyard and Recreational Area

G.6.1   Description

The recreational area serves as a retention basin for the courtyard. Thus, all rain on them acts as a grassy area and infiltrates at a rate of 0.35 in/hr. Its surface is a minimum of 1 ft below surrounding areas and underlain with a foot of gravel. It accepts all runoff from the courtyard and itself. Surface and underground storage are enough to retain all runoff up to and including a 100-year, 24-hour storm. Thus, all runoff and potential pollutants are retained on site.

G.6.2   Inflow Hydrographs

With a type B soil, the NRCS curve number is 61. Total area of the inner courtyard and the recreational area is 10.30 ac. Tc is estimated as 86.1 minutes or 1.44 hours. Use Tc as 1.25 hours to place greater stress on the recreational area.

G.6.2.1   2-Year Storm

Ia/P was used as 0.3 hours to place greater stress on it. Peak runoff is 4.83 cfs at hour 12.2.

G.6.2.2   10-Year Storm

Ia/P was used as 0.1 hours to place greater stress on it. Peak rate is 7.77 cfs at hour 13.0.

G.6.2.3   100-Year Storm

Ia/P was used as 0.1 hours to place greater stress on it. Peak rate is 16.48 cfs at hour 13.0.

G.6.3   Depth-Storage Calculations

Its area is 272 × 344 ft with an area of 93,570 ft^2 or 2.15 ac with 5:1 side slopes. It is a minimum of 1-ft deep (see Fig. G.4) and underlain with 1.5 ft of gravel. Depth-volume calculations total 92,350 ft^3 or 2.12 AF. Runoff volume is 10.3 × 3.3/12 or 2.83 AF. This area temporarily stores all runoff from a 100-year, 24-hour storm on its surface and underground.

G.6.4   Depth-Outflow Calculations

Outflow from this recreational area is infiltration into the earth itself. Since this area is 272 × 344 ft and the soil infiltration rate is 0.35 in/hr, outflow rate is:

$$Q = 0.35 \text{ in/hr} \times 1 \text{ ft}/12 \text{ in} \times 272 \text{ ft} \times 344 \text{ ft} \times 1 \text{ cfs-hr}/3{,}600 \text{ ft}^3 = 0.76 \text{ cfs}$$

Thus, 0.76 cfs infiltrates into the earth. This rate is more than sufficient to drain water off these various playing surfaces in sufficient time for them to dry out without unduly delaying play.

G.6.5   Routing Curve Calculations

The routing curve was not plotted because infiltration is constant at 0.76 cfs.

G.6.6   Hydrograph Routing

G.6.6.1   2-Year Storm

Results indicate that maximum depth is only 0.10 ft deep in the gravel. Runoff is infiltrated into the soil and gravel, so no runoff ponds on the surface in the recreational area.

G.6.6.2   10-Year Storm

Results indicate that maximum depth is only 0.86 ft deep in the gravel. Runoff is infiltrated into the soil and gravel, so no runoff pounds on the surface in the recreational area.

G.6.6.3   100-Year Storm

Peak inflow rate is 16.48 cfs at hour 13.0. Peak outflow rate is 0.76 cfs at hour 21.0, a delay of 8.0 hours. Results indicate that maximum depth on the surface in the recreational area is 0.82 ft during this 100-year, 24-hour event. Water is on the surface for 20.8 hours, hour 13.6 to 34.4, in the recreational area. Gravel is completely drained in another 16 hours. This is adequate time for all pollutants to be either trapped on the surface or in the gravel. Thus, in the central courtyard and recreational area, no runoff or pollutants exit the site during a 100-year, 24-hour storm event.

G.7   Summary

This 30-acre site was to be developed first with three 2-story office buildings as portrayed in Fig. G.2. About half the site was utilized for surface parking with its impervious area. A 21-in RCP served as an outflow structure from a detention basin. Peak-flow rates were reduced and runoff flowed through this 21-in RCP to an existing storm sewer. Since this detention basin would normally be dry, little, if anything, was done to enhance stormwater runoff quality. Two grassy areas were equipped with picnic tables, benches, and some recreation equipment.

This site was then revised to include eight 2-story office buildings with two-story underground garages. Intensive greenroofs, rain gardens, and a recreational area handled all runoff from these buildings on-site during a 100-year, 24-hour storm event. Building space was about tripled from 460,800 to 1,228,800 ft². Both construction costs and income increased but all runoff and pollutants were retained on-site.

Moreover, half the site was used for grassy areas with picnic tables, benches, shrubs, trees, gazebos, separate jogging trails and pathways, plus a large recreational area that included tennis, basketball, and badminton courts plus areas for basketball hoops and horseshoe pits. Its surface was covered with porous asphalt and grass. This recreational area served as a detention basin with all runoff being infiltrated. No runoff or pollutants left the site, relieving stress on the local storm sewer system. Employees could exit and enter their cars in a dry location.

Also, flooding and pollution regulations were met since all runoff and pollutants were retained on-site. Depending on the local stormwater ordinance, most, if not all, stormwater fees would be waived. This would save the owner additional thousands of dollars annually. The existing storm sewer system would be capable of conveying additional inflow—if the existing inlets throughout the system were large enough.

Industrial Site

H.1 Introduction

Figure H.1 shows an 11.0-ac industrial site. A 340 by 660-ft building contains 5.15 ac with an extensive greenroof with grass, flowers, and areas with tables and chairs for coffee breaks and lunch. Driveways, parking, and outdoor storage area are paved with porous concrete underlain with gravel. A rain garden receives runoff from the offices area and surrounding grass.

The bioswale receives runoff from the production area roof plus rain on the bioswale. The porous concrete areas receive rainfall through the concrete into 1-ft of gravel with a porosity of 40 percent. All runoff drains into the soil from a 100-year, 24-hour storm event. Water flows from the north to the south side of the site at a 0.5 percent slope.

Soil type is hydrologic soil group (HSG) B and the 100-year, 24-hour storm is 6.5 in.

H.2 Roof

The several layers of the reinforced greenroof include a concrete slab and a layer of 10 in gravel. With a porosity of 40 percent, gravel on the roof temporarily stores about $340 \times 660 \times 10 \times 0.4/12$ or 74,800 ft³ of water. Various types of sedums and flowering grasses and plants yield a variety of colors on the roof and help to reduce the energy needed to heat and cool the building. Six inches of amended soil retains 2 in. of water. The 40 percent porosity of the gravel reduces and delays runoff from the roof. A vertical concrete wall on the roof separates runoff between the production space and offices.

H.3 Production Space Roof

With a rainfall of 6.5 in., runoff from the production space is as follows. With 2.0 in. of rain retained in the grass, net rainfall into the gravel is 4.5 in. Over its area, this is $340 \times 480 \times 4.5/12$ is 61,200 ft³ of water. Ten inches of gravel can temporarily store about $340 \times 480 \times 10 \times 0.4/12$ is 54,400 ft³ of water.

H.3.1 Time of Concentration

Assume water flows diagonally across the roof to the bioswale via three downspouts. This flow length is $((80)^2 + (340)^2))^{0.5}$ or 350 ft. Time of concentration (Tc) for this portion of the roof is estimated as 54.6 minutes or 0.91 hour. Use a Tc of 1.0 hour.

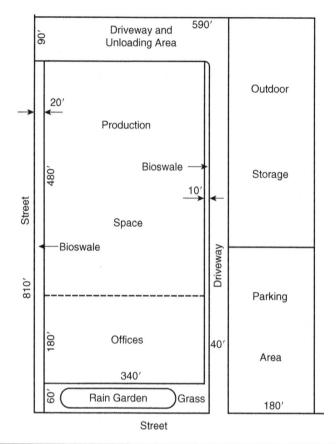

FIGURE H.1 Plan view of industrial site.

H.3.2 100-Year Inflow Hydrograph

Roof area is 340×480 is 163,200 ft^2 or 3.75 ac or 0.00586 mi^2. Peak of the inflow hydrograph is 9.40 cfs at hour 12.8.

H.3.3 Depth-Storage Calculations

Storage volume in the 10-in. of gravel is determined every 0.1-ft totaling 54,180 ft^3.

H.3.4 Depth-Outflow Calculations

The outflow structure is three 1.5-in. square downspouts. Depth-outflow calculations each 0.1-ft were made. Peak outflow rate from the gravel to the downspouts is 0.21 cfs.

H.3.5 Routing Curve Calculations

Routing curve calculations were made and are plotted in Fig. H.2.

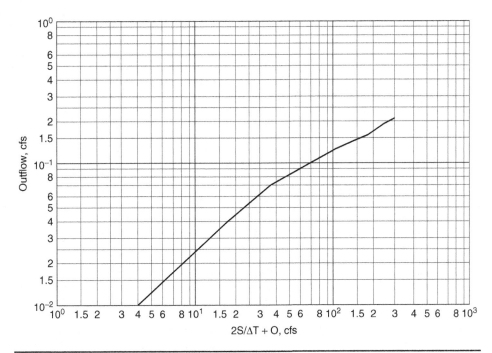

Figure H.2 Routing curve for production space roof.

H.3.6 100-Year Hydrograph Routing

The 100-year hydrograph was routed. Peak inflow of 9.40 cfs at hour 12.8 is reduced to 0.20 cfs at hour 24.0. Maximum depth in the gravel is 0.78 ft and is completely drained in 10 days from the rain's end. Since the downspouts' outflow peak into the bioswale is about 1 day, rainfall peak onto the bioswale occurs several hours previously.

H.4 Offices Area Roof

Assuming a net rainfall of 4.5 in, rainfall onto the office's area roof will be as follows. Over its area, there is $180 \times 340 \times 4.5/12$ or about 22,950 ft^3 of rainfall. With 10 in. of gravel on the roof, $180 \times 340 \times 10 \times 0.4/12$ is 20,400 ft^3 of temporary storage volume. A downspout from the roof conveys runoff into the rain garden.

H.4.1 Time of Concentration

Assume water flows from the concrete wall diagonally across the roof to the middle of the rain garden. This length of flow is $((180)^2 + (170)^2))^{0.5}$ or 248 ft. Tc for this portion of the roof is estimated as 36.1 minutes or 0.60 hour. Use a Tc of 0.50 hour.

H.4.2 100-Year Inflow Hydrograph

Roof area is 180×340 is 61,200 ft^2 or 1.40 ac or 0.00219 mi^2. Inflow hydrograph values were estimated with a peak of 5.22 cfs at hour 12.4.

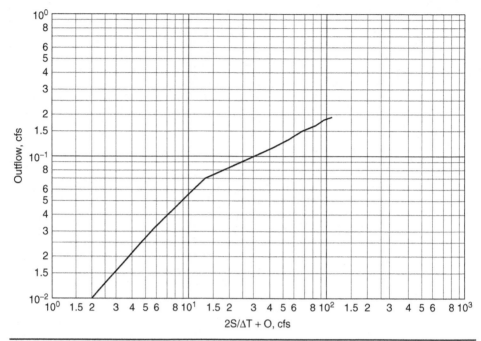

Figure **H.3**   Routing curve for offices area roof.

H.4.3   Depth-Storage Calculations

Depth-storage calculations were made each 0.1 ft. Total storage volume in the gravel is 20,390 ft^3.

H.4.4   Depth-Outflow Calculations

The outflow structure is a 2.5-in. square downspout. Depth-outflow calculations for this outlet were made with a peak outflow rate from the 10 in. of gravel of 0.19 cfs.

H.4.5   Routing Curve Calculations

Routing curve calculations were made and plotted in Fig. H.3.

H.4.6   100-Year Hydrograph Routing

The 100-year, 24-hour runoff hydrograph calculations were made. The peak-inflow rate of 5.22 at hour 12.4 was reduced to just 0.17 cfs at hour 18.7, a delay of 6.3 hours. Maximum depth in the 1.0 ft of gravel is 0.66 ft. The gravel is drained in about 5.0 days.

H.5   Bioswale

Rainfall into the swale is $20 \times 720 \times 6.5/12$ or 7,800 ft^3 plus runoff from the production-area roof. It is composed of amended soil and vegetation underlain with 1.0 ft of gravel. It has a 10-ft bottom width with 5:1 side slopes, is 1.0 ft deep, and a length of 700 ft.

Intermediate gravel berms along its length maximizes storage volume and slows runoff into the gravel beneath the engineered surface soil. Storage within it is about

$(1 \times (10 + 20)/2) \times 710$ or $10{,}650$ ft^3. Storage in the gravel below the swale is $20 \times 710 \times 1 \times 0.4$ or $5{,}680$ ft^3. Total volume of storage in and beneath the swale is $16{,}330$ ft^3, sufficient to contain the immediate runoff from the bioswale and delayed runoff from the production space roof during the 100-year, 24-hour storm event.

H.5.1   Time of Concentration

Water flows from the bioswale's upstream end just south of the driveway and unloading area to the bioswale's downstream end, a distance of 710 ft. Tc for the bioswale is 33.5 minutes or 0.56 hour. Use a Tc of 0.50 hour.

H.5.2   100-Year Inflow Hydrograph

Area of the bioswale is 20×720 is $14{,}400$ ft^2 or 0.33 ac divided by 640 is 0.000516 mi^2. Curve number is 61 from TR-55 (USDA, 1986). The inflow hydrograph peak is 0.63 cfs at hour 12.4. The outflow hydrograph from the production space roof is added to this inflow hydrograph.

H.5.3   Depth-Storage Calculations

Depth-storage calculations were made. Total volume of the 1.0 ft of gravel and the 1.0-ft deep bioswale is $16{,}120$ ft^3.

H.5.4   Depth-Outflow Calculations

Outflow is infiltration into the earth itself. Minimum infiltration rate for this B-type soil is 0.35 in/h. Flow rate is:

$$Q = 20 \times 720 \times 0.35/12/3{,}600 = 0.12 \text{ cfs}$$

H.5.5   Routing Curve Calculations

The calculations were made and no need to plot the curve since flow is constant at 0.12 cfs.

H.5.6   100-Year Hydrograph Routing

The inflow hydrograph to the grass bioswale from rain falling directly on it plus the outflow hydrograph from the production space roof were added together and routed.

The peak inflow rate of 0.69 cfs at hour 12.4 is reduced to 0.12 cfs at hour 51.0, a delay of 38.6 hours. Maximum depth in the bioswale is 0.41 ft. Water is on the surface from hour 26.4 to hour 56.0, a total of 23.6 hours during the 100-year storm. The gravel is completely drained in another 1.1 days.

H.5.7   Summary of Production Space Roof and Bioswale

The surface swale and gravel beneath it are both 1.0 ft deep. An interesting aspect is that with temporary detention on the roof and subsequent disparity in the timing of the two peak flows, the surface swale has water 0.41 deep in it for 1 day after the 100-year, 24-hour storm event. Gravel contains water in it for 2.4 days before it is empty again after the rain stops.

H.6   Rain Garden

CN is 61 and runoff is 2.3 in. Rain garden area is $350 \times 60 + 660 \times 10$ or $27{,}600$ ft^2 or 0.63 ac or 0.00099 mi^2. Total runoff volume is $27{,}600 \times 2.3/12$ or $5{,}290$ ft^3 plus runoff from the offices area roof. Its bottom width is 40 ft with 5:1 side slopes, is 1.0 ft deep, and is

290 ft long. Storage volume within the bioswale will be about $(1 \times (40 + 50)/2) \times 295$ or about 13,280 ft³. Storage volume within the 6 in of gravel below it is $60 \times 350 \times 0.50 \times 0.4 = 4,200$ ft³. Thus, total storage volume in and beneath the rain garden is 17,480 ft³. It must detain runoff during the 100-year, 24-hour storm event from the offices area roof, rain garden, and grass on the building's east side. Outflow from the rain garden is infiltration into the soil.

H.6.1   Time of Concentration

Runoff flows from the 10 ft width of grass's upstream end just south of the driveway and unloading area to the rain garden, a distance of 760 ft. Tc for this swale into the rain garden is estimated as 40.9 minutes or 0.68 hour. Use a Tc of 0.75 hour. Peak runoff from the offices area roof is delayed for another several hours.

H.6.2   100-Year Inflow Hydrograph

Area of the rain garden and surrounding grass is 0.00099 mi². Curve number is 61. Peak of the inflow hydrograph is 0.97 cfs at hour 12.6. To this hydrograph must be added the outflow hydrograph from the office area roof.

H.6.3   Depth-Storage Calculations

Depth-storage calculations were made. Total volume of the 6 in. of gravel and the 1.0-ft deep rain garden is 17,480 ft³.

H.6.4   Depth-Outflow Calculations

The outflow structure is the native soil under the rain garden. Minimum infiltration rate for this B type soil is 0.35 in/h. Flow rate is:

$$Q = 60 \times 350 \times 0.35/12/3,600 = 0.17 \text{ cfs}$$

H.6.5   Routing Curve Calculations

Routing curve calculations were made, but the routing curve does not need to be plotted because the outflow rate is a constant 0.17 cfs.

H.6.6   100-Year Hydrograph Routing

The inflow hydrograph from rain falling directly on it plus the outflow hydrograph from the offices area roof are added. Then routing calculations were made for this combined 100-year, 24-hour runoff hydrograph. The peak inflow rate of 1.10 cfs at hour 12.6 is reduced to 0.17 cfs at hour 25.0, a delay of 12.4 hours. Maximum surface depth is 0.06 ft and remains on the surface from hour 18.4 to hour 33.2, a total of 14.8 hours. The gravel is completely drained at hour 48.0.

H.7   Porous Concrete Areas

Total area of porous concrete is $90 \times 370 + 220 \times 810$ or 211,500 ft³. A rainfall of 6.5 in. will infiltrate the porous pavement and store in the 1.0 ft of gravel beneath the pavement. Total inflow volume will be $211,500 \times (6.5 - 0.041)/12 = 113,840$ ft³. Storage volume within the 1.0-ft of gravel below the pavement is $211,500 \times 1 \times 0.4$ or 84,600 ft³.

H.7.1   Time of Concentration

Tc is the time rain infiltrates into the porous pavement. Use Tc of 6.0 min or 0.10 hour.

H.7.2   100-Year Inflow Hydrograph

Pavement area is 211,500 ft^2 or 4.86 ac or 0.00759 mi^2. Its CN is 98. The inflow hydrograph peak is 49.07 cfs at hour 12.1.

H.7.3   Depth-Storage Calculations

Depth-storage calculations were made and total volume of the foot of gravel is 84,600 ft^3.

H.7.4   Depth-Outflow Calculations

Minimum infiltration rate for this B-type soil is 0.35 in/h. Flow rate is:

$$Q = 211,500 \times 0.35/12/3,600 = 1.71 \text{ cfs}$$

H.7.5   Routing Curve Calculations

Routing curve calculations were made but does not need to be plotted because the outflow rate is a constant 1.71 cfs.

H.7.6   100-Year Hydrograph Routing

Routing calculations for this storm were made. The peak-inflow rate of 49.07 cfs at hour 12.1 is reduced to 1.71 cfs at hour 14.2, a delay of 2.1 hours. Maximum depth in the 1.0 ft of gravel is 0.79 ft. The gravel is completely drained at hour 31.0.

H.7.7   Summary of Porous Pavement

Temporary storage in the 1.0 ft of gravel below the porous pavement is 0.79 ft deep at hour 14.2 during the 100-year, 24-hour storm and drains completely by hour 31.0. The additional storage in the gravel could be utilized in later years if it becomes partially plugged.

H.8   Summary of the Industrial Site

The building, driveways, outdoor storage area, and parking lot cover 91 percent of the site. The 5.15-ac building has a greenroof with 10 in. of gravel as one of its layers. The 4.86 ac of porous concrete are underlain with 1.0-ft of gravel. This combination plus a bioswale and a rain garden infiltrate the entire runoff from the 100-year, 24-hour storm containing 6.5 in. of rain.

The greenroof and rain garden add color and interest to the site. Porous concrete pavement ensures that no one gets his/her feet wet during any rain event because all rain immediately infiltrates, leaving only a wet pavement. The absense of any runoff allows an existing storm sewer system to intercept runoff from a larger storm event. Any pollutants originating on or brought onto the site remain on site either by infiltration or from removal by maintenance crews.

Potential Source Control Best Management Practices

1. **Control the Use and Disposal of Fertilizers, Pesticides, and Herbicides**
   a. Use native vegetation in public spaces to reduce the need for herbicides/pesticides.
   b. Review correct certification procedures/requirements for high-volume users (e.g., commercial applicators, public agency personnel) and make recommendations for expanding programs to provide better education about stormwater quality impacts, as necessary.
   c. Continue to collect stormwater samples for pesticide/herbicide analyses in an effort to define the local problem with these pollutants.
   d. Review current public education programs related to low volume use of pesticides/herbicides/fertilizers (e.g., household use) and make recommendations for supplementing the program with education regarding stormwater quality impacts and use of non-polluting alternate products.
   e. Evaluate existing O&M (and/or landscape management) programs for public rights-of–way and public-drainage channels and ensure that these programs limit the discharge of pollutants from pesticides/herbicides/fertilizers in runoff.
   f. Restrict use of pesticides, herbicides, and fertilizers (e.g., regulate the sale of household pesticide/fertilizer products).
   g. Educate the manufacturers and distributors of pesticides, herbicides, and fertilizers about stormwater quality impacts and encourage them to educate the public about the proper use and management of the products.
   h. In cooperation with the State Department of Agriculture, Natural Resources Conservation Service (NRCS), and others, develop maps of appropriate fertilizer types and appropriate rates and combine with educational programs.
   i. Establish planting/landscape policies for specified districts and/or for specified land uses which encourage use of vegetation, either indigenous or imported that are self-sustainable without the human applications of pesticides, herbicides, or fertilizers.
   j. Develop a program to educate architects, landscape architects, and engineers about development design practices that reduce the need for pesticides, herbicides, and fertilizers during project life.

2. **Control Littering and Improper Solid Waste Disposal Practices**
   a. Educate the public regarding the stormwater pollution impacts that result from littering practices (e.g., dumping yard debris in drainage channels).
   b. Promote public involvement in "Keep Watershed Clean" campaigns in "adopt-a-creek" programs for specific waterways.
   c. Provide, collect, and maintain more litter receptacles in strategic public areas and during major public events.
   d. Work with citizen action programs to facilitate efforts to reduce littering.
   e. Strengthen enforcement of existing regulations which provide legal authority to control littering.
   f. Strengthen enforcement of existing regulations which provide legal authority to control improper disposal of potentially harmful solid waste material into the storm drainage system.
   g. Review and revise as necessary, existing solid waste management programs (e.g., reduce, recycle, and control trash and yard debris) to take stormwater quality into account.
   h. Provide free pickup and disposal for leaf and yard debris.
   i. Facilitate efforts to report illegal dumping, illicit connections, and other incidents. Work with citizen action groups: consider operation of a telephone "hotline" for citizens to report incidents; and post signs at areas where illegal dumping may occur that encourages citizens to report them.
   j. Educate residents about the advantages of composting and about proper composting techniques.
   k. Provide neighborhood compost sites, city operated or citizen operated. Post instruction signs and/or organize loose volunteer organizations to instruct residents and to maintain the site.
   l. Develop methods to encourage inter-agency cooperation on illegal dumping problems. Clarify responsibilities where necessary.

3. **Prevent the Dumping of Pollutants into Storm Sewers and Drainage Channels**
   a. Educate regarding the impacts that result when oil, antifreeze, pesticides, herbicides, paints, solvents, or other potentially harmful chemicals are dumped into storm sewers or drainage channels.
   b. Label (stencil) storm drain inlets and provide signs along the banks of drainage channels and creeks explaining the environmental impacts of dumping wastes. Encourage volunteer assistance.
   c. Continue to support programs which provide convenient means for people to properly dispose of oil, antifreeze, pesticides, herbicides, paints, solvents, or other potentially harmful chemicals and waste materials (recycle if possible).
   d. Clarify and strengthen enforcement of existing regulations which give co-applicants and/or USA legal authority to prevent and eliminate the improper disposal of pollutants into storm sewers and drainage channels, including illicit connections and illegal dumping.
   e. Develop and implement an aggressive field program to search for, detect, and prevent dumping or routine discharging of pollutants into storm sewers and drainage channels.

f. Develop a program for public agencies to ensure appropriate disposal of used automobile fluids.

g. Develop a program for clean up after structural fires and vehicular accidents to prevent pollutants and debris from being washed into the storm drain system.

4. Spill Prevention and Response

a. Review, and improve existing spill prevention and response programs. The objective is to eliminate pollutant spills on roads or open space, where they are washed into storm drains or waterways.

b. Support and cooperate with existing programs by others to ensure that *private* trucks hauling materials do not leak, spill, or otherwise release contaminants onto roadways or open spaces where they may be washed into storm drains or waterways.

c. Review current procedures and develop and implement a program to ensure that *municipal* trucks hauling materials do not leak, spill, or otherwise release contaminants onto roadways or open spaces where they may be washed into storm drains or waterways. Consider educational activities as appropriate.

d. Educate owners and operators of trucks about the impacts of leaks, spills, and other releases from bulk materials during transportation (pollution effects of materials that are spilled onto roadways or other open spaces being washed into storm drains or waterways).

5. Eliminate Cross Connections between Sanitary and Process Wastewater Systems and the Storm Sewer System

a. Develop and implement an aggressive field program to search, detect, and control illicit connections (including connections from sanitary sewers and commercial/industrial wasewater sewers). The field program should include procedures to enforce existing regulations which give the legal authority to eliminate illicit discharges.

b. Include bacteria as an analytic parameter in the illicit discharge field screening and investigation program, to detect and limit illicit connections and leaks from sanitary sewers.

c. Coordinate new/existing investigative efforts (e.g., field-screening programs, sampling programs, in-line sewer inspections, and routine maintenance activities), with an information management system (e.g., databases, mapping) to facilitate detection/elimination of illicit discharges.

6. Control of Industrial/Commercial Activities

a. Develop and implement a maintenance program for oil/water separators.

b. Research, strengthen (if necessary), and enforce regulations and building codes to require water quality controls in areas which are significant sources (e.g., gas stations, automotive repair shops, commercial/industrial facilities, parking areas, and food service establishments).

c. Educate commercial/industrial sector regarding the effective use of "housekeeping" practices to reduce pollutants. Include the use of absorbents, cleaning compounds, oil/grease traps, and other techniques for controlling pollutants from gas stations, automotive repair shops, commercial/industrial facilities, parking areas, and food service establishments.

d. Develop technical guidance which will facilitate compliance with regulations requiring water quality controls (oil/grease traps, plate separators, synthetic absorbent media, grassy swales) for commercial/industrial facilities.

e. Develop and enforce requirements (e.g., structural controls, best management practices (BMPs), connection requirements, developing and redeveloping practices) for existing and developing industrial/commercial areas.

f. Educate regarding the need to keep rainfall and runoff from contacting potential contaminants. Describe typical examples of the problem and practical solutions.

g. Develop a program to coordinate with the Department of Environmental Quality's (DEQs) industrial permitting program.

h. Develop and implement regulations which require landowners and/or tenants to provide covers (e.g., roofs, tarps) to keep rain out of areas that contain contaminants (e.g., chemical storage areas, waste-storage areas, contaminated industrial areas), and keep runoff from draining through areas that contain contaminants.

i. Develop and implement an aggressive field program to search for, detect, and correct situations where rainfall and/or runoff presently contact potential contaminants.

j. Develop a program which addresses the National Pollutant Discharge Elimination System (NPDES) regulation requiring monitoring of stormwater discharges from industrial facilities.

k. Coordinate with the EPA/DEQ to be sure that all potential water-quality impacts are adequately considered at the time NPDES permits are issued for any discharges to storm sewers or drainage channels. Where feasible, require monitoring of all pertinent constituents as a permit.

l. Develop a program to encourage businesses (such as auto supply outlets and service stations) to provide collection services for used automotive fluids. Regulate to assure adequate disposal.

7. Control Leaks from Gasoline, Fuel Oil, Chemical Storage Tanks, and Automobiles

a. Educate regarding environmental impacts resulting from leaks and spills from gasoline, fuel oil, and chemical tanks (above and below ground).

b. Coordinate with efforts (by others) to identify the implementation of existing regulations calling for new improved tank designs (e.g., double wall, monitoring facilities), an aggressive self-monitoring program to be conducted by landowners and tenants, and implement a strategically focused spot-check program to search for, identify, test, and control leaking storage tanks.

c. Develop a program to encourage conversion from oil to natural gas or electric for home heating.

d. Educate regarding: the need to itensify vehicle inspection and maintain efforts to reduce leakage of oil, antifreeze, hydraulic fluid, etc.

8. Control Pollutants from Construction Sites

a. Improve channel construction and maintenance activities so that erosion control is emphasized.

b. Implement an erosion-control program including erosion control requirements/ policy area-wide, incorporating new techniques/practices, increasing inspections and enforcement activities, and educating affected groups (architects, engineers, contractors, and public agency personnel).

c. Take a more active role in the National Environmental Protection Act (NEPA) process to better address the topics of erosion potential, proposed erosion/sediment control plans, proposed inspection programs, related environmental impacts, and enforceable mitigation measures minimizing environmental impacts.

d. Establish incentives and/or requirements for contractors/developers to ensure that potential damages from erosion and sediment deposition are addressed and paid for.

e. Work with the state (DEQ) and play a larger role in implementing and supervising state requirements for erosion control at construction sites.

f. Implement a comprehensive control program which addresses discharge of all other pollutants from construction sites, which include expanding existing regulations/policy area-wide, incorporating new techniques/practices, increasing inspection and enforcement activities, and educating all affected groups (architects, engineers, contractors, and public agency personnel).

9. **Control Erosion on Lands Other Than Construction Sites (Forest Practices, Agricultural Land, etc.)**

a. Coordinate with the Soil Conservation Service (SCS and local resource conservation programs to support their activities to control erosion and sedimentation problems).

b. Request that the Department of Agriculture (DA) begin educating farmers, ranchers, and other managers of agricultural and/or open space lands regarding the need for practical methods for sediment control and erosion control.

c. Develop and implement programs to actively search for, identify, evaluate, and prioritize erosion problems on undeveloped land, parkland, and agricultural land.

d. Develop and implement programs to work with landowners, tenants, and/or public agencies to apply practical erosion-control and sediment-control practices.

e. Request that the DA begin educating managers and users of parkland and open-space lands regarding the need to establish and enforce practical, site-specific regulations to control off-trail activities.

f. Develop and implement practical programs for revegetating and otherwise restoring eroding areas (e.g., areas damaged by fires, overgrazing, landslides, improper tillage, and off-road vehicle use).

g. Research, strengthen (if necessary), and enforce regulations to require erosion control and/or hydraulic-runoff controls on privately owned vacant or neglected land.

10. **Control Human and Animal Wastes**

a. Implement a program to monitor septic tanks and cesspools. The program should identify failing systems and require repair or replacement of failing systems.

b. Implement and enforce leash laws and pet-waste cleanup ordinances in selected public-use areas.

c. Coordinate with the Department of Agriculture to control discharge of wastes into the storm sewer system from confined animal feeding operations (CAFO).

d. Educate the public through brochures/other means regarding the need to clean up and properly dispose of pet wastes and the adverse water-quality impacts of feeding ducks in local waterways.

e. Provide informational signs and dispense animal waste litter bags in parks and other areas.

f. Educate regarding the need for proper management of wastes from suburban livestock (e.g., horses, chickens) and agricultural operations in the watershed.

g. Restrict livestock from entering stream channels or damaging vulnerable stream bank areas.

11. **Control Airborne Particles**
    a. Encourage and cooperate with the State's program to control vehicle emissions through licensing smog test requirements.

    b. Encourage and cooperate with programs that seek to reduce particulate atmospheric emissions of pollutants from individual, public, commercial, industrial sources, and outdoor burning.

    c. Encourage and cooperate with the state's program to control and minimize atmospheric emissions from fireplaces and wood-burning stoves, and to educate the public about the regulations.

    d. Encourage and cooperate with programs that seek to reduce automobile use by various means (e.g., ride sharing, carpooling, public transportation, and human-powered transportation).

    e. Educate regarding the relationship between air pollution and stormwater quality problems. Coordinate with and obtain information from air quality agencies.

    f. Cooperate with public transportation agencies, public agency motorpools, and/or public works departments to provide effective air pollution controls on publicly owned vehicles and motorized equipment, and/or to use alternate clean-burning fuels where possible.

12. **Operation/Maintenance of Stormwater Drainage and Quality Facilities**
    a. Keep up-to-date inventories and maps of the storm sewer system. Include mapping of storm drainage amenities such as grassy swales and detention basins. Consider use of a computerized mapping system such as a graphical information system (GIS).

    b. Analyze sediments from stormwater facilities to determine whether or not there may be any disposal problems related to the sediments (i.e., do sediments contain hazardous contaminants?).

    c. Develop a comprehensive O&M plan for all public stormwater facilities (new and existing) which maximizes water-quality benefits while maintaining flood capacity. Incorporate methods to evaluate effectiveness. Provide a means of recording the observation of field inspections and maintenance personnel, and procedures for transmitting this information to the appropriate department/agency, so the information can be used to locate and eliminate the pollutant source(s).

d. Require O&M plans at permitting time for stormwater facilities related to new private development and redevelopment/retrofitting. Conduct inspections and follow-up regularly after construction to ensure that approved O&M plans are being followed.

e. Require O&M plans for existing private stormwater facilities where feasible.

f. Develop and implement an aggressive field program to search for, test, remove, and properly dispose of sediment deposits (in drainage channels, streams, and stormwater storage/retention basins) containing relatively high concentrations of pollutants. Find places to dispose of sediments.

g. Remove (or perforate) paved bottoms of drainage channels, where practical, to maximize infiltration and reduce peak rates.

h. Conside increasing frequency of cleaning out inlets, catch basins, storm sewers, pump stations, channels, and stormwater retention basins in areas where sediment and/or debris tend to accumulate. Implement approved programs where appropriate.

i. Research methods of economically and safely detoxifiying high concentrations of pollutants at locations where they accumulate, e.g., detention basins, drainage channels, sedimentation trapping facilities, etc. Current information on bioremediation and solar detoxification may offer some solutions to the disposal of wastes issue.

j. Determine the effectiveness of retrofitting selected storm sewers, sanitary sewers, and portions of the existing publically owned treatment works (POTWs) allowing the plants to receive and treat runoff from small storms and strategic portions of large storms (e.g., "first flush").

k. Determine the feasibilitiy of retrofitting existing drainage and flood control facilities (e.g., storm drain inlets, detention basins, drainage channels) to function as water quality facilities. Retrofits could include installation of in-line sediment trap devices, detention facilities, or wetland/riparion vegetation. If feasible, develop a plan and implement.

13. Maintenance of Streets and Paved Areas

a. Provide incentives to property owners to optimize and intensify street sweeping of private parking lots and other paved private areas.

b. Develop and implement intensified street sweeping programs in strategic locations (e.g., central business districts, shopping malls, major parking lots, industrial areas) and/or at strategic times (e.g., following extended periods of dry weather).

c. Improve street sweeping on a watershed-wide basis (e.g., sweep more areas, more frequently).

d. Investigate effectiveness of using street flushers to reduce pollutants in runoff by researching what is done in other locales. Make recommendations to co-applicants/cities if appropriate.

e. Review existing street-design standards with respect to water quality (e.g., sloped medians may increase infiltration and enhance water quality) and modify as appropriate. Need to keep ground water impacts in mind.

f. Evaluate ways that transporation authorities (e.g., Department of Transportation (DOT), others) can reduce pollutant discharges associated with their road/highway maintenance and rehabilitation operations. Implement improved programs where appropriate.

g. Where appropriate, establish requirements for proper maintenance and cleaning of paved surfaces on private property. Enforce requirements.

14. **Control Roadway Deicing Materials**
   a. Prohibit the use of salt for deicing activities.

   b. Develop improved strategies for applying deicing materials, limiting material discharge to the storm sewer system.

   c. Establish a program to pick up (and properly reuse) sand that is distributed on streets for deicing purposes.

   d. Develop and implement programs for proper storage of deicing materials to prevent materials from entering the storm sewer system (e.g., providing roofs and covers for stockpiled materials).

   e. Examine the sources of sand used for deicing (e.g., review purchasing specifications), ascertaining if the material components are acceptable for discharge to storm sewer systems (depending on source, some sands could contain organic pollutants, nutrients, and heavy metals).

15 **Reduce the Volume of Post-Construction Site Runoff Which Enters Storm Sewers**
   a. Research, strengthen (if necessary), implement, and enforce regulations which give jurisdictions the legal authority to require site drainage designs and systems which minimize the total volume of runoff and the peak rate of runoff from new construction, where local conditions permit.

   b. Require new commercial, industrial, institutional, and major multifamily residential building complexes to have drainage facilities that incorporate on-site detention, to assure that neither the total volume of runoff nor the peak rate of discharge to the storm sewer system or drainage channels are increased.

   c. Require new public- and private-sector developments to make significant use of design techniques for new buildings, landscaping, recreation areas, walkways, and parking areas to maximize detention.

   d. Educate regarding need to minimize the total runoff volume and the peak runoff rate from a given area. Describe basic principles and suggest practical alternative means to enhance surface detention.

   e. Require (or consider providing incentives for) control of peak rate and total volume (e.g., impervious area recharge rebates).

   f. Provide education and guidance encouraging engineers, architects, and building departments to implement systems which temporarily detain rainfall peaks on rooftops and/or in detention facilities, minimizing the peak rate of discharge to the storm sewer system or drainage channels.

   g. Establish and enforce legal authority to prohibit new direct connections (e.g., roof drains to storm systems) and which require retrofitting existing buildings where practical.

h. Require (or provide incentives for) roof drains not be connected directly to storm systems.

i. Educate regarding the need to minimize the total roof drain runoff volume contributing directly to both storm sewers and drainage channels. Describe basic principles and suggest practical alternatives to minimize their peak rate of discharge.

16. Provide Sedimentation Facilities to Remove Pollutants

a. Develop a program encouraging public/private agencies to research new, more economical methods of improving stormwater quality, i.e., compost filters and other experimental techniques.

b. Require (or consider providing incentives for) owners of selected private areas (parking areas, vacant land, etc.) to route runoff through appropriate control structures: sedimentation basins, grassy swales, oil traps/separators (which provide cleaning), or others.

c. Require (or consider providing incentives for) use of treatment-based controls, detention basins.

d. Develop a program (possibly financial rewards/public recognition) to encourage individuals and inventors to create efficient, cost effective treatment techniques to replace some of the expensive devices and O&M practices now being employed, e.g., oil/water separators, sedimentation manholes, etc.

17. Provide Infiltration Basins to Remove Pollutants

a. Determine the effectiveness and potential negative environmental impacts of installing facilities on public property to encourage infiltration of runoff from neighborhood streets (e.g., use street circles, cul-de-sacs, and other curbside areas as infiltration areas or dry wells). Consider possible conflicts with existing policies and regulations. If feasible, install and maintain the facilities.

18. Employ Vegetation/Employ Natural Resources to Remove Pollutants from Storm Runoff

a. Develop a program to provide incentives to property owners who protect natural areas on their property considered to have valuable water-quality characteristics.

b. Develop a pilot project for multiple objective stream restoration. Emphasize water quality, habitat, education, and recreation. Measure water-quality changes.

c. Support government and community tree-planting programs.

19. Requirements and Design Standards for New Development and Redevelopment

a. Review and modify existing design standards (for flood control and water-quality facilities) to improve water quality. Research methods being used in other areas and develop applicable regional criteria.

b. Develop a program to educate public agency personnel (responsible for plan checking and permit issuance) and engineers/architects of new and/or existing design standards and structural techniques that reduce negative water-quality impacts to streams and the storm system.

c. Review and modify the current planning procedures utilized by the agencies for review of new development and significant redevelopment. Create new/improved checklists and routing procedures where needed. Educate all affected public agency personnel about the improvements.

d. Measure/evaluate post-construction compliance with water-quality objectives through a program of follow-up inspections and enforcement of performance standards for all stormwater quality improvement facilities in new construction or major redevelopment.

20. Comprehensive and Research-Type Controls

a. Determine the feasibility of building, establishing, and maintaining new water-quality facilities (e.g., detention basins, manmade/natural wetlands, etc.). If feasible, develop a plan for conducting the work and implement the plan. Include evaluation methods to assess the facility's performance.

b. Develop a pilot program for a selected small well-defined subbasin (which is representative of a typical urbanized watershed) to implement every applicable aspect of the stormwater management plan (SWMP) and conduct monitoring to assess effectiveness of the overall pilot program.

Manning's Roughness Coefficients, n^a

Description	Manning's n^b
I. Closed Conduits:	
A. Concrete pipe	0.011–0.013
B. Corrugated-metal pipe or pipe-arch	
1. 2/3 by 1/2-in corrugation (riveted pipe)^c	
a. Plain or fully coated	0.024
b. Paved invert (values are for 25 percent and 50 percent of circumference)	
(1) flow full depth	0.021–0.018
(2) flow 0.8 depth	0.021–0.016
(3) flow 0.6 depth	0.019–0.013
2. 6- by 2-in corrugation (field bolted)	0.030
C. Vitrified clay pipe	0.012–0.014
D. Cast-iron pipe, uncoated	0.013
E. Steel pipe	0.009–0.011
F. Brick	0.014–0.017
G. Monolithic concrete:	
1. Wood forms, rough	0.015–0.017
2. Wood forms, smooth	0.012–0.014
3. Steel forms	0.012–0.013
H. Cemented rubble masonry walls:	
1. Concrete floor and top	0.017–0.022
2. Natural floor	0.019–0.025
I. Laminated treated wood	0.015–0.017
J. Vitrified clay liner plates	0.015

[a]Estimates are by Bureau of Public Roads unless otherwise noted and are for straight alignment. A small increase of value of n may be made for other channel alignments.
[b]Ranges of sections I through III are for good to fair construction. For poor quality construction, use larger values of n.
[c]*Friction Factors in Corrugated Metal Pipe*, by M. J. Webster and L. R. Metcalf, Corps of Engineers, Department of the Army; published in Journal of the Hydraulics Division, Proceedings of the American Society of Civil Engineers, Vol. 85, No. HY9, Sep. 1959, Paper No. 2148, pp. 35–67.

II. Open Channels, Lined[d] (straight alignment)[e]

A. Concrete, with surfaces as indicated:

1. Formed, no finish	0.013–0.017
2. Trowel finish	0.012–0.014
3. Float finish	0.013–0.015
4. Float finish, some gravel on bottom	0.015–0.017
5. Gunite, good section	0.016–0.019
6. Gunite, wavy section	0.018–0.022

B. Concrete, bottom float finish, sides as indicated:

1. Dressed stone in mortar	0.015–0.017
2. Random stone in mortar	0.017–0.020
3. Cement rubble masonry	0.020–0.025
4. Cement rubble masonry, plastered	0.016–0.020
5. Dry rubble (riprap)	0.020–0.030

C. Gravel bottom, sides as indicated

1. Formed concrete	0.017–0.020
2. Random stone in mortar	0.020–0.023
3. Dry rubble (riprap)	0.023–0.033

D. Brick	0.014–0.017

E. Asphalt:

1. Smooth	0.013
2. Rough	0.016

F. Wood, planed, clean	0.011–0.013

G. Concrete-lined excavated rock:

1. Good section	0.017–0.020
2. Irregular section	0.022–0.027

III. Open Channels, Excavated[d] (straight alignment, natural lining)[e]

A. Earth, uniform section:

1. Clean, recently completed	0.016–0.018
2. Clean, after weathering	0.018–0.020
3. With sho+rt grass, few weeds	0.022–0.027
4. In gravely soil, uniform section, clean	0.022–0.025

B. Earth, fairly uniform section:

1. No vegetation	0.022–0.025
2. Grass, some weeds	0.025–0.030
3. Dense weeds or aquatic plants in deep channels	0.030–0.035
4. Sides clean, gravel bottom	0.025–0.030
5. Sides clean, cobble bottom	0.030–0.040

C. Dragline excavated or dredged:

---

[d]For important work and where accurate determination of water profiles is necessary, the designer is urged to consult the following references and to select *n* by comparison of the specific conditions with the channels tested: *Flow of Water in Irrigation and Similar Channels*, by F. C. Scobey, Division of Irrigation, Soil Conservation Service, U.S. Department of Agriculture, Tech. Bull. No. 652, Feb. 1939; and *Flow of Water in Drainage Channels*, by C. E. Ramser, Division of Agricultural Engineering, Bureau of Public Roads, U.S. Department of Agriculture, Tech. Bull. No. 129, Nov. 1929.

[e]*Handbook of Channel for Soil and Water Conservation*, prepared by the Stillwater Outdoor Hydraulic Laboratory in cooperation with the Oklahoma Agricultural Experiment Station; published by the Soil Conservation Service, U.S. Department of Agriculture, Publ. No. SCS-TP-61, Mar. 1947, rev. June 1954.

1. No vegetation	0.028–0.033
2. Light brush on banks	0.035–0.050

D. Rock:
1. Based on design section   0.035
2. Based on actual mean section:
   a. Smooth and uniform   0.035–0.040
   b. Jagged and irregular   0.040–0.045

E. Channels not maintained, weeds and brush uncut:
1. Dense weeds, high as flow depth   0.080–0.120
2. Clean bottom, brush on sides   0.050–0.080
3. Clean bottom, brush on sides, highest stage of flow   0.070–0.110
4. Dense brush, high stage   0.100–0.140

IV. Highway Channels and Swales with Maintained Vegetation [f,g]

(values shown are for velocities of 2 and 6 fps)

A. Depth of flow up to 0.7 ft
1. Bermudagrass, Kentucky bluegrass, buffalograss:
   a. Mowed to 2 in   0.070–0.045
   b. Length 4–6 in   0.090–0.050
2. Good stand, any grass
   a. Length about 12 in   0.180–0.090
   b. Length about 24 in   0.300–0.150
3. Fair stand, any grass
   a. Length about 12 in   0.140–0.080
   b. Length about 24 in   0.250–0.130

B. Depth of flow 0.7–1.5 ft
1. Bermudagrass, Kentucky bluegrass, buffalograss:
   a. Mowed to 2 in   0.050–0.035
   b. Length 4–6 in   0.060–0.040
2. Good stand, any grass
   a. Length about 12 in   0.120–0.070
   b. Length about 24 in   0.200–0.100
3. Fair stand, any grass
   a. Length about 12 in   0.100–0.060
   b. Length about 24 in   0.170–0.090

V. Street and expressway gutters:
A. Concrete gutter, trowel finish   0.012
B. Asphalt pavement:
1. Smooth texture   0.013
2. Rough texture   0.016

---

[f]*Flow of Water in Channels Protected by Vegetative Linings*, by W. O. Ree and V. J. Palmer, Division of Drainage and Water Control Research, Soil Conservation Service, U.S. Department of Agriculture, Tech. Bull. No. 967, February 1949.

[g]For calculation of stage or discharge in natural stream channels, it is recommended that the designer consult the local District Office of the Surface Water Branch of the U.S. Geological Survey, to obtain data regarding values of *n* applicable to streams of any specific locality. Where this procedure is not followed, the table may be used as a guide. The values of *n* tabulated have been derived from data reported by C. E. Ramser (see endnote d) and from other incomplete data.

    C. Concrete gutter with asphalt pavement:
       1. Smooth   0.013
       2. Rough   0.015
    D. Concrete pavement:
       1. Float finish   0.014
       2. Broom finish   0.016
    E. For gutters with small slopes where sediment
       may accumulate, increase above values of n by   0.020

VI. Natural stream channels[h]

    A. Minor streams[i] (surface width at flood stage less than 100 ft)
       1. Fairly regular section
          a. Some grass and weeds, little or no brush   0.030–0.035
          b. Dense growth of weeds, depth of flow materially
             greater than weed height   0.035–0.050
          c. Some weeds, light brush on bank   0.035–0.050
          d. Some weeds, heavy brush on banks   0.050–0.070
          e. Some weeds, dense willows on banks   0.060–0.080
          f. For trees within channel, with branches submerged
             at high stage, increase all above values by   0.010–0.020
       2. Irregular sections, with pools, slight channel meander,
          increase values in 1a–e about   0.010–0.020
       3. Mountain streams, no vegetation in channel,
            banks usually steep, trees and brush along
            banks submerged at high state
          a. Bottom of gravel, cobbles, and few boulders   0.040–0.050
          b. Bottom of cobbles, with large boulders   0.050–0.070
    B. Flood plains (adjacent to natural streams)
       1. Pasture, no brush
          a. Short grass   0.030–0.035
          b. High grass   0.035–0.050
       2. Cultivated areas
          a. No crop   0.030–0.040
          b. Mature row crops   0.035–0.045
          c. Mature field crops   0.040–0.050
       3. Heavy weeds, scattered brush   0.050–0.070
       4. Light brush and trees[j]
          a. Winter   0.050–0.060
          b. Summer   0.060–0.080

---

[h]The tentative values of n cited are principally derived from measurements made on fairly short but straight reaches of natural streams. Where slopes calculated from flood elevations along a considerable length of channel, involving meanders and bends, are to be used in velocity calculations by the Manning formula, the value of n must be increased to provide for the additional loss of energy caused by bends. The increase may be in the range of 3 to 15 percent.

[i]The presence of foliage on trees and brush under flood stage will materially increase the value of n. Therefore, roughness coefficients for vegetation in leaf will be larger than for bare branches. For trees in channel or on banks, and for brush on banks where submergence of branches increases with depth of flow, n will increase with rising stage.

     5. Medium to dense brush
       a. Winter   0.070–0.110
       b. Summer   0.100–0.160
     6. Dense willows, summer, not bent over by current   0.150–0.200
     7. Cleared land with tree stumps, 100–150/ac
       a. No sprouts   0.040–0.050
       b. With heavy growth of sprouts   0.060–0.080
     8. Heavy stand of timber, a few down trees,
       little undergrowth
       a. Flood depth below branches   0.100–0.120
       b. Flood depth reaches branches   0.120–0.160

C. Major streams (surface width at flood stage more than 100 ft)
   Roughness coefficient is usually less than for minor
   streams of similar description on account of less
   effective resistance offered by irregular banks or
   vegetation on banks. Values of n may be reduced
   somewhat. Follow recommendations of note g if possible.
   The value of *n* for larger streams of most regular sections,
   with no boulders or brush, may be in the range of from: 0.028–0.033

APPENDIX K

References

American Society of Civil Engineers; Water Environment Federation (1992). *Design and Construction of Urban Stormwater Management Systems*; ASCE Manuals and Reports of Engineering Practice No. 77; WEF Manual of Practice No. FD-20; American Society of Civil Engineers: New York.

Bossy, H.G. (1961). "Hydraulics of Conventional Highway Culverts"; a paper presented at the Tenth National Conference, Hydraulics Division, American Society of Civil Engineers.

Brater, Ernest F. and Horace Williams King (1976). *Handbook of Hydraulics, 6th Ed.*; McGraw-Hill: New York.

Chow, Ven Te (1959). *Open-Channel Hydraulics*; McGraw-Hill: New York.

Commonwealth of Virginia (Revised 2002). *Drainage Manual*; Commonwealth of Virginia, Virginia Department of Transportation, Location and Design Division, Hydraulics Section.

FAA (1970). *Airport Drainage; Avisory Circular, 150-5320-5B*; Federal Aviation Administration, U.S. Department of Transportation, Washington, D.C.

Federal DOT (1984). *Drainage of Highway Pavements*; Hydraulic Engineering Circular No. 12; U.S. Department of Transportation, Federal Highway Administration, Hydraulics Branch, Bridge Division, Office of Engineering: Washington, D.C.

Federal Register, Environmental Protection Agency (1990). Storm Water Phase I Final Rule, 40 CFR Parts 122, 123, 124, Federal Register, September 1990: Washington, D.C.

Federal Register, Environmental Protection Agency (1999). Storm Water Phase II Final Rule, 40 CFR Parts 9, 122, 123, 124, Federal Register, October 29, 1999: Washington, D.C.

FHWA (1980). *Design Charts for Open Channel Flow*; Hydraulic Design Series (HDS) No. 3, Federal Highway Administration, U.S. Department of Transportation, Washinton, D.C.

French, John L. (Unknown). "Hydraulic Characteristics of Commonly Used Pipe Entrances"; Report No. 4444; National Bureau of Standards.

Harrison, Rick. (2008). *Prefurbia: Reinventing the Suburbs: From Distainable to Sustainable*; Sustainable Land Development International: Dubuque, Iowa.

Kerby, W. S. (1959). Time of Concentration for Overland Flow; *Civil Engineering*; Vol. 29, March, 1959, 174

Kuichling, Emil (1889). Relationship Between Rainfall and Discharge of Sewers in Populous Districts; *Trans. ASCE*; Vol. 20, pages 1–56.

Linsley, Ray K. et al. (1975). *Hydrology for Engineers*; McGraw-Hill: New York.

Manning, Robert (1891). "On the Flow of Water in Open Channels and Pipes"; *Transactions, Institution of Civil Engineers of Ireland*; Vol. 20, pp. 161–207: Dublin, Ireland.

McCuen, Richard H. (1989). *Hydrologic Analysis and Design*; Prentice Hall: Englewood Cliffs, New Jersey.

Metcalf and Eddy, Inc.; University of Florida and Water Resources Engineers, Inc. (1971). *Storm Water Management Model, Volume 1—Final Report*; EPA Report 11024DOC07/71 (NTIS PB-203289); Environmental Protection Agency: Washington, D.C.

National Weather Service (1973). *Precipitation Frequency Atlas of the Western United States, Atlas 2*; National Oceanic and Atmospheric Administration, National Weather Service: Silver Springs, Maryland.

National Weather Service (1983). *Hydrometerological Report No. 55, PMP Estimates, United States Between the Continental Divide and the 103rd Meridian*; National Oceanic and Atmospheric Administration, National Weather Service: Silver Springs, Maryland.

National Weather Service (2003). *Precipitation Frequency Atlas of the Western United States, Atlas 14*; National Oceanic and Atmospheric Administration, National Weather Service: Silver Springs, Maryland.

NOAA (1977). *Hydro-35*; National Oceanic and Atmospheric Administration, National Weather Service: Silver Springs, Maryland.

Ragan, R. M and J. O. Duru (1972). Kinematic Wave Nomograph for Times of Concentration; *Proc. Am. So. of Civ. Engrs, Jour. Hyd. Div.*; 98(10): 1165–1171.

Rain Garden Network (2008a). *All About Rain Gardens*; www.raingardennetwork/about.htm (accessed September 22, 2008).

Rain Garden Network (2008b). *Benefits of Rain Gardens*; www.raingardennetwork/benefits.htm (accessed September 22, 2008).

Rossmiller, Ronald L. (2008). *Calculating Water Surface Profiles*; Short Course Notes Taught at the University of Wisconsin, Engineering Professional Development Department: Madison

Stoll, Garner and Gil W. Rossmiller (2003). *Be Unique: A Model for Anti-Monotony in Residential Development*; American Planning Association, Zoning News Magazine (now Zoning Practice); October, 2003, 73–76.

Sustainable Land Development International (2010). *Strategic Objectives and Guiding Principles*; http//www.sldi.org/index.php?option=com_content&task=view&id=44&Itemid=57 (accessed January 5, 2012).

Sustainable Land Development Today (2008). *SLDI in Focus*: www.SLDTonline.com/1/2008, pp. 46–47.

Urban Design Tools, Low Impact Development (2011a). *Bioretention*; http//www.lid-stormwater.net/biotrans_specs.htm (accessed September 29, 2011).

Urban Design Tools, Low Impact Development (2011b). *Permeable Pavers*; http//www.lid-stormwater.net/permpavers.benefits.htm (accessed September 29, 2011).

U.S. Corps of Engineers (1981). *HEC-1, Flood Hydrology Package*; Hydrologic Engineering Center, United States Corps of Engineers: Davis, California.

U.S. Department of Agriculture (1984). *Computer Program for Project Formulation*; *Technical Release 20*; U.S. Department of Agriculture, Soil Conservation Sevice: Washington, D.C.

U.S. Department of Agriculture (1986). *Urban Hydrology for Small Watersheds; Technical Release 55, 2nd Ed.*; U.S. Department of Agriculture, Natural Resources Conservation Service, Engineering Division: Washington, D.C.

U.S. Department of Agriculture (2008). *Bioswales*; http//www.mt.usda.nrcs.gov/technical/ecs/water/lid/bioswales.html (accessed September 25, 2008).

U.S. Corps of Engineers (1981). *HEC-1 Flood Hydrograph Package-Users Manual*; U.S. Army Corps of Engineers, Hydrologic Engineering Center: Davis, California.

U.S. Department of Commerce (Reprinted 1963). *Rainfall Frequency Atlas of the United States*; Technical Paper No. 40 ; U.S. Department of Commerce, Weather Bureau: Washington, D.C.

U.S. Department of the Interior (Reprinted 1976). *Measurement of Peak Discharge at Dams by Indirect Method, Application of Hydraulics, Book 3, Chapter A5*; Techniques of Water Resources Investigations of the USGS, U.S. Department of the Interior, United States Geological Survey: Washington, D.C.

U.S. Department of the Interior (Revised Reprint 1977). *Design of Small Dams*; U.S. Department of the Interior, Bureau of Reclamation: Washington, D.C.

U.S. Department of Transportation (1965). *Design of Roadside Drainage Channels*; Hydraulic Design Series No. 4; U.S. Department of Transportation, Federal Highway Administration, Hydraulics Branch, Bridge Division, Office of Engineering: Washington, D.C.

U.S. Department of Transportation (Reprinted 1977). *Hydraulic Charts for the Selection of Highway Culverts*; Hydraulic Engineering Circular No. 5; U.S. Department of Transportation, Federal Highway Administration, Hydraulics Branch, Bridge Division, Office of Engineering: Washington, D.C.

U.S. Department of Transportation (2008). *Stormwater Best Management Practices in an Ultra-Urban Setting: Selection and Monitoring: Dry an Wet Vegetated Swales*; http//www.fhwa.dot.gov/environment/ultraurb/3fs4.htm (accessed October 10, 2008).

U.S. Department of Transportation (2008). *Stormwater Best Management Practices in an Ultra-Urban Setting: Selection and Monitoring: Porous Pavements*; http//www.fhwa.dot.gov/environment/ultraurb/3fs15.htm (accessed October 10, 2008).

U.S. Environmental Protection Agency (1983). *Results of the Nationwide Urban Runoff Program*, Volume 1; Final Report; U.S. Environmental Protection Agency, Water Planning Division: Washington, D.C.

U.S. Environmental Protection Agency (2009a). *National Pollution Discharge Elimination System (NPDES)*; http//cfpub.epa.gov/npdes/stormwater/menuofbmps/index.cfm?action=browse&Rbutton=d (accessed April 1, 2009).

U.S. Environmental Protection Agency (2009b). *National Pollution Discharge Elimination System (NPDES)*; http//cfpub.epa.gov/npdes/stormwater/menuofbmps/index.cfm?action=browse&Rbutton=d (accessed April 6, 2009).

U.S. Environmental Protection Agency (2009c). *National Pollution Discharge Elimination System (NPDES): Wet-Ponds*; http//cfpub.epa.gov/npdes/stormwater/menuofbmps/index.cfm?action=browse &Rbutton=d (accessed April 6, 2009).

U.S. Environmental Protection Agency (2009d). *National Pollution Discharge Elimination System (NPDES)*; http//cfpub.epa.gov/npdes/stormwater/menuofbmps/index.cfm?action=browse&Rbutton=d (accessed April 5, 2009).

U.S. Environmental Protection Agency (2009e). *National Pollution Discharge Elimination System (NPDES): Porous Pavements*; http//cfpub.epa.gov/npdes/stormwater/menuofbmps/index.cfm?action=browse& Rbutton=d (accessed April 1, 2009).

U.S. Environmental Protection Agency (2009f). *National Pollution Discharge Elimination System (NPDES): Riparian/Forested Buffer*; http//cfpub.epa.gov/npdes/stormwater/menuofbmps/index.cfm?action =browse&Rbutton=d (accessed April 6, 2009).

U.S. Soil Conservation Service (1984). *Technical Release 20, Computer Program for Project Formulation*; Soil Conservation Service, United States Department of Agriculture: Washington, D.C.

Weather Bureau (1934). *Monthly Weather Review: Methods and Results of Definite Rain Measurements, Vol. 62, pages 5–7*; Weather Bureau, United States Department of Commerce: Washington, D.C.

Weather Bureau (1955). *Rainfall Intensity-Duration-Frequency Curves; Technical Paper No. 25*; Weather Bureau, U.S. Department of Commerce, Washington, D.C.

Weather Bureau (1961). *Rainfall Frequency Atlas of the United States; Technical Paper No. 40*; Weather Bureau, U.S. Department of Commerce, Washington, D.C.

UW-Extension (2008a). *Protecting Our Waters: Buffers*; http//www.clean-waters.uwex.edu/plan/buffers. htm (accessed September 26, 2008).

UW-Extension (2008b). *Protecting Our Waters: Street Trees*; http//www.clean-waters.uwex.edu/plan/ streetrees.htm (accessed September 26, 2008).

Wikipedia (2011). *Rain Guage*; http//en.wikipedia.org/wiki/Rain_guage (accessed December 21, 2011).

Index

Note: Page numbers followed by "*f*" and "*t*" indicate figures and tables respectively.

CPSIA information can be obtained at www.ICGtesting.com
Printed in the USA
BVOW08*0754181214

379315BV00008B/9/P